Environmental Science and Engineering

The ultimate goal of this series is to contribute to the protection of our environment, which calls for both profound research and the ongoing development of solutions and measurements by experts in the field. Accordingly, the series promotes not only a deeper understanding of environmental processes and the evaluation of management strategies, but also design and technology aimed at improving environmental quality. Books focusing on the former are published in the subseries Environmental Science, those focusing on the latter in the subseries Environmental Engineering.

More information about this series at http://www.springer.com/series/7487

Sotaro Higuchi

Municipal Solid Waste Landfill Technology in Japan

Springer

Sotaro Higuchi
Fukuoka University
Fukuoka city
Fukuoka prefecture, Japan

ISSN 1863-5520 ISSN 1863-5539 (electronic)
Environmental Science and Engineering
ISBN 978-981-16-2733-0 ISBN 978-981-16-2734-7 (eBook)
https://doi.org/10.1007/978-981-16-2734-7

Translation from the Japanese language edition: Saishu Shobun Gijutsu by Sotaro Higuchi, © Sotaro Higuchi 2018. Published by Yakubo Printing. All Rights Reserved.

This Springer imprint is published by the registered company Springer Nature Singapore Pte Ltd.
The registered company address is: 152 Beach Road, #21-01/04 Gateway East, Singapore 189721, Singapore

Preface

The fundamental principle of waste management in Japan is to suppress the generation of waste and try to do the recycling and to landfill the residues after the volume reduction and detoxification of the discharged waste by intermediate processing such as incineration. It is because Japan is small in size, and it is tough to secure the site for landfill disposal. For this reason, ahead of other countries, Japan introduced incinerators for MSW (municipal solid waste) and became a country that has a world incineration power. On the other hand, regarding landfill technology, the semi-aerobic landfill took a central role in the 1960s when the diffusion of the incinerator is less than 50%, and the landfill site is utilized as a bioreactor. With the spread of incinerators, the landfill technology focusing on incineration residues and "cover type landfill site (closed system landfill site)" has been developed and spread. Besides, there are some unique and unprecedented landfill technologies, such as the existence of a sea surface disposal site (landfill the waste left on the water surface) and the absence of capping regulations after the landfill.

From these facts, the technology for MSW landfill site can be said as a technology that has an internationally Galapagos syndrome in a sense.

On the other hand, gas recovery by whole quantity landfill or anaerobic landfill is still the mainstream in the case of the MSW landfill technology in other foreign countries. Although the unique technology, "semi-aerobic landfill" (Fukuoka method), has been transferred to foreign countries in recent years, there is no experience in applying the whole quantity landfill to semi-aerobic landfill. In other words, Japan introduced the incinerator in advanced of the world, so it does not have the accumulation technology of the whole quantity landfill. Under such circumstances, the number of countries which introduced the incinerator began to increase, and a landfill waste quality similar to the Japanese style in the 1970s began to appear. It is expected that a landfill site similar to the Japanese style at present will also appear in the future.

Under these circumstances, the MSW landfill technology in Japan is thought to be considered as the technology in the world at the forefront of the future.

Therefore, by organizing the technologies according to the methods (whole quantity landfill, introduction process of intermediate processing such as incineration and the like, completely introduced the intermediate processing such as incineration and the like) of MSW management, we can show our experience of MSW landfill technology to the foreign countries. Besides, the future of Japan's landfill technology, where incineration is at the center of waste management, can be pointed out.

Fukuoka, Japan Sotaro Higuchi

Contents

About the Author

Sotaro Higuchi
Experience
He was born in 1949 in Fukuoka City. In 1972, he graduated from the Faculty of Engineering, Fukuoka University. He joined Japan Engineering Consultants Inc. in 1997 and took a doctoral degree at Kyushu University. He was appointed as a professor of the Graduate School of Engineering, Fukuoka University, in 2000. In 2008, he was appointed as the director of the Institute of Resource Recycling and Environmental Pollution Control System, Fukuoka University. In 2020, he was appointed as a special professor, Emeritus Professor, Fukuoka University. His research fields include environmental engineering and waste management engineering. His research works are as follows:

Journal Sketch of 100 Days in Xi'an. (Syonensha, 1991).
Design and Construction of MSW Landfill Site. (Nippo, 1993).
Waste Management. (Shared writing) (Chuohoki Publishing, 1995).
Waste Handbook. (Shared writing) (Ohmsha, 1996).
Issues of Groundwater, and Countermeasures. (Shared writing) (Kankyo Shimbunsha, 1998).
Technical Standard for MSW in German. (Joint supervision) (NTS, 1998).
Technical Standard for Special Waste in German. (Joint supervision) (NTS, 1998).
Design and Construction of MSW Landfill Site. (Chinese version) (Tongji University, 2000).
MSW Landfill Site in Japan 2000. (Shared writing) (Kankyosangyo Shimbunsha, 2000).

Guidebook of Leachate Treatment Technology. (Supervision) (Kankyosangyo Shimbunsha, 2001).
Handbook of Landfill Regeneration Technology. (Supervision) (Kajima Institute Publishing, 2005).
2008/6 Appointed as the director of the Institute of Resource Recycling and Environmental Pollution Control System, Fukuoka University.
2020/4 Appointed as a Special Professor, Emeritus Professor, Fukuoka University.

Chapter 1
The Transition of Landfill Technologies

1.1 Social Situation and Landfill Disposal Volume Over Landfill

Figure 1.1 shows the detail amount of municipal waste emissions and disposal every five years. According to it, the amount of emissions has increased until 2000, reaching a peak of about 50 million tons annually and then turned to decline.

Meanwhile, about 40% of the municipal waste emission was already incinerated in 1965, and the rate reached about 80% after 1990. As approximately 80% of the municipal waste emission was a combustible waste, the incineration rate was 50% in 1965. The incineration rate reached almost 100% around 1990 as envisioned. 10% of incinerated waste was landfilled as incineration residue. The proportion of direct landfill is decreasing year by year, as 47.6% in 1965, 19.9% in 1990, 5.9% in 2000, and 1.5% in 2010. It is because the municipalities promoted the construction of a recycling center, recycling treatment, and disposal reduction. Based on the assumption that incineration rate reached almost 100% in 1990, direct landfill waste before 1990 includes combustible waste, incombustible waste oversized waste while combustible waste has virtually disappeared after 1990 as inferred.

Figure 1.1 also shows revision time of legislation and standard, social background, and so on. Late 1960–1970s was the period of high economic growth, increasing waste became a social problem, and then in 1970 "Waste Management and Public Cleaning Law" (hereinafter referred to as Waste Management Law) was established.

In 1977, the Ministry of Health and Welfare and Prime Minister's office decided "Structural and maintenance standards for Municipal Solid Waste (MSW) Landfill Sites" (hereinafter referred to as Joint Orders).

Gas emission regulations (nitrogen oxides, hydrogen chloride gas) in 1977 and 1979 are cited as they are deeply involved in the landfill sites.

Furthermore, "Explanations of Guidelines on MSW Landfill Site" indicates that 25% of the subsidy should be issued for the construction of landfill sites for municipal waste in principal in 1978. The storage and processing function was defined as the functions of landfill sites, purification function by "semi-aerobic landfill" was

S. Higuchi, *Municipal Solid Waste Landfill Technology in Japan*, Environmental Science and Engineering, https://doi.org/10.1007/978-981-16-2734-7_1

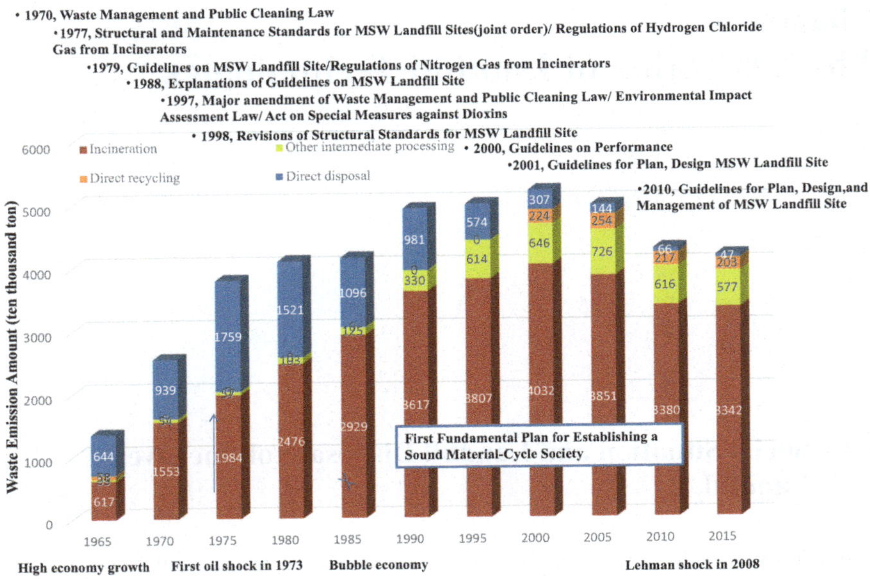

Fig. 1.1 Waste emission amount and social situation (Higuchi S 2015)

introduced and "semi-aerobic landfill" became the standard landfill technology in Japan.

In the 1980s, the spread of incineration facilities further accelerated, and fly ash mixing into landfills also began to increase. Under such circumstances, the "Explanations of Guidelines on MSW Landfill Site" was revised in 1989 (Hereinafter referred to as Explanations of Revision Guidelines). In this revision, the introduction of lime soda method as a countermeasure against calcium scale and inorganic salt was added. Thereafter, conflicts with the residents concerning the construction and maintenance of the landfill sites occurred frequently in various places due to their anxiety of groundwater contamination caused by breakage of the seepage control work and dioxins discharged by incineration. Municipalities and business operators have increased double seepage control work to ensure safety.

In 1995, Ministry of Health and Welfare obliged municipalities to add double seepage control work as a requirement for giving subsidies of the landfill site. In 1997 "Waste Management Law" was revised, and "Act on Special Measures against Dioxins" was enacted. 1997 is a major turning period concerning waste treatment system in Japan. External circumstances of landfill sites also changed significantly, and regulations were strengthened as the revision of "Waste Management Law," the establishment of "Environmental Impact Assessment Law" and enact of "Act on Special Measures against Dioxins," which was an obligation of "Living Environment Survey." In the following 1998, Joint Order was revised first in 21 years in accordance with the revision of "Waste Management Law," and for structural standards, double seepage control work, leachate adjusting facilities, and groundwater collecting facilities were obliged. For Maintenance standards, groundwater monitoring facilities, deployment inspections at stable industrial waste landfill sites were obliged. In addition, as structural standards, etc. of completed landfills with abolition criteria were substantially revised, double seepage control work, removal technology of chloride ions in leachate, separation and decomposition technology of dioxins, etc. were additionally reinforced. In 2010, "Guidelines for Plan, Design, and Management of MSW Landfill Site" was revised with maintenance and management technology added, and it is still used now.

In the twenty-first century, as resource depletion and environmental pollutions, sustainable society is called for, and Japan has also adopted the 3R policy, and in 1990, municipal and industrial waste landfilled was 110 million tons and the target volume would reduce by 25%, 27 million tons by 2010 according to "The Fundamental Plan for Establishing a Sound Material-Cycle Society." For municipal waste, municipalities stopped direct landfill of incombustible garbage and oversized waste and promoted the construction of intermediate processing and recycling to reduce volume. For industrial waste, emission companies and others worked on unused resource recycling and achieved it by public and private sectors working together (Fig. 1.2).

However, here incineration ash, etc. refer to intermediate processing residue of fly ash, bottom ash, and incombustible and oversized waste). Through these processes, the volume of landfill waste was greatly reduced, and the quality of it also fluctuated.

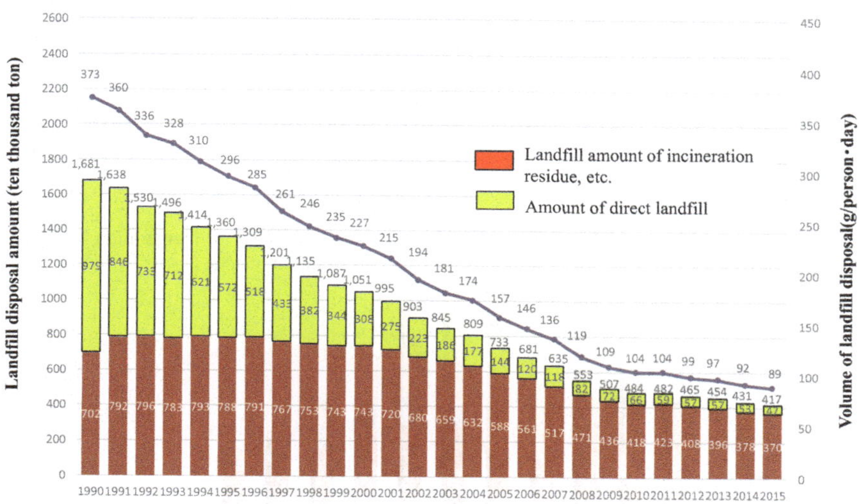

Fig. 1.2 Transition of landfill disposal amount (Waste Recycling Office, Ministry of Environment 2012, 2017)

1.2 The Transition of Landfill Waste

Diffusion of incineration facilities and recycling centers, and strengthening emission regulations of incineration facilities, etc. have led to changes in landfill waste materials. Figure 1.3 shows the type of transition of municipal landfill waste (*Since there is no data of incineration residues before 1980, 10% of incinerating waste is assumed. Besides, incineration residue contains intermediate processing residue). The direct landfill of combustible waste ceased around 1990, and direct landfill of incombustible waste also declined sharply after 1990. Most of the waste in landfill sites was incineration residue and intermediate processing residue of incombustible waste.

The reason why direct landfill has declined sharply in the 20 years since 1990 is thought to be the result of municipalities establishing recycling centers, recycling incombustible and oversized waste, and sorting incineration.

Figure 1.4 shows subdivided landfill ratio from 1965 to 2015.

From this, the proportion of incineration residue in landfill waste rapidly increases and accounts for 75.9% in 2015. Combining with 12.8% of crushed incombustible residue together, intermediate processing residue reaches 88.7%.

It is thought that combustible waste generates from incombustible and oversized waste as the establishment of the recycling centers mentioned above, and fly ash increases as a sophisticated treatment of gas emission.

For these reasons, it is inferred that quality of landfill waste has been rapidly mineralized due to incineration residue and incombustible waste residue for 25 years since 1990, and the landfill environment is becoming increasingly alkaline and high salted.

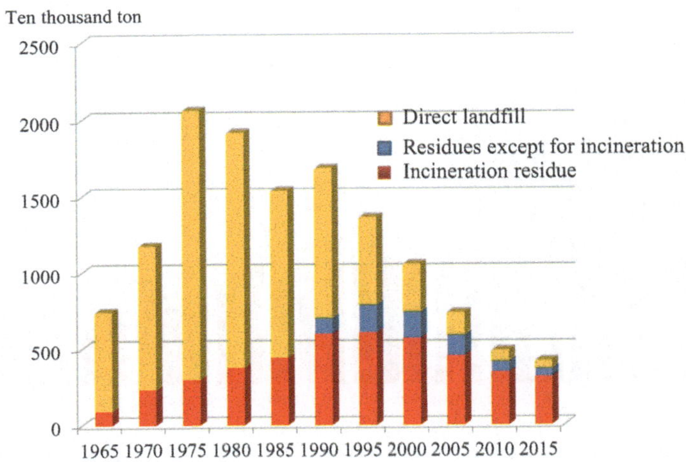

Fig. 1.3 Transition of landfill disposal types (Waste Recycling Office, Ministry of Environment 2017; Environmental Industry Newspaper Company 2014)

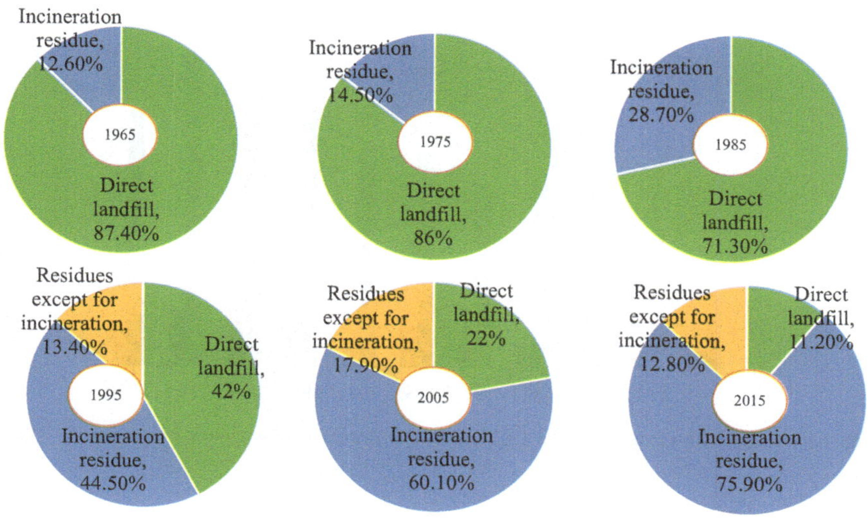

Fig. 1.4 Transition of landfill disposal amount (Waste Recycling Office, Ministry of Environment 2017; Environmental Industry Newspaper Company 2014)

1.3 The Transition of Researches in the Field of Landfill

It seems that there was hardly any waste-related researches in the 1960s. The publication situation of waste-related papers and landfill site related papers is shown in Fig. 1.5 in the Table of Contents of Japan Society of Civil Engineering (JSCE) annual academic lecture paper collection since 1970. About 10 waste-related researches

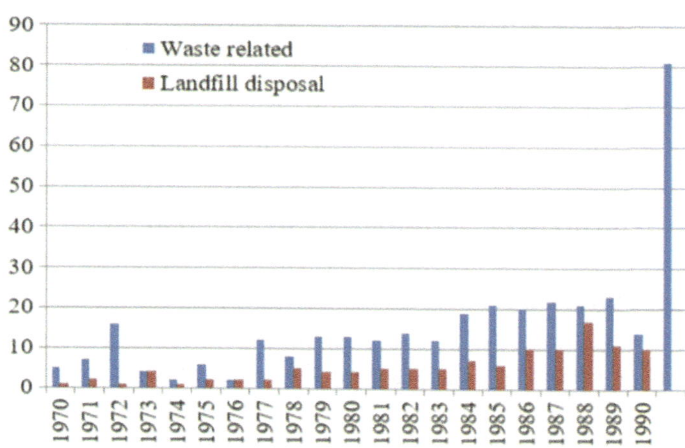

Fig. 1.5 Numbers of presentations in JWMA

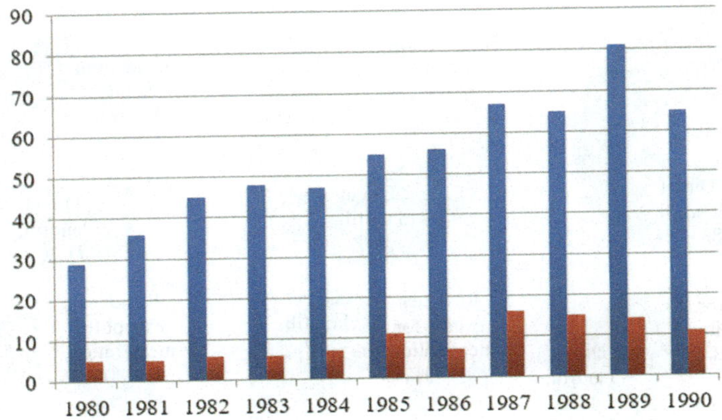

Fig. 1.6 Numbers of presentations in JSCE annual academic lecture paper collections

and 1–3 landfill related researches published per year in the 1970s. The contents of waste-related researches were mainly about radioactive waste disposal, quality data of waste, and location theory of waste disposal facilities. Landfill related researches were concentrated on landfill characteristics of kitchen waste and semi-aerobic landfill. In the late 1970s, research cases increased along with the social problems caused by waste, and incineration related researches began. When Japan Waste Management Association (JWMA) held a research presentation seminar in 1980, the number of researches increased at once. 29 waste-related researches were presented in the 1st seminar, of which the number of landfill related presentations was 5, intermediate processing related was 10. After that, it continued to increase as shown in Fig. 1.6, and the number of researches and papers published also increased in a stroke as the establishment of Japan Society of Waste Management (now is called Japan Society of Material Cycles and Waste Management) in 1990. Figure 1.6 shows the number of published papers at the JWMA research presentations from 1980 to 1990 when the Japan Society of Waste Management (JSWM). The total number of presentations in 2016 was 128, of which landfill related was 20.

1.4 The Transition of Technologies by the Year

1.4.1 The 1960s

The high economic growth began in Japan. The amount of waste in the urban areas increased sharply while many municipalities landfilled all waste caused a severe shortage of landfill sites in big cities. Sea surface disposal sites were required, and incinerators were introduced. Landfilling techniques had not been established, the situations close to open dumping show problems such as odors, insects, crows, and water pollution due to high concentration organic leachate, etc. are appeared. According to the date from Ministry of Health and Welfare, the total collected amount of municipal waste in 1965 was 13.54 million ton, among which the incineration treatment was 6.44 million t, and the incineration rate was 47.6% (Japan Waste Management Association 1989). Figure 1.7 shows the landfill situation of the landfill site at that time.

Leachate treatment technologies had not been established, but treatments such as sprinkling filtration method to deal with high concentration BOD, nitrogen, etc. had been conducted on trial (Fig. 1.8). Masataka Hanashima of Fukuoka University proposed a semi-aerobic landfill concept, and it was adopted in Fukuoka city after the demonstration experiment, and put to practical use. At that time, the landfill

Fig. 1.7 Landfill sites in 1960s

Fig. 1.8 Leachate treatment by sprinkling filtration

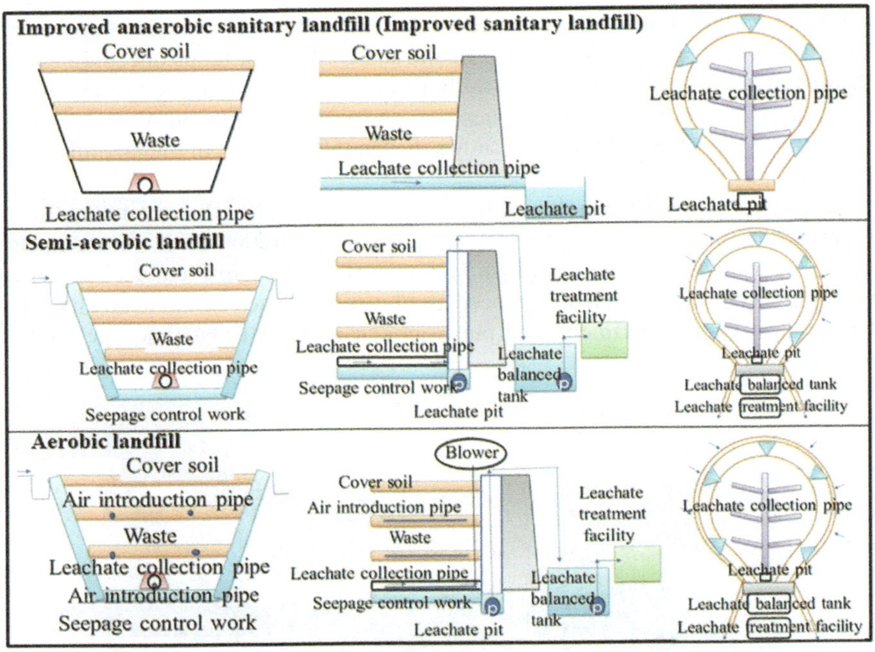

Fig. 1.9 Classifications of landfills

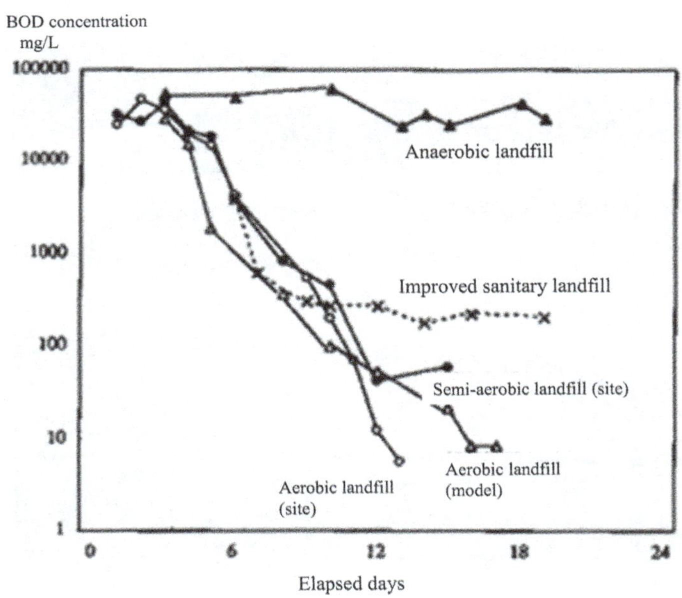

Fig. 1.10 Transition of BOD in landfill by time (Hanashima M 1985)

was classified as "aerobic landfill," "semi-aerobic landfill," and "anaerobic sanitary landfill" based on the presence of air in the landfill layer (Figs. 1.9 and 1.10).

In the semi-aerobic landfill, air flows into the landfill layer from the end of the bottom collecting pipe by natural aeration, through collecting pipe holes, air goes inside of the landfill layer, the landfill layer is kept in aerobic condition, and the organic substance is decomposed. Because of aerobic decomposition, the temperature in the layer rises.

On the other hand, insolubilization of heavy metals by sulfides is carried out in anaerobic landfill layer where cannot get the air, and methane gas (CH_4) which is lighter than air is generated by anaerobic decomposition of organic substance. The inflow of air into the landfill layer is caused by heat raised by aerobic decomposition of organic substance and CH_4 generated by anaerobic decomposition released from the gas venting facilities or the drainages into the atmosphere from the upper part of the landfill layer, and then negative pressure is generated in the landfill layer against the atmosphere.

Therefore, air flows into the landfill layer from the open end of the water collecting pipe, and the aerobic region and the anaerobic region coexist in the landfill layer. The stabilization progresses due to the interaction between them. Since it stabilizes very early compared to the anaerobic landfill structure, it was named "semi-aerobic landfill." Figure 1.11 is an example of a conceptual diagram of a semi-aerobic landfill. The end of collecting pipe is opened, and leachate flows into the water intake pit.

The water intake pit is equipped with a pump that sends leachate to the leachate adjusting facility. The operation and stop of the pump are controlled by the water gauge installed in the pit, but the starting water level must be set lower than the bottom of the collecting pipe. If the starting water level is set above the bottom of the collecting pipe, the water collecting pipe would be sealed with water so that the inflow of air is blocked, and the inside of the landfill layer would become anaerobic.

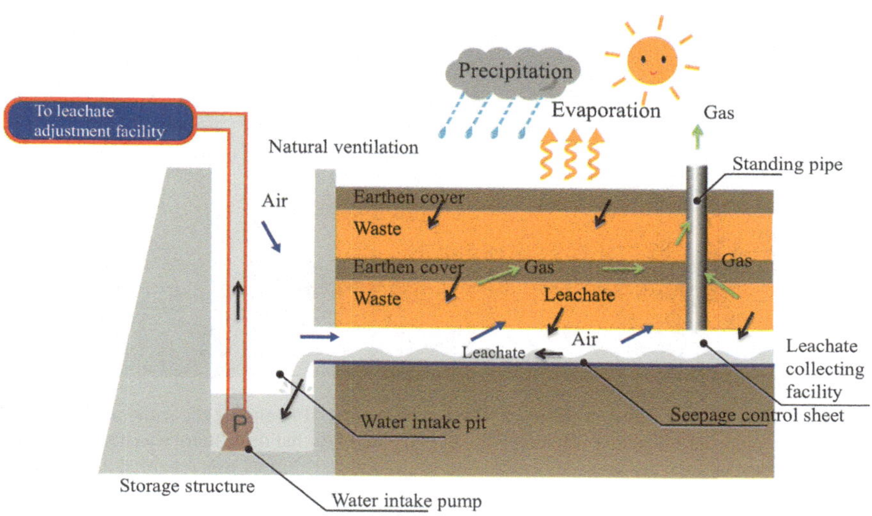

Fig. 1.11 Conceptual diagram of a semi-aerobic landfill

There is also a method that makes the collection pipe penetrate through the storage structure, and the leachate flows down to the water intake pit by gravity. In that case, water tightness maintenance of the stored structure should be taken, and an on/off valve should be installed at the end of the water collecting pipe to counter heavy rain.

1.4.2 The 1970s

In 1970, the Waste Management Law was enacted. Structural and maintenance standards were established as technical standards for MSW landfill sites. In 1978, "Explanations of Guidelines on MSW Landfill Sites" indicates that 25% of the subsidy should be issued for the construction of landfill sites for municipal waste in principal. In it, the storage and processing function was defined as the functions of landfill sites. Purification function by "semi-aerobic landfill" was introduced and "semi-aerobic landfill" became the standard landfill technology in Japan (Fig. 1.12).

Particularly it was stated that the water collecting pipe, as the heart of "semi-aerobic landfill," should be secured a large cross-section to ensure the amount of ventilation. On the other hand, since gas emission regulation was formulated at almost the same time. The proportion of fly ash to the disposal sites started to increase, and the effect of it was not mentioned. In addition, the seepage control work, groundwater collection drainage pipes, and leachate adjustment facilities which are now obliged to be installed were not clearly stated. Especially leachate adjusting facility was recommended to have a treatment capacity for several days of leachate, because of the high ratio of direct ration and its water retention capacity. There was also a survey report stating that leachate adjustment facility was unnecessary. On the other

Fig. 1.12 Disposal site constructed by semi-aerobic landfill. It is almost the same as the present structure except for the absence of the seepage control sheet. At that time, there was no standard for seepage control and the permeability coefficient of ground, 10^{-5} cm/s was regarded as a measure of impermeable ground

Fig. 1.13 Tower type sprinkling filter (appearance)

hand, the shortage of the landfill site became increasingly serious in urban areas, and the construction of the sea surface disposal sites represented by the Phoenix Project increased. Although the incineration rate was about 50%, leachate and gas concentrations tend to be reduced, but water pollution and malodor remained a major social problem.

Water quality items subject to leachate treatment were BOD, COD, and T-N. Countermeasures against agricultural damage caused by nitrogen (straighthead phenomenon due to overgrowth) became a subject of water quality management. Plastic filter material was used to reduce weight in the leachate treatment method, and ground-based tower type sprinkling filters were constructed in Kobe, Fukuoka, Sendai city, etc. to improve the processing efficiency (Figs. 1.13 and 1.14). In addition, the Lagoon Treatment method spread mainly in the north of Kanto (Fig. 1.15).

1.4.3 The 1980s

The spread of incinerators was remarkable, and the incineration rate reached 90%. As a result, the BOD, COD and T-N concentrations in the leachate sharply decreased, while high concentration inorganic salt problems such as equipment scaling by calcium ions in leachate, water treatment disorders due to chlorine ions, agricultural damage, etc. became manifest. Salt countermeasure technology was developed since the seepage control sheet began to be laid in landfill sites (Fig. 1.16).

Fig. 1.14 Tower type sprinkling filters (filter material)

Fig. 1.15 Lagoon treatment method

Under such circumstances, "Explanations of Guidelines on MSW Landfill Sites" was revised. In this revision, it was stated that leachate treatment and landfill management that required consideration of upstream systems such as exhaust gas treatment method of incineration facilities are necessary, and introduction of lime soda method as a measure against calcium scale and inorganic salt was added. The necessity of

Fig. 1.16 Landfill site with seepage control work

leachate adjusting facilities was clarified, and it was recommended to determine its capacity by calculating the water balance with leachate treatment facilities.

Biological treatment and physiochemical treatment were also organized as leachate treatment methods, and the standard of raw water quality of landfill waste, mainly kitchen and incombustible waste was indicated.

For seepage control work, "For landfill sites, seepage control work must be established to prevent leachate contaminate public water areas and groundwater," "In order to keep its function, groundwater collection and drainage facilities are necessary to set," and seepage control work is classified into vertical and surface types. As a guide "~ if the permeability coefficient of soil ground is larger than 10^{-5} cm/s, or if the Lugeon value of bedrock is larger than 5–10, seepage control work should be established in principle." are described.

Synthetic rubber, synthetic resin, asphalt, earth lining, and pavement facing have been introduced for surface seepage control work.

For synthetic resin sheet, "soft polyvinyl chloride is common. There are also sheets of ethylene-vinyl acetate, polyethylene and the like." are stated.

Under the circumstance, conflicts with the residents concerning the construction and maintenance of the landfill sites are frequently occurred in various places due to their anxiety of groundwater contamination caused by breakage of the seepage control work and dioxins discharged by incineration. Municipalities and business operators have increased double seepage control work to ensure safety due to the background. Besides securing landfill sites, the 1980s is said to the dawn of the development of new landfill technologies accompanying the change of landfill waste materials because of semi-aerobic landfill method which was the basic landfill technology until then.

Fig. 1.17 First covered-type landfill site

1.4.4 The 1990s

The incineration rate of combustible waste is almost 100%. Landfill waste materials at the municipal waste landfill sites has reached nearly the present standard and melting treatment of fly ash began to be implemented. The concern of contamination of groundwater caused by breakage of seepage control work etc. became a social problem, and in 1998, reinforcement of structural and maintenance standards was carried out, and double seepage control work was obligated. In addition, abolition criteria were enacted. The first facility of the covered-type landfill site (closed system disposal site) was constructed (Fig. 1.17). The subsidy system was changed to a grant system, and the "MSW landfill Site Performance Guidelines" was established.

1.4.5 The 2000s –Present

In 2000, instead of "Explanations of Guidelines on MSW Landfill Sites," "Planning and Design Guidelines for Landfill Site Maintenance" was issued. As the volume of landfill waste sharply declined, and the capacity of landfill sites became smaller, so the constructions of covered-type landfill sites increased rapidly. In addition, the reverse osmosis membrane method for leachate treatment and desalination treatment equipment of electrodialysis membrane method have been increasingly introduced. Target items of leachate treatment were focused on low concentrations of BOD, COD, T-N, SS, calcium ions, and chlorine ions. And the importance of maintenance became more important as abortion criteria were enacted in 1998. In 2010 "Planning and Design Guidelines for Landfill Site Maintenance" has been revised to "Planning, Design, and Management Guidelines of Landfill Site Maintenance."

References

Environmental Industry Newspaper Company (2014) Yearbook on waste, 2014 edition

Hanashima M (1985) Research on the aerobic landfill of waste. Kyushu University Thesis

Higuchi S (2015) Technical transformation and future trend of landfill site. J Mater Cycles Waste Management 26(1):3–11

Japan Waste Management Association (1989) Edited by Environment Maintenance Division, Water Environment Department, Ministry of Health and Welfare. Waste in Japan 89, appended chart

Waste Recycling Office, Ministry of Environment (2012) Emission and processing status of municipal waste

Waste Recycling Office, Ministry of Environment (2017) Emission and processing status of municipal waste

Chapter 2
Issues and Measures for Landfill Technologies

The landfill technologies have developed with the changes of landfill waste materials, which changed with the upstream parts of a process, that is, collection process or intermediate processing system. However, the landfill technologies so far were problem-solving types which solve a problem after it was noticed. For example, scale often occurred in equipment used for leachate treatment, after analysis, it is caused by the rising of Ca^{2+} in leachate. To remove Ca^{2+} in leachate, the lime soda method was developed. It can be said it was symptomatic and post-treatment correspondence. The reason was that each process independently worked to improve the technologies from the same standpoint and did not consider the influence on downstream processes. There was no system that overlooked the whole. It is no exaggeration to say that even now the most issues of landfill technologies are related to it. In this chapter, we will also consider these cases and countermeasures for solving problems.

2.1 Changes in Landfill Waste Materials

2.1.1 Quality of Waste Carried to Landfill Sites

Reasons for the changes in waste materials were discussed in Chap. 1. Due to the spread of intermediate processing facilities, the main types of waste landfilled have been changed to incineration residue or crushed incombustible residue. Reasons are (1) legal regulations such as Containers and Packaging Recycling Law and Home appliance Recycling Law, (2) changes of sorted collection and treatment policy, (3) improvement of incineration efficiency and waste gas treatment, (4) advancement of recycling facilities. Concerning incineration facilities, the purpose of the incineration facility has traditionally been placed on reducing the waste volume and rendering the emission gas harmless, and in recent years, as the heat recovery added and strengthened self-regulation of emission gas by municipalities, ignition loss decreased, and content of chlorine and calcium tend to increase. Regarding recycling facilities,

S. Higuchi, *Municipal Solid Waste Landfill Technology in Japan*, Environmental Science and Engineering, https://doi.org/10.1007/978-981-16-2734-7_2

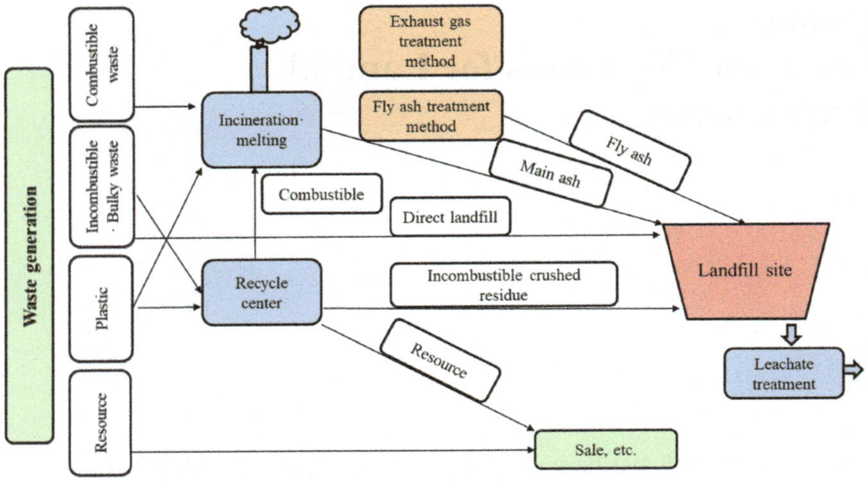

Fig. 2.1 Route of waste treatment

as the improvement of crushing and sorting accuracy, combustible and recyclable waste decreased, and the amount of landfill waste also decreased. Figure 2.1 shows the route of waste treatment. Table 2.1 shows the composition of incineration residue at present, Tables 2.2 and 2.3 show composition of crushed incombustible waste and elution test, Fig. 2.2 shows photos of crushed incombustible waste by composition.

K city in Table 2.1 uses stoker furnace, and emission gas treatment is a dry type, with highly reactive lime as chemicals. F city uses stoker furnace, and the system is a wet type with chemical agent NaOH. And high concentrations of Cl and Ca are contained in fly ash of incineration residue.

The residue from the recycling center, which accounts for about 15% of landfill waste is poor in the universality of quality because crushed incombustible waste has characteristics in physical composition and eluate depending on the collection, crushing and sorting method.

Particularly, incombustible or bulky waste will not be mixed before crushing, so the uniformities of the quality of crushed incombustible waste materials is also indeterminate. In addition, BOD and COD are also detected at relatively high concentrations from the elution test, and mercury and lead are sometimes detected although at low concentration.

It is predicted to be caused by fluorescent tubes or the bases of electronic toys, etc. However, the home appliance may also be mixed.

Table 2.1 Composition of incineration residue (Higuchi S 2013)

Item	Analysis items	Unit	Lower limit of determination	Results of analysis			
				Fly ash in City K	Main ash in City K	Fly ash in City F	Main ash in City F
Fly ash and main ash content analysis items	COD_{Mn}	mg/kg	50	3150	4000	1080	3320
	COD_{Cr}	mg/kg	50	29,500	22,800	35,600	25,300
	T-P	mg/kg	1	5700	2140	4070	3050
	Ca	mg/kg	0.1	276,000	162,000	104,000	174,000
	Cs	mg/kg	0.1	3.9	0.6	5.5	0.4
	Pb	mg/kg	0.1	2100	1420	5600	1140
	K	mg/kg	0.1	41,800	7120	117,000	5920
	S	mg/kg	10	25,400	4840	46,500	2280
	Cl	mg/kg	10	201,000	9320	265,000	14,800
	Mg	mg/kg	10	5460	10,100	6930	10,490
	Cu	mg/kg	1	490	1890	1400	3260
	Zn	mg/kg	1	12,500	3680	32,200	2750
	pH	–	–	12.2 (16 °C)	12.5 (20 °C)	6.7 (18 °C)	12.5 (21 °C)
	Moisture content	%	0.1	0.5	20.1	1.2	23.3

Table 2.2 Composition of crushed incombustible waste (Lei et al. 2014; Study Group on Salt Recycling System 2004)

	City A	City B	City C
Waste plastic	61	13.2	53.5
Foam polystyrene	0.4	–	0.7
Glass ceramics	18	42.6	24.6
Metal	5	30.5	18.9
Waste wood	6	1.5	1.6
Paper	1	–	0.2
Fiber	2	0.9	–
Others (earth and sand)	6.6	11.3	0.5

Unit wt%

2.1.2 The Necessity of Grasping Landfill Waste Materials

Generally speaking of waste materials, physical composition, chemical composition, element analysis are carried out, facilities of incineration and exhaust gas treatment equipment are planned at planning design of the incineration facilities.

However, there are still few cases of investigating landfill waste materials at planning and design period of landfill sites. Even when investigating, there are many

Table 2.3 Elution test of crushed incombustible waste (Lei et al. 2014)

	Unit	City A	City B	City C
pH	–	9.3	8.4	8.5
BOD	mg/L	210	240	39
COD	mg/L	230	230	72
T-N	mg/L	18	14	5.5
Cl	mg/L	60	46	8.5
Pb	mg/L	<0.01	0.22	<0.01
Hg	mg/L	0.002	<0.0005	<0.005

Fig. 2.2 Crushed incombustible waste by composition (City A)

cases where planning and design are advanced qualitatively based on the type and its landfill ratio. This is partly due to the landfill sites are typical nuisance facilities, the extreme difficulty to secure the site, the chemical stabilization of the landfill layer or fail to consider the abolition of sites. From these facts, it is important to grasp the method of intermediate processing, the exhaust gas regulation value, and the type and amount of chemicals used in processes.

In the case of incineration facility, quality change of incineration residue can be minimized by mixing in pits, but in the case of crushing and sorting facility there is no mixing process, so the quality of crushed incombustible residue is easy to fluctuate. Furthermore, in the case where home appliances or the like are mixed, mercury or lead may be detected in the leachate although it is a trace amount because it is not subjected to thermal decomposition, so the sorting work is considered to be high importance. Figure 2.3 conceptually shows the transition of landfill waste materials in Japan.

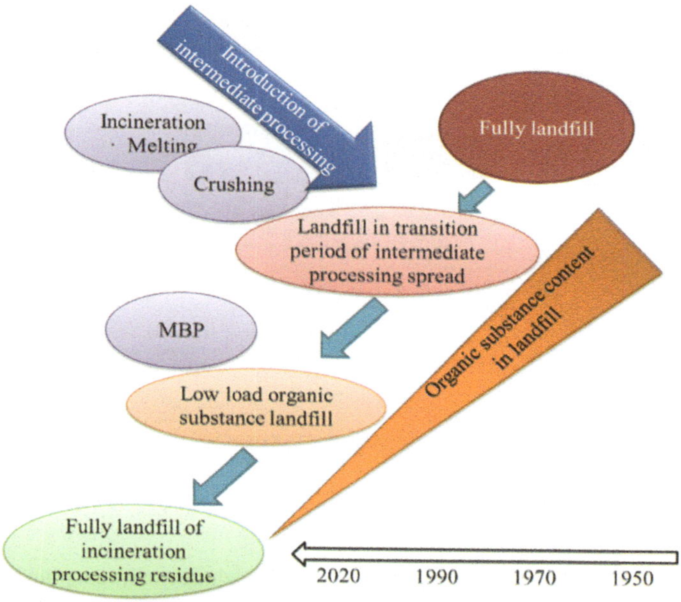

Fig. 2.3 Conceptual diagram of landfill waste quality transition

Prior to the high economic growth period, waste was almost fully landfilled, the organic substance content in landfill waste was high. BOD, COD, and T-N in leachate of landfill site were high concentrations and organic pollution in public water area became a problem. To deal with these problems, intermediate processing facility mainly centered on incineration facility was introduced from the 1960s, and after the spread transition period, the incineration rate of combustible waste became nearly 100% in the 1990s.

As a result, the proportion of incineration residues in landfill waste increased, landfill waste is transitioned to which mainly composed of inorganic substance, and organic substance content of it decreased. In the landfill site, the concentration of organic substance in leachate declined, instead Ca^{2+} and Cl^- concentrations rose and the landfill management and leachate treatment methods significantly changed.

MBP in the figure is an abbreviation for Mechanical Biological Pretreatment. In the 1990s, when the generation of dioxins accompanying incineration became a social problem, facilities that aim for volume reduction and inactivation of landfill waste became popular other than incineration in Germany. This is made up of thorough crushing sorting and recycling, gas recovery from organic waste such as kitchen waste, composting technology of residue and was also introduced to Japan. It is methane fermentation and energy recovery by mixing treatment of excretion and organic waste, and it has become popular as biomass utilization technology.

Figure 2.4 is a conceptual diagram for understanding the landfill waste materials from the current waste treatment system in Japan. It is a necessary diagram for

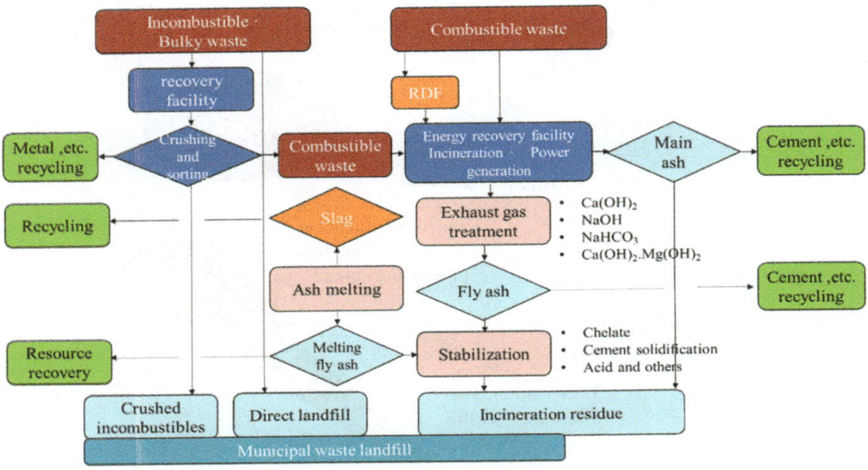

Fig. 2.4 Conceptual diagram for understanding landfill waste quality

establishing appropriate landfill management and leachate treatment plan by grasping what kind of intermediate processing has been carried and what chemicals have been added when waste is delivered to the landfill site. Regarding incombustible waste and bulky waste, since the collection methods and definitions are different for each municipality, the properties of crushed incombustible waste are also different. Since the composition of the crushed incombustible residue and elution test results described above have large seasonal and daily variations, it is desirable to collect as much data as possible for the longest possible time. In the absence of data, it is one way to get samples from a municipality that requires a similar sorting and crushing plant.

Regarding incineration residue

① Method of Incineration and melting (stoker furnace, fluidized bed furnace, etc.)
② Design value, regulated value and actual operating value of exhaust gas treatment
③ Types of chemicals and blowing equivalent used for exhaust gas treatment (lime, highly reactive lime, sodium bicarbonate, dolomite hydroxide, etc.)
④ For stabilization treatment methods of fly ash and using chemicals: chemical types (organic or inorganic), its addition rate and moisture content when added (It is desirable to sample the bottom ash and fly ash from the incineration/power generation facility during operation and perform composition analysis)
⑤ When organic chelate is used for stabilization treatment, the composition of organic chelate has large influence especially on COD and T-N composition, so it is desirable to obtain a standard composition from the manufacturer and analyze
⑥ Composition of bottom ash and elution test value

Regarding crushed incombustible residues

① Classification and types of incombustible waste
② Types of the crushed sorting facility Types of the crushed sorting facility
③ Composition and elution test value of crushed incombustible residue

When there is direct landfill waste

① Types of direct landfill waste
② Properties

It is possible to grasp the properties of the landfill waste and to make it the basic information of the landfill plan and leachate treatment plan.

In the case of industrial waste, it is important to consider an appropriate treatment method according to Waste Data Sheet (WDS) and Material safety data sheet (MSDS).

2.2 Issue of High Concentration Inorganic Salt

Since the 1980s, incineration facilities became popular and the regulation of HCl gas in exhaust gases became full scale, Ca and Cl contents in the fly ash increased, and the Ca^{2+} and Cl^{-} concentrations in leachate also increased. F At first, the problem of adhesion of calcium scale to agitators and submersible pumps constituting the leachate treatment facility became obvious. Then the Cl^{-} concentration rose, causing corrosion of equipment and inhibition of nitrification in the biological nitrification-denitrification step. The reason is when chlorine-containing refuse is burned and thermal decomposed, HCl gas is generated. Since HCl gas is hazardous, in the incineration facility, exhaust gas is treated with dry or semi-dry type in which Ca $(OH)_2$ is blown in and neutralized as calcium chloride (hereinafter referred to as $CaCl_2$) and collect as fly ash with a dust collector, while exhaust gas is treated with a wet type in which NaOH is blown and neutralized as sodium chloride (hereinafter referred to as NaCl).

Wet type was adopted in some metropolis because of high processing efficiency but high cost of chemicals, while dry/semi-dry type was adopted in many cities as the cheap price of chemicals.

Dry/semi-dry type

$$2HCL + Ca(OH)_2 \rightarrow CaCl_2 + 2H_2O$$

$$Ca(OH)_2 + SO_2 + 1/2O_2 \rightarrow CaSO_4 + H_2O$$

Wet type

$$HCL + NaOH \rightarrow NaCl + H_2O$$

$$2NaOH + SO_2 \rightarrow Na_2SO_3 + H_2O$$

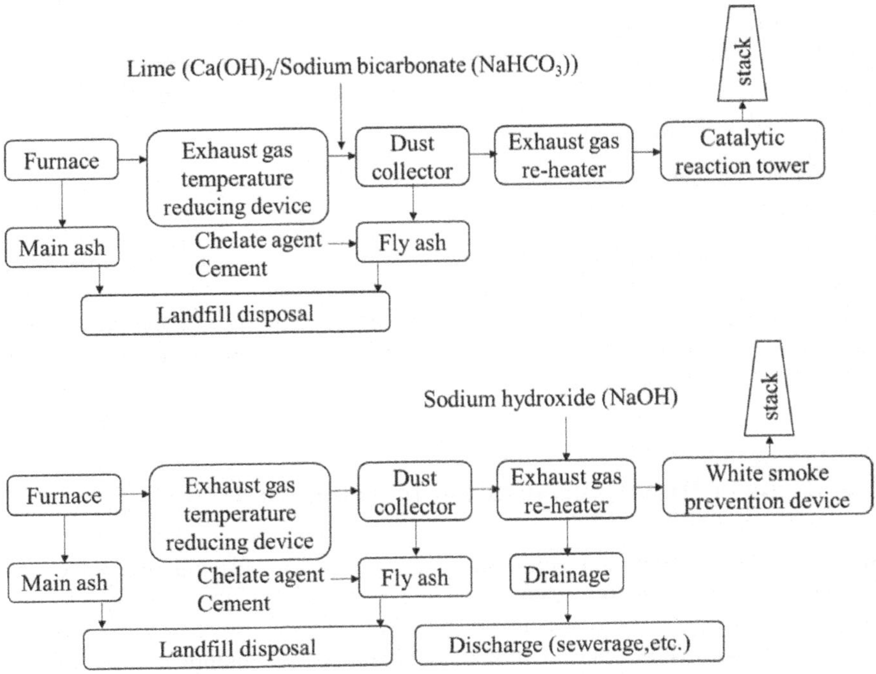

Fig. 2.5 Exhaust gas treatment and fly ash stabilization in incineration facility

Besides, $CaCl_2$ treated by dry type is finally landfilled as fly ash, so high concentration inorganic salt problem developed. Meanwhile, in the wet type, since NaOH is blown into the exhaust gas after capturing fly ash with the dust collector, almost all municipalities adopting the wet type discharged drainage into the sewer as scrubber waste or released into the sea after treatment of regulated substances, the inorganic salt problem did not become big. Figure 2.5 shows the flow of dry/semi-dry type and wet type of ash treatment.

2.2.1 *Issue of Calcium Scale*

When exhaust gas is treated by a dry or semi-dry type in an incineration facility, it is collected by a dust collector and landfilled, so it contains $CaCl_2$ and unreacted lime in fly ash. They are dissolved as Ca^{2+} and Cl^- in leachate by precipitation. Ca^{2+} in the leachate precipitates and solidifies as calcium carbonate (hereafter $CaCO_3$) when it is oxidized, $CaCO_3$ scale is attached to the pump, agitator, piping which constitutes the water collecting pipe and leachate treatment facility, causing piping blockage, pump failure, etc. Figure 2.6 shows a case of Ca scale adhering to an agitator.

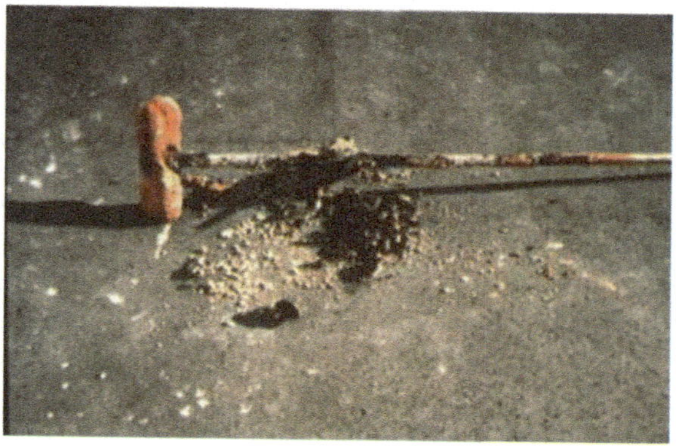

Fig. 2.6 Ca scale adhering to an agitator

As a pretreatment of a leachate treatment facility in the 1980s, scale dispersant and Ca removal method by lime soda method were developed and set up.

$$Ca + Na_2CO_3 \rightarrow CaCO_3 + 2Na$$

In addition, due to the Lime Soda Act, a large amount of sludge is generated, the landfill method to distribute lime soda (sodium carbonate) directly to the landfill site to suppress the Ca^{2+} concentration has also been carried out. On the other hand, when exhaust gas is treated by wet type in an incineration facility, since 2–3 equivalents of power lime is sprayed as the gas-solid reaction, unreacted lime increases and also the amount of fly ash. Highly reactive lime which increases the reaction efficiency by finely grinding lime has been used. Figure 2.7 shows an example of calcium removal flow in leachate treatment.

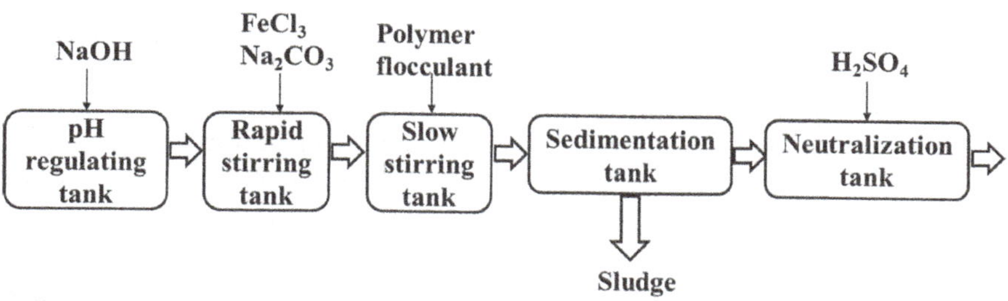

Fig. 2.7 Calcium removal flow in the leachate treatment

2.2.2 *Issue of Chloride*

The chloride problem in landfill site is caused by $CaCl_2$, a by-product in the fly ash accompanying HCl gas treatment in incineration facility.

$CaCl_2$ dissolves as Ca^{2+} and Cl^- in leachate due to precipitations. As Cl^- in leachate is not a substance regulated by the Water Pollution Prevention Act, it is generally not subject to leachate treatment. However, high concentrations cause metal corrosion and biological treatment in the leachate treatment process, especially nitrification inhibition during nitrogen treatment. Besides, when it is discharged to the public water area, if it is used for agricultural water, it will affect the agriculture industry.

Generally, it is said that problems occur when it reaches 500 mg/L for paddy rice and 300 mg/L for strawberry. Figure 2.8 is a case of paddy rice got damaged where water containing Cl^- at a high concentration was discharged and when the Cl^- concentration in agricultural water exceeds 500 mg/L, it withered. The same situation happened for strawberry when it reached 300 mg/L.

Conventionally, the influence of leachate on agriculture was nitrogen, and when the amount of nitrogen in agricultural water exceeded 1.0 mg/L, the water would be over nutrition, causing the straighthead phenomenon (the plant grows tall but does not bear fruit), which became a problem.

In addition to the problem of nitrogen, the problem of Cl^- occurred. Besides, Cl^- causes metal corrosion problems in leachate treatment facilities using iron, etc. Figure 2.9 is an example of corrosion of a steel pipe used in the leachate treatment facility.

To avoid the influence of Cl^- damage, desalination treatment by reverse osmosis membrane treatment (RO) or electrodialysis membrane treatment (ED) has been carried out. However, due to the desalination process, Concentrate will be produced, which will be dried and filled in flexible container pack, and then stored or processed in consignment to industrial waste landfill sites which is not sea surface landfill or where is no salt regulation. That is, in landfill site which is the final runner of waste treatment, inappropriate materials which couldn't be processed generated. Since these unfit matters are not harmful substances prescribed in the Waste Disposal Law, development of a proper process method or recycling technology of by-product salt is required.

Normal paddy rice after rice planting Start to wither due to salt damage Withered rice

Fig. 2.8 Paddy rice damage caused by salt damage

Fig. 2.9 Metal corrosion due to salt damage

2.2.3 *Generation of By-Product Salt*

2.2.3.1 Generation Source

The generation source of the by-product salt, which is the cause of the inorganic salt problem, is shown in Fig. 2.10. The by-product salt is mainly condensed salt generated as a result of desalination treatment of leachate at landfill sites that has fly ash landfilled after the dry type exhaust gas treatment. In the case of the wet type exhaust gas, more are discharged as drainage to the sewage treatment facility and then to sea from the sewer terminal treatment facility. The part is treated by wash scrubber, then dried and recovered. It also occurs when water washing desalination is performed at the time of cement recycling of incineration residues and resource recycling. The sources are incineration residues, that is, incineration facilities.

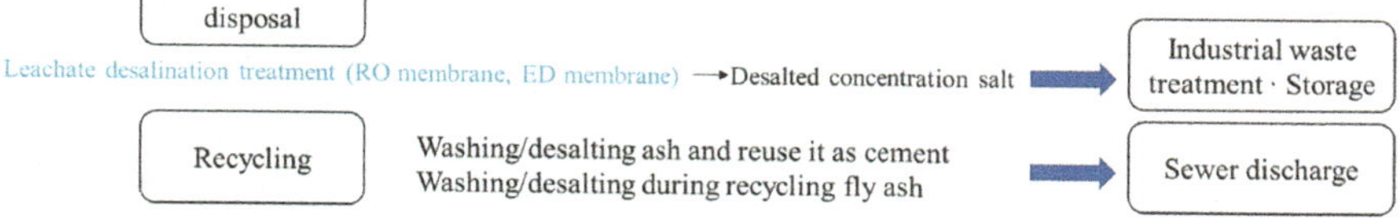

Fig. 2.10 Source of by-product salt (Higuchi S 2011)

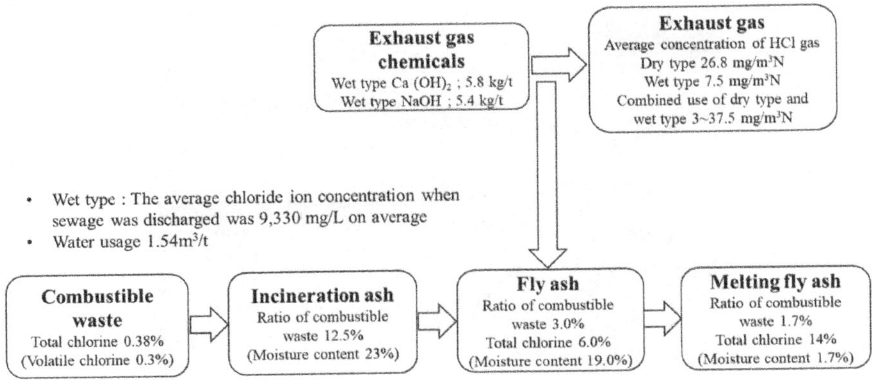

Fig. 2.11 Actual state of incineration facilities (Higuchi S 2011)

2.2.3.2 Actual Status of Incineration Facilities

In order to grasp the actual state of fly ash and bottom ash discharged from the incinerators or status of by-product salt released into the air, questionnaires were sent to 111 incineration facilities which generate by-product salt (Tohoku region excluded). The survey was conducted by post and received responses from 78 facilities and the response rate was 70%. And we got the average data from 78 responses. The result is shown in Fig. 2.11. In incineration/melting treatment, the slaked lime used for the dry desalination treatment was 5.8 kg per ton of waste, and the amount of the caustic soda used in the wet desalination treatment was 5.4 kg per ton of waste. The average concentration of HCl gas in the dry type was 26.8 mg/m^{3}N, and in wet type was 7.5 mg/m^{3}N. The total chlorine content in the combustible waste was 0.38%, incineration ash incidence was 12.5, and 3.0% for fly ash. 1.7% for melting fly ash.

2.2.3.3 Quantity of Generation

Referring to (2) Actual status of incineration facilities, the amount of by-product salt generated from combustible waste was calculated. The amount of municipal waste discharged in 2013 (Heisei 25) is 42.29 million tons, of which the incineration amount of combustible waste is about 33.73 million t, including the incineration of process residues to about 35.15 million t, finally, 3.33 million t of incineration residues are landfilled. Combustible waste contains 149 thousand t of Cl. Among it, Cl discharged to sewage as wet type spray covers about 20%, 21,000 t. The amount of Cl contained in the incineration residue is presumed to be about 123,000 tons per year, and about 337 t of Cl per day is landfilled. Cl is generated from waste (for example, vinyl chloride and food waste, etc.), and from incineration ash (such as bottom ash and the like). Incineration ash (bottom ash), fly ash, melting fly ash, the spray is discharged as desalination residue from dedicated devices for removing acid

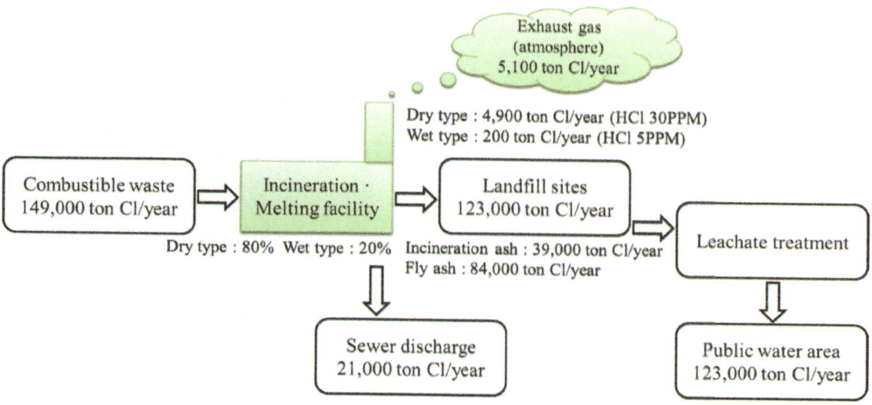

Fig. 2.12 By-product salt production from incineration facilities (2013) (Higuchi S 2011)

gas after two-stage bag filter. On the other hand, Cl discharged into the air as exhaust gas is estimated to be 5100 tons/year. However, the amount is decreasing year by year due to strengthening regulations on exhaust gas by municipalities. In the opposite, it is inferred that the amount of fly ash landfilled Cl increases. The landfilled Cl is discharged to the public water area as leachate, and a part is collected and stored as desalted concentrated water. The material balance of Cl caused by incineration of waste is shown in Fig. 2.12.

2.2.3.4 Desalination Treatment and By-Product Salt of Leachate

Leachate treatment facilities incorporating the desalination process are gradually increasing. According to the results of the questionnaire conducted by the authors (In November 2010, we sent questionnaires to 27 facilities (2%) producing by-product salt out of 1343 leachate treatment facilities in the whole country, responses were received from 16 facilities, the response rate was 59%. However, 7 of the 16 facilities had no by-product salt generation facilities running due to stop or abolishment of desalination equipment). Figure 2.13 shows the proportion of the facilities by type of landfill sites. Figure 2.14 shows the ratio of facilities by processed water discharge destinations, and Fig. 2.15 shows the ratio of facilities by desalination treatment methods.

The controlled landfill site was classified into open type and closed type, which covered 82% and 18% respectively. In the case of discharge destination, the percentage of "discharge" and "not discharge" are totally the same. Rivers and oceans were considered as discharge destinations, in this survey, all were discharged to rivers. In the case of "not discharge", it would be reused at landfill sites or the facilities installed at the landfill sites. It was clear that 80% was used as cooling water at the incineration facility and 20% was used for water sprinkling at landfill sites. Besides,

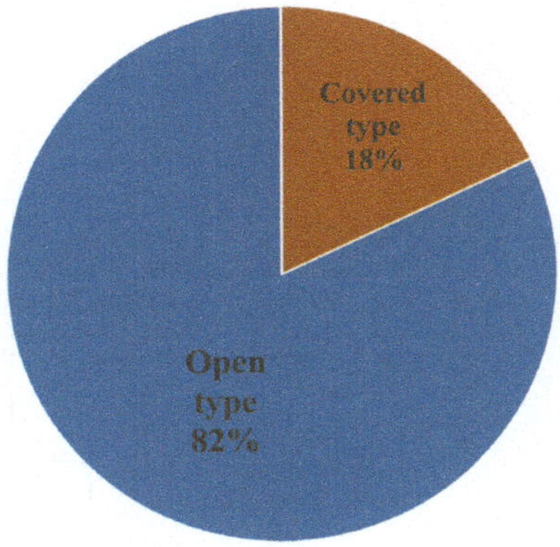

Fig. 2.13 Types of landfill site (Higuchi S 2011)

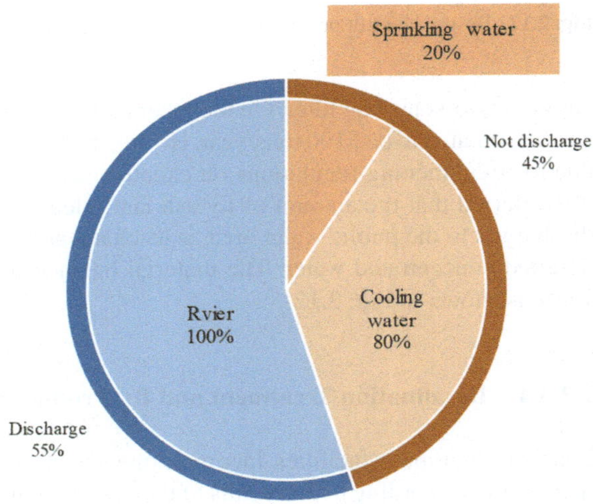

Fig. 2.14 Discharge destinations (Higuchi S 2011)

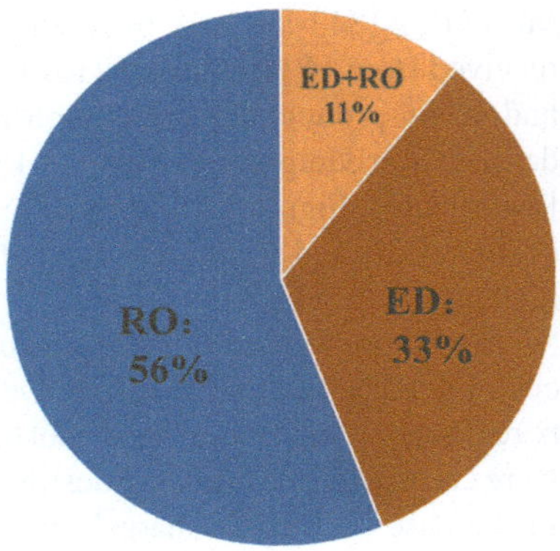

Fig. 2.15 Desalination treatment methods (Higuchi S 2011)

when it is classified by desalination treatment methods, RO method occupied more than half, ED method accounted for 30%, and the remaining 10% used RO and ED together to desalinize. Results on processing method and disposal cost of by-product salt are shown in Fig. 2.16.

As processing methods of by-product salt, there are landfill, reuse, and storage. Since the treatment as municipal and industrial waste is landfill, 87% in total and 832 t a year is landfilled as disposal.

The storage volume is about 18 t, 2%; which only 11% is being reused at present. It was confirmed that it was reused for weed inhibitors, anti-freezing agent, leather tanning industry. Regarding the storage of by-product salt, although the amount is small, 3 out of 16 facilities are conducting, and it seems that they themselves store it.

Fig. 2.16 Processing method and cost of by-product salt (Higuchi S 2011)

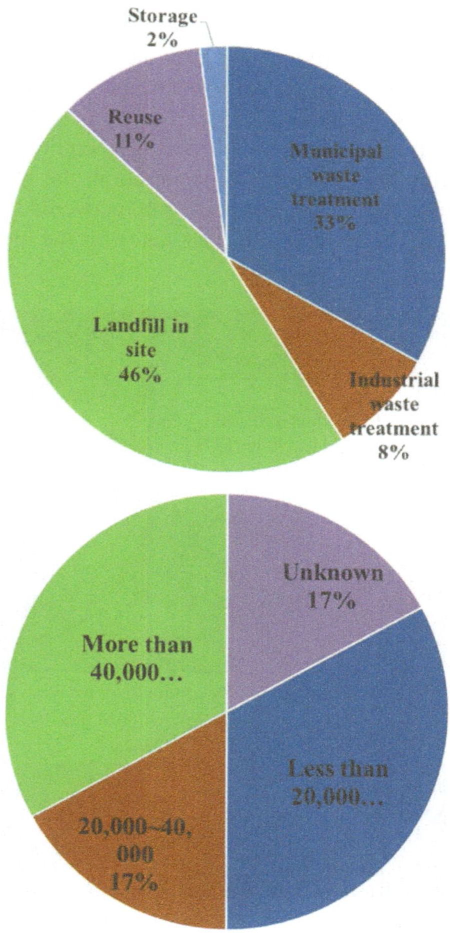

As for the disposal cost of by-product salt disposal, the difference is large from 15,000 yen per t to 63,000 yen. There is a big difference depending on the shipping cost, consignee, etc.

2.2.4 Composition of By-Product Salt

The by-product salt differs for each waste disposal process. Figure 2.17 shows the by-product salt. Table 2.4 shows the composition of by-product salt by different sources. When comparing the compositions of by-product salt by different sources and natural salt, the former has higher contents of K and Ca than the latter. K is estimated to be contained in the fly ash or bottom ash after being concentrated at the time of incineration since K is large in the waste of the body of animals and plants.

The main cause of more Ca is the use of $Ca(OH)_2$ for exhaust gas treatment, but for by-product salt after leachate desalination treatment, since Ca separation is performed as pretreatment in the case of electrodialysis membrane (ED), there is little Ca in it.

In the case of reverse osmosis membrane (RO), since it is often carried out directly without performing Ca separation, the by-product salt contains much Ca. On the other hand, the by-product salt recovered from the facility using sodium bicarbonate ($NaHCO_3$: sodium bicarbonate) for the incineration gas treatment contains less Ca

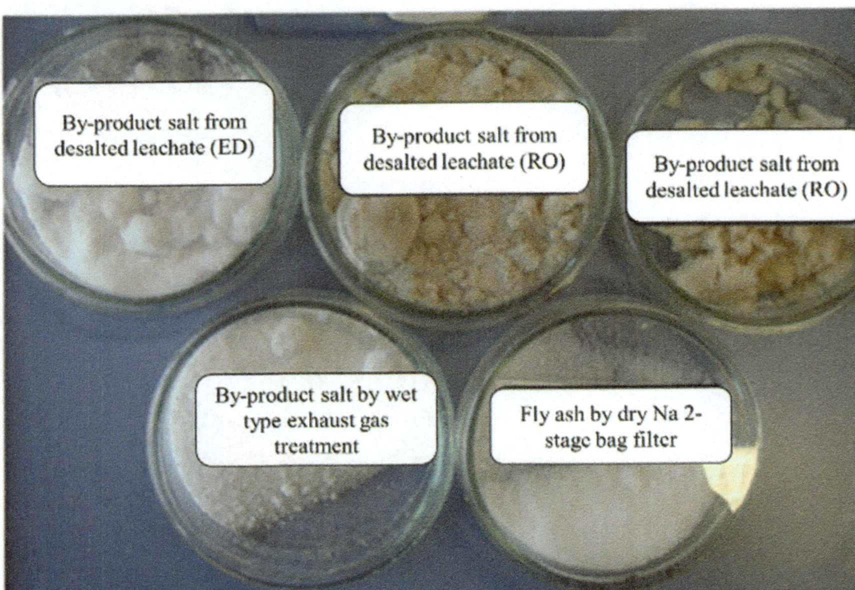

Fig. 2.17 Appearances of by-product salt by source (Higuchi S 2011)

Table 2.4 Compositions of by-product salt by source (Higuchi S 2011)

Item	Electrodialysis membrane	Reverse osmosis membrane	Wet type exhaust gas treatment	Dry type sodium salt	Natural salt	Testing methods
Moisture	9.8 wt%	10.2 wt%	1.2 wt%	0.9 wt%	–	–
Ca	475 mg/kg	61,000 mg/kg	3900 mg/kg	244 mg/kg	330 mg/kg	Standard addition method
Mg	3400 mg/kg	654 mg/kg	12 mg/kg	500 mg/kg	180 mg/kg	ICP atomic emission spectroscopy
Na	30.4 wt%	18.4 wt%	39.2 wt%	40.5 wt%	39 wt%	
K	5.4 wt%	12.1 wt%	0.05 wt%	0.02 wt%	0.024 wt%	
Mn	Below 2 mg/kg	Below 2 mg/kg	Below 2 mg/kg	Below 2 mg/kg	0.5 mg/kg	
Al	Below 5 mg/kg	7 mg/kg	5 mg/kg	Below 2 mg/kg	0.7 mg/kg	
Si	53 mg/kg	6 mg/kg	9 mg/kg	Below 5 mg/kg	–	
I	–	9 mg/kg	132 mg/kg	Below 5 mg/kg	–	
Cl	46.9 wt%	47 wt%	51.9 wt%	60.4 wt%	60 wt%	Automatic titration
SO_4^{2-}	41 g/kg	16 g/kg	99 g/kg	0.42 g/kg	0.88 g/kg	Ion chromato
NO^{3-}	1.3 g/kg	1.2 g/kg	0.22 g/kg	Below 0.05 g/kg	–	
Hg	Below 0.01 mg/kg	Below 0.01 mg/kg	Below 0.01 mg/kg	Below 0.01 mg/kg	–	Hydride generation gold amalgam
						Nondispersive two-beam bundle cold atomic absorption method

because the agent is Na type. It is necessary to reuse by-product salt according to the sources and compositions.

2.2.5 *Disposal and Recycling of By-Product Salt*

According to the questionnaires above, by-product salt is disposed of as industrial waste, storage, and reuse such as weed inhibitors, anti-freezing agent, leather tanning, etc. However, there are few cases.

One method of by-product salt recycling is to use it in the soda industry, but soda industrial products are used in a wide range from chemistry to food and the imported salt cost is low, so they are reluctant to use by-product salt. The by-product salt in landfill sites is in the form of calcium chloride, and generally, calcium removal equipment is installed in the leachate treatment facility, so for Ca, Mg, Na, etc., there is little difference between natural and by-product salt, but a big difference in the content of K and SO_4.

Furthermore, it can be said that the presence of heavy metals such as Pb, Cr, Cd or trace dioxins, is the feature of by-product salt from landfill sites. On the other hand, NaCl recovered by the wet type exhaust gas is closer in composition to the natural salt but often contains more I and SO_4.

Considering the possible use of by-product salt, in addition to the soda industry, at present, it can be used as road anti-freezing agent, aluminum refining flux agent, a sterilizing agent with sodium hypochlorite (NaClO.)

2.2.5.1 Use for the Soda Industry

The Soda Industry is the place where major salt is used in Japan. In the use of by-product salt in the soda industry, particularly in the electrolytic soda industry, it is taken up in the form of solidified salt and then electrolysis of the water-soluble salt to make products. If by-product salt can be applied to the existing production line, it is possible to sufficiently consume the production amount. However, in the existing electrolytic soda system, there is a possibility that the amount of generated sludge will increase due to contamination of Mg, Ca, SO_4 or heavy metal raw materials, and even the system itself will not function due to blockage, etc. So, it is necessary to improve the quality of advanced by-product salt. In addition, there are several methods for electrolysis, so it is also necessary to consider a method that is advantageous to the properties of by-product salt. When considering the electrolytic soda system (eco alkali) using only by-product salt as raw material, it is necessary to secure regenerated salt that is stable both quantitatively and qualitatively. Besides, it is limited to uses that do not enter the food system, such as a neutralizing agent or BDF catalyst.

2.2.5.2 Use for Aluminum Refining

In order to increase the purity of molten aluminum, by-product salt is added. However, the demand in Japan is low, it is difficult to make it as a universal salt reuse system.

2.2.5.3 Use for the Leather Industry

At an early stage of processing leather, a large amount of highly-concentrated salt water is used. Although there is little demand, the reuse system may be possible to be established depending on geographical conditions.

2.2.5.4 Diversion to Road Anti-freezing Agent

Although it is difficult for the anti-freezing agent market to consume all the regenerated salt production, it is an effective element of establishment of by-product salt utilization system since there is a universal market in various parts of the country including northern Japan. As a point to be noted, when sprayed outdoors, trace metals in the by-product salt may affect the vegetation around the road, so the quality that fully considering environmental impact is required.

2.2.5.5 Use of Water Treatment Sterilizing Agent

Sodium hypochlorite (NaClO) is produced by a non-diaphragm electrolytic method or the like and used as a sterilizing agent in a sewage terminal treatment plant or the like. There is also sewage terminal treatment plant that uses NaClO produced from seawater, which has the lowest risk both technically and environmentally, but the disadvantage is the amount injected as a sterilizing agent increases because of low production concentration.

It can be said that it is the most feasible usage to remove the administrative barriers of waste field and sewerage field where by-product salt occurs. Based on these facts, we will describe the use of by-product salts for road anti-freezing agent and water treatment sterilizing agent, which are the most realistic at present.

① Road anti-freezing agent

A. Quality standard

Road anti-freezing agent is used as $CaCl_2$, NaCl, $MgCl_2$ or mixtures thereof in snowy areas. Quality is determined by the road administrator. Table 2.5 lists chloride anti-freezing specifications made by the Ministry of Land, Infrastructure, Transport, and Tourism (MLIT).

Based on these, Salt Recycle System Association (SR) sets the compositional standards for using by-product salt as road anti-freezing agent as shown in Table 2.6.

B. Biotoxicity

As the antifreezing agent spreads by-product salt in the general environment, when it is diffused by precipitation, there is concern about salt damage and the effect on and living things. As shown in Table 2.5, MLIT provides standard values for acute oral toxicity test of mouse, eye irritation test of rabbit, skin irritation test of rabbit, acute toxicity test of Japanese medaka. The results of the toxicity tests conducted by the author using rats, medaka, plants (Komatsuna, a leafy green vegetable) on different by-product salts of different sources shown in Fig. 2.17 were as follows.

- Rat: In acute oral toxicity test with rats, no death or other abnormalities were found in all by-product salt, and no toxicity was observed (Fig. 2.18).

Table 2.5 Composition of by-product salt by source (Study Group on Salt Recycling System 2004)

1. The quality standard of chloride road anti-freezing agent

(1) Composition

- Component: magnesium chloride and sodium chloride. A compound or mixture of calcium chloride and sodium chloride. (Anti-corrosion agent is contained in the case of anti-corrosion)
- Composition: magnesium chloride and sodium chloride. A compound or mixture of calcium chloride and sodium chloride. The mixing rate is 4.6~6.4 (fixed ratio). (Submission of ingredient list)
- Particle size: the residual weight left on 5 mm sieve is less than 2%, the residual weight passed through 5 mm sieve and left on 1.2 mm sieve is more than 90%, the passed weight on 0.6 mm sieve is less than 2%. The standard of sample is 250 g.
- pH: 7~10. (10% of aqueous solutions)
- Volume of insoluble: less than 1%.

(2) Safety

- The component (5WN% of aqueous solutions) in chloride road anti-freezing agent is within the standard value of water pollution prevention law.
- Values of chloride road anti-freezing agent meet the standard value in the following test.

Acute oral toxicity test on mice	LD50 > 2,000 mg/kg
Eye irritation test on rabbits	Below middle degree
Skin irritation test on rabbits	Below middle degree
Acute toxicity test on Himedaka	LD50 > 2,000 mg/L

- Submission of a product data sheet.

(3) Corrosion resistance test (Dried repeat test, including 3% of aqueous solutions, sodium chloride)

- Conduct a corrosion resistance test in Industrial Research Institute of Hokkaido.

(4) Submission of certificates

- Submit test lists, etc. (for test lists except for component list, it is valid as long as the components and standard do not change) to prove (1) and (2). The test method basically conforms to JIS, etc.

(others)

- In the case of workability and storage quality, prevent the generation of dust and odor that hinders the operation during installation of spraying machine.
- Prevent the changes in consolidation and fusion, etc. that hinder the operation.

Table 2.6 Proposed quality standards for by-product salt as road anti-freezing agent (SR) (Study Group on Salt Recycling System 2004)

	Essential point	Item	Standard value	Note
Recommended quality	Salt content	Na, Cl	Above 75%	
		Na, K, Ca, Cl	Above 90%	
	Insoluble content	Insoluble content	Below 1%	
	Permissible limit of toxic substance in elution test	Cadmium and its compounds	Cadmium is below 0.1 mg/L	
		Cyanide compound	Cyanide is below 1 mg/L	
		Organic phosphorous compound	Below 1 mg/L	
		Lead and its compounds	Lead is below 0.1 mg/L	
		Hexavalent chromium and its compounds	Hexavalent chromium is below 0.5 mg/L	
		Arsenic and its compounds	Arsenic is below 0.1 mg/L	
		Mercury and alkyl mercury and other mercury compounds	Mercury is below 0.005 mg/L	
		Alkyl mercury	Not detected	
		PCB	Below 0.003 mg/L	
		Thiuram	Below 0.06 mg/L	
		Simazine	Below 0.03 mg/L	
		Thiobencarb	Below 0.2 mg/L	
		Selenium and its compounds	Selenium is below 0.1 mg/L	
Reference quality		Boron and its compounds	Boron is below 10 mg/L	
		Fluorine and its compounds	Fluorine is below 8 mg/L	
		Ammonia, ammonium compound, nitrous acid compound and nitric acid compound	Ammonia nitrogen multiplied by 0.4, total amount of nitrite nitrogen and nitrate nitrogen is below 100 mg/L	
		COD	Below 160 mg/L	
	Toxic substance	Dioxins	Below 5 pg-TEQ/g	
	Others	Moisture content	Below 5%	

(continued)

Table 2.6 (continued)

	Essential point	Item	Standard value	Note
		Particle size, consolidation state	No hindrance to the operation of scattering snow and ice	

Notes (1) Salt content is set according to *Testing Method for Salt(second edition)* published by The Salt Industry Center of Japan in Jan 2002
(2) Moisture content is set according to loss on drying test in *Testing Method for Salt (second edition)* published by The Salt Industry Center of Japan in Jan 2002
(3) The permissible limits of toxic substance in elution test is based on the method for measuring toxic substance specified in 10% concentration aqueous solution by Water Pollution Control Law
(4) For reference quality, appropriate settings should be made in accordance with the application method, etc.

Fig. 2.18 Toxicity test by rat (Higuchi S 2011)

- Medaka: In the fish acute toxicity test, there were differences in toxicity to medaka with 5 types of by-product salt, and the order of toxicity was leachate ED concentrated by-product salt (0.39%) > leachate RO concentrated by-product salt (0.84%) > leachate RO concentrated by-product salt-2 (1.28%) > dry type Na fly ash by-product salt (1.81%) >wet type gas cleaning by-product salt (2.13%).

In order to investigate the differences from commercial salt (NaCl), a simple exposure test of NaCl was carried out. As a result, the LC50 of NaCl was 1.6–1.7% (Probit method cannot calculate). Therefore, it was shown that wet gas cleaning by-product salt and dry type Na fly ash by-product salt are less toxic than NaCl, and leachate by-product salt is more toxic than NaCl.

Besides, it is said that freshwater fish cannot adjust osmotic pressure in dense salt concentration like seawater, which leads to death. The electric conductivity (EC value) of each by-product salt is almost the same value at the same concentration, and the EC value is greatly different (salt concentration differs) in each LC50. For by-product salt, death is observed even at low salt concentration, it is inferred that differences in each LC50 are not due to a malfunction of osmotic pressure adjustment but caused by some toxicity difference.

- In the early life stage toxicity test of fish as well, the toxicity of the medaka eggs and larvae just after hatching was different by by-product salt. The influence on hatching, that is, the effect on eggs can be expressed only within the range of the LC50 values of leachate ED concentrated by-product salt and leachate RO concentrated by-product salt in the range of 0.8–1.6% and 1.6–3.2%. The order of toxicity is leachate ED concentrated by-product salt > Leachate RO concentrated by-product salt-2 > leachate RO concentrated by-product salt-1 > dry type Na fly ash by-product salt> wet type gas cleaning by-product salt.
- Influence on survival days—or the effect on hatched larvae is considered to be toxic in the order below: leachate ED concentrated by-product salt > leachate RO concentrated by-product salt-1 > leachate RO concentrated by-product salt-2 > dry type Na fly ash >wet type gas cleaning by-product salt.

Wet gas cleaning by-product salt is less toxic than NaCl; dry type Na fly ash by-product salt and leachate RO concentrated by-product salt-2 are of the same toxicity as NaCl; leachate RO concentrated by-product salt-1 and leachate RO concentrated-2 are more toxic than NaCl (Fig. 2.19).

Fig. 2.19 Toxicity test by medaka (Higuchi S 2011)

- Vegetation: seeds of Komatsuna were planted in Bayer pot and water with different salt concentrations including by-product salt was observed. In cultivation tests concerning germination and growth of seedlings, somewhat differences were found in influence on germination growth. Regarding germination, the toxic order is leachate ED concentrated by-product salt from LC50 ≈ dry type Na fly ash by-product salt > wet type gas cleaning by-product salt>leachate RO concentrated by-product salt > leachate RO concentrated by-product salt-2. Also, leachate RO concentrated by-product salt-2 and dry type Na fly ash by-product salt had the same effect, while wet type gas cleaning by-product salt, leachate RO concentrated by-product salt-1 and the group of leachate RO concentrated by-product salt-2 were weaker than leachate ED concentrated by-product salt, but the same degree of influence was seen. At the end of the growth survey test, the influence order was dry type Na fly ash by-product salt > leachate ED concentrated by-product salt > leachate water RO concentrated by-product salt-1 > leachate water RO concentrated by-product salt-2>wet type gas cleaning by-product salt. The influence of wet type gas cleaning by-product salt on the growth of plants was less than half as compared with dry type Na fly ash by-product salt, leachate ED concentrated by-product salt and SD which were relatively toxic. Regarding salt damage of plants, EC of water-saturated soil where growth is started to be inhibited is set to 0.9–1.5 mS/cm for weak salt tolerance (Saga Agricultural Technology Support Center, Agriculture, Forestry and Fisheries, Commerce and Industry headquarters, Saga Prefecture). In this test, soil EC was 1.5 mS/cm or more in the 1.5% concentration group for each by-product salt at the end. Therefore, from the beginning of the test, it is considered that growth inhibition occurred due to high EC in the high concentration zone. In the phytotoxicity test using Neubaiel as in this test, because it uses a closed pot that does not have a drainage port, there is a decline in EC due to the growth of plants, but the decrease due to irrigation does not occur and the influence of growth inhibition is considered to be sustained. In the case of general arable land, it is inferred that EC value will decrease due to leaching caused by precipitation, etc. And the influence of inhibition will decrease (Fig. 2.20).

As mentioned above, no acute oral toxicity was observed in all by-product salts, but some influences were observed on fish and land plants, and the degree of influence varied depending on by-product salt. In this test, the influence of by-product gas cleaning by-product salt on both fish and plants is small, but the leachate-based concentrated by-product salt showed slightly stronger toxicity than incineration system by-product salt.

Dry type Na fly ash by-product salt has a larger influence on fish than the wet type gas washing by-product salt but was comparable or smaller than leachate RO concentrated by-product salt-2, whereas for plants was obviously more toxic than concentrated by-product salt-2. As described above, the degree of influence on living things by by-product salt was different, and furthermore, it was recognized that the degree of toxicity to animals and plants may be different. For these facts, even though the by-product salt is used as it is, which meets the soil environmental standards, and

Fig. 2.20 Toxicity test by vegetation (Higuchi S 2011)

may not cause problems, but it is desirable to carry out a biotoxicity test and purify it according to the place of use.

③ Water treatment sterilizing agent

As the by-product salt approaches the composition close to natural salt when purified, it can be used as soda industrial salt. However, since soda industrial products are used in a wide variety of fields, especially when it is assumed to be used in the food field, as the source of by-product salt reuse is accompanied by waste disposal, no matter how purified it is, there are things that directly touch people or things that are likely to enter the body, which is emotionally resistant to acceptance. Furthermore, although the amount of by-product salt generated differs depending on the size of the incineration facility and leachate treatment facility at the landfill site, it is not economic as the amount generated from one place is small, while the transportation expenses to the soda factory are high. Therefore, there is a method to generate NaClO that can be recycled on site and used as water treatment sterilizing agents.

A. Eco-hypochlorite

NaClO can be produced from salt. As shown in Table 2.4, when comparing natural salt and by-product salt, more Ca, K, etc. are contained in the by-product salt. As for Ca, it can be separated by pretreatment, but economic separation method of K has not been established now. For that reason, the generated NaClO contains potassium hypochlorite (KClO). Therefore, it cannot be distributed as an ordinary product according to the JIS standard. However, since it has sufficient sterilization effect, it is called Eco-hypochlorite (Hereafter, Eco-hypo). The effective chlorine amount of Eco-hypo varies depending on the manufacturing methods, and it is 0.1–6.0%, which is lower than that commercial hypochlorite of 12%.

B. A production method of Eco-hypo

Eco-hypo can be produced by non-diaphragm electrolysis or diaphragm electrolysis. Eco-hypo generating apparatus using seawater non-diaphragm electrolysis has many achievements in ship field, and also introduced in sewage treatment plants. In addition, the apparatus is also used for sterilization by dissolving natural salt to produce NaClO in water treatment plants and pools, which is technically established. As for by-product salt, the authors have conducted experiments using by-product salts of various sources and are almost established (Higuchi 2011; Takahi et al. 2015). Figure 2.21 shows a schematic diagram of the apparatus.

When passing salt water with by-product salt dissolved, Cl_2 is generated on the anode side and NaOH on the cathode side, and Eco-hypo is produced.

Figure 2.22 shows the outline of diaphragm electrolysis. When adding salt water from the anode side and NaOH from the cathode side, Cl, NaOH, H are generated from each reaction. At this time, since Na^+ moves from the anode side through the ion-exchange membrane to the cathode side, high concentration NaOH is generated.

Cl and NaOH are reacted outside the electrolytic cell, which becomes a mechanism to generate eco-hypo. Since high concentration eco-hypo is produced, it can be used as a substitute for commercial NaClO. For by-product salt, the authors succeeded in generating 5–6% eco-hypo by experiment using by-product salt of various sources, but due to voltage rise caused by scale formation on the ion exchange membrane, it stopped for a short time, has not reached the practical use level. Besides, since the ion exchange membrane used for diaphragm is easily deteriorated and expensive, it cannot be used unless impurities are sufficiently removed.

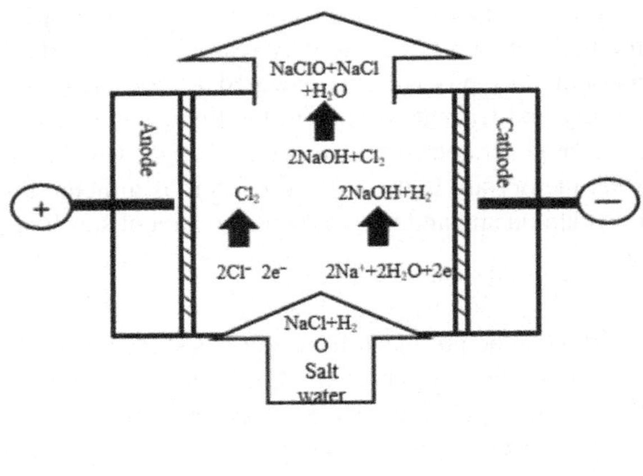

Fig. 2.21 Schematic diagram of non-diaphragm electrolysis method (Higuchi S 2011)

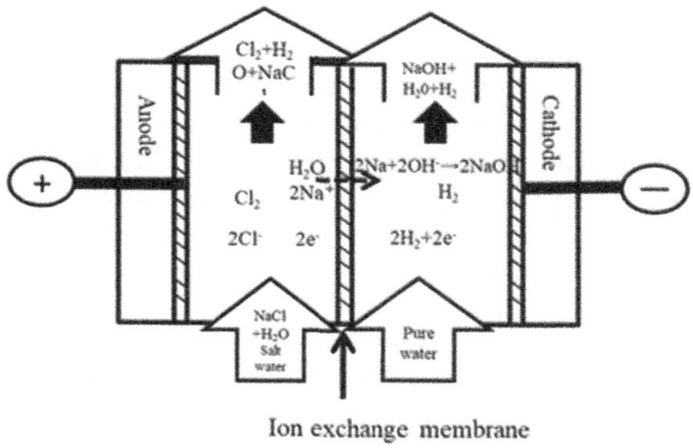

Fig. 2.22 Schematic diagram of diaphragm electrolysis (Higuchi S 2011)

C. Example of eco-hypo generation (salt water production from dried by-product salt)

Table 2.7 shows the case of concentration salt recovered by electrodialysis membrane (ED) and reverse osmosis membrane (RO) as by-product salt accompanying leachate desalination treatment at landfill sites, and eco-hypo produced by non-diaphragm electrolytic using dry by-product salt recovered by exhaust gas treatment of incineration facility,dry sodium method 2-stage bag filter.

In the by-product salt accompanying leachate desalination by ED and RO, salt water of 3% concentration was made from the dry by-product salt as the experimental condition, then eco-hypo was generated by the non-diaphragm electrolysis method. As a result, the effective chlorine concentration was at about 5000 ppm at

Table 2.7 Ecology secondary generation by non-diaphragm electrolytic method (Higuchi S 2011) fixed

By-product salt occurrence place	By-product salt	Salt water concentration (wt%)	Ca concentration (mg/L)	Effective chlorine concentration (mg/L)	Voltage (V)
Leachate desalting concentration dried salt	ED (Electrodialysis membrane)	3	3.2	6500–6900	4.1–4.4
Leachate desalting concentration dried salt	RO (Reverse osmosis membrane)	3	25.9	3700–5200	4.1–4.6
Incineration facility	$NaHCO_3$	5	15	3300–4900	3.3– 3.6
Incineration facility	$NaHCO_3$	10	30.5	4400–4900	3.7–4.0

the maximum, and high concentration eco-hypo which exceeded the expected 1000–2000 ppm was obtained. At the Cl^- concentration of 3%, it was confirmed that when the Ca^{2+} concentration was set to 100 mg/L or less, the electrolysis voltage did not rise, and continuous operation was possible. In the by-product salt produced by dry type exhaust gas treatment by sodium, when the Cl^- concentration was set to 5%, the eco-hypo with a maximum effective chlorine concentration of 6.900 ppm was obtained, when the electrolysis voltage did not rise, and continuous operation was possible. In the case where the Cl^- concentration was set to 10%, eco-hypo of about 4500 ppm was obtained.

D. Example of eco-hypo generation (Generated from desalted concentrate)

In the desalination treatment step in the leachate treatment, the concentrate is dried by a drum dryer or the like, and then stored or consigned, but in the processing step, drying concentrated water consumes much heavy oil and electricity. In the desalination treatment step in the leachate treatment, the concentrate is dried by a drum dryer or the like, and then stored or consigned, but in the processing step, drying concentrated water consumes much heavy oil and electricity. Therefore, the experiment was conducted in which the concentrate was applied directly to a non-diaphragm electrolytic apparatus. As by-product salt by RO was originally RO concentrate with a concentration of 3 to 5%, and to obtain the same result as Table 2.7 as expected, using non-diaphragm electrolytic method with concentrate from high salt concentrate by ED, effective Cl^- 9000 ppm and eco-hypo with high concentration could be obtained. At that time, Ca^{2+} was 79 mg/L, which allowed long-term continuous operation for one month (Table 2.8).

E. Self-oxidation of Eco-hypo

When a sterilizing agent is used in the process of eco-hypo production, where to set the production facility depends on the distance between the generation place of by-product salt and the use place of eco-hypo. The effective chlorine content of eco-hypo is 0.3–0.9%, which is lower than 12% of commercial hypochlorite. Therefore, when eco-hypo is generated at the place of production, the cost for transportation increases. Because when the use place is far away, it is needed to dry, solidify and transport it, it may be more efficient to generate it at the use place. Figure 2.23 shows the measurement of the self-oxidation rate by leaving the produced eco-hypo.

Table 2.8 Eco-hypo by direct electrolysis of ED concentrate (Tsuboi T et al. 2015)

By-product salt occurrence place	By-product salt	Salt water concentration (wt%)	Ca^{2+} (mg/l)	Effective chlorine concentration (mg/l)	Voltage (V)
Leachate desalted concentrate	ED (Electrodialysis membrane)	8.2	79	8900–9100	4.1–4.4

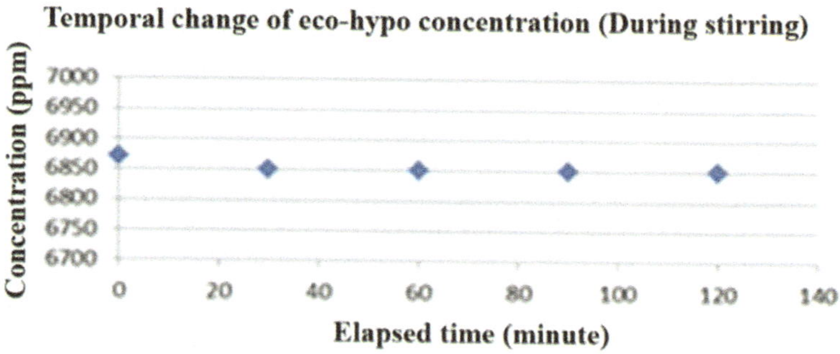

Fig. 2.23 Concentration change for one week (Leave in nature state) (Tsuboi T et al. 2015)

Next, self-oxidation prevention experiments were carried out by injecting alkali and raising the pH in order to minimize auto oxidation of eco-hypo generated on the assumption that the place of ecological sub-production and the place of use are different. Figure 2.23 shows the concentration change due to natural leaving.

Figure 2.23 Concentration change for one week (Leave in nature state) (Tsuboi T et al. 2015)

Eco-hypo gradually self-oxidizes even at normal temperature, but the oxidation process is accelerated by temperature rise and ultraviolet rays due to sunlight.

In this experiment, the influence of sunlight on concentration and pH of eco-hypo was verified by keeping generated eco-hypo outdoors for one week. The condition was filling the eco-hypo in a 250 ml transparent container with a lid and leaving it at the place where sunlight hit (Fig. 2.23).

As a result, it was found that both effective chlorine and pH decreased significantly due to temperature rise and ultraviolet rays caused by sunlight. The effective chlorine concentration of eco-hypo on the 7th day of the experiment was 1032 ppm, which was about 6000 ppm lower than the concentration on the first day. Also, during this experiment, the average temperature in the week was 23 °C, and the weather was temporarily cloudy but basically sunny. Next, NaOH was added to prevent self-oxidation by pH adjustment. Agitation was carried out at 340 rpm for 2 h assuming vibration agitation during transportation of Eco-hypo. The result is shown in Fig. 2.24. From Fig. 2.24, it was found that even under stirring for 2 h under the condition of pH 12, almost no self-oxidation was observed, and by adjusting pH the effect of transportation of several tens of kilometers could be resolved.

2.2.5.6 The Sterilization Effect of Eco-hypo

In order to confirm the sterilization effect of Eco-hypo, quality analysis and sterilization effect experiment was conducted. Eco-hypo used in the experiment for the sterilization effect experiment was from by-product salt produced by ED.

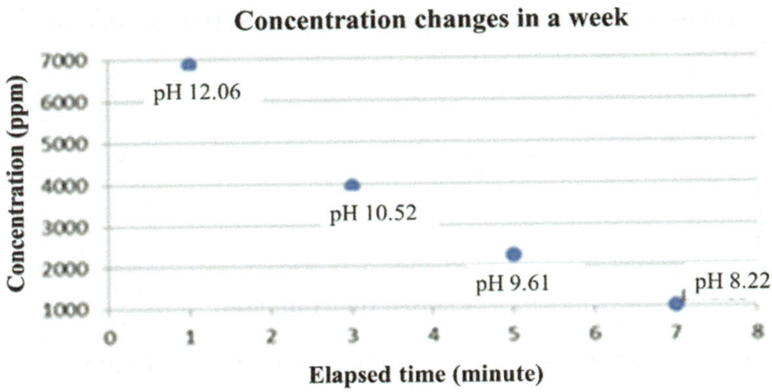

Fig. 2.24 Change of effective chlorine by time (Tsuboi T et al. 2015)

Table 2.9 Quality of eco-hypo (Higuchi S 2011; Ushikoshi et al. 2013)

Item	Unit	Eco-hypo	Commercial hypo
Effective chlorine	W/V%	0.378 (0.360)	0.376 (0.373)
Appearance	W/V%	Faint pale yellow transparent liquid	Faint pale yellow transparent liquid
Density	–	1.048	1.008
Free alkali	W/W%	<0.04	<0.04
Bromic acid	mg/kg	200	<3
Chloric acid	mg/kg	<300	1100
Sodium chloride	W/V%	6.31 (6.01)	0.25 (0.25)

① Quality analysis of Eco-hypo

Table 2.9 shows the quality analysis result of Eco-hypo. It was found that this Eco-hypo contains high concentrations of bromic acid and sodium chloride and does not satisfy the quality of the third grade of product II of the Japan Water Works Association Standards and cannot be treated as a reagent for tap water. In addition, chloric acid was detected in the hypo on the market, but this is because ① about 3 months have passed since purchase ② because the stock solution of effective chlorine 8.206 w/v% was diluted, and then hypochlorous acid was decomposed to chloric acid.

② Impurity (heavy metals) analysis of Eco-hypo

It is thought eco-hypo may contain impurities. Here, we assume the sterilizing agent at processing facilities such as Industrial wastewater, sewage drainage, landfill site leachate, etc. as an application of Eco-hypo, so heavy metals were picked up and analyzed based on "JIS K 0102 (2008) factory wastewater test method." The result is shown in Table 2.10. Eco-hypo was equal to commercial hypo.

Table 2.10 Impurity analysis of eco-hypo (Lei et al. 2014; Tuboi T et al. 2015)

Analysis item		Eco-hypo	Commercial hypo
Cadmium	mg/kg	<0.5	<0.5
Lead	mg/kg	<0.5	<0.5
Hexavalent chromium	mg/kg	<1	<1
Arsenic	mg/kg	<0.5	<0.5
Mercury	mg/kg	<0.01	<0.01
Alkyl mercury	mg/kg	<0.01	<0.01
Selenium	mg/kg	<0.5	<0.5
Boron	mg/kg	<1	<1
Flurine	mg/kg	<1	<1
Cu	mg/kg	<1	<1
Zn	mg/kg	<1	<1
Soluble iron	mg/kg	<1	<1
Soluble manganese	mg/kg	<1	<1
Chromium	mg/kg	<1	<1

③ Effect of sterilizing agent

Three conditions shown in Table 2.11 were set for the test liquid. Prior to the experiment, a preliminary experiment was performed to calculate the Cl consumption of each test liquid.

Commercial hypo was gradually added to 5 ml test liquid and Cl consumption was estimated from the relationship between the amount of Cl added and the amount of residual chlorine (Fig. 2.25). From the result of the preliminary experiment, the amount of Cl added to 100 ml test liquid was determined so that the residual chlorine concentration after hypo addition was within the range of 0.1–1.0 mL in the experiment, 100 ml of each test liquid was added in a sterilization container, and eco-hypo or commercially available hypo was added to the test liquid step wisely. Eco-hypo or a commercially available hypo was added so that the residual chlorine concentration in the test liquid after the addition was within the range of 0.1–1.0 mL. Since the test liquid itself contained COD, the amount of Cl consumption in the test liquid was grasped in the preliminary experiment and then the addition amount

Table 2.11 Test liquid condition of sterilization effect (Lei et al. 2014; Tsuboi et al. 2015)

Type	Details of test solution
Test liquid A	Treated water of sewage treatment facility in City A before sterilization
Test liquid B	Treated water of sewage treatment facility in City B before sterilization
Test liquid C	Penetration water of private stable waste disposal site in City C

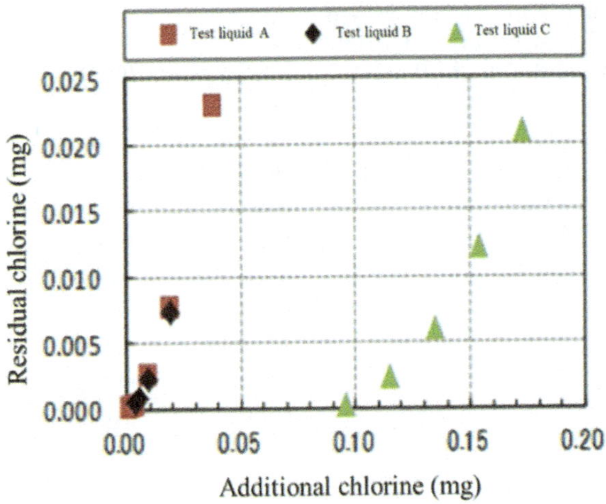

Fig. 2.25 Relationship between added Cl and residual chlorine (Lei et al. 2014; Tsuboi et al. 2015)

was determined. After gently stirring the test liquid with added hypo, analysis of the number of general bacteria and coliform bacilli was performed, and residual chlorine was measured with the remaining test liquid.

Experimental results for confirming the effect as a sterilizing agent are shown in Fig. 2.26. For samples added together with Eco hypo and Commercially Available hypo, the coliform bacilli were hardly detected, which was much lower than the average of 3000 cells/mL, which is the emission standard of coliform bacilli. The effect of eco-hypo as a sterilizing agent was almost equal to that of commercially available hypo.

2.2.5.7 Proposed By-Product Salt Standards for Eco-hypo Recycling

In case of generating eco-hypo with by-product salt by non-diaphragm electrolytic method, it is necessary to set the Ca^{2+} concentration of salt water in the electrolysis tank to 100 mg/L or less as one criterion. For items other than Ca^{2+}, the by-product salt standard proposed by SR is shown in Table 2.12.

2.2.5.8 Feasibility Study for Use as a Sterilizing Agent in Sewage Terminal Treatment Facility

It was assumed that by-product salt generated from the landfill site leachate possessed by the municipalities was used as a sterilizing agent of the sewage treatment plant possessed by the same municipality with the non-diaphragm electrolytic method. Feasibility study (hereinafter referred to as FS) was performed. The conceptual diagram is shown in Fig. 2.27.

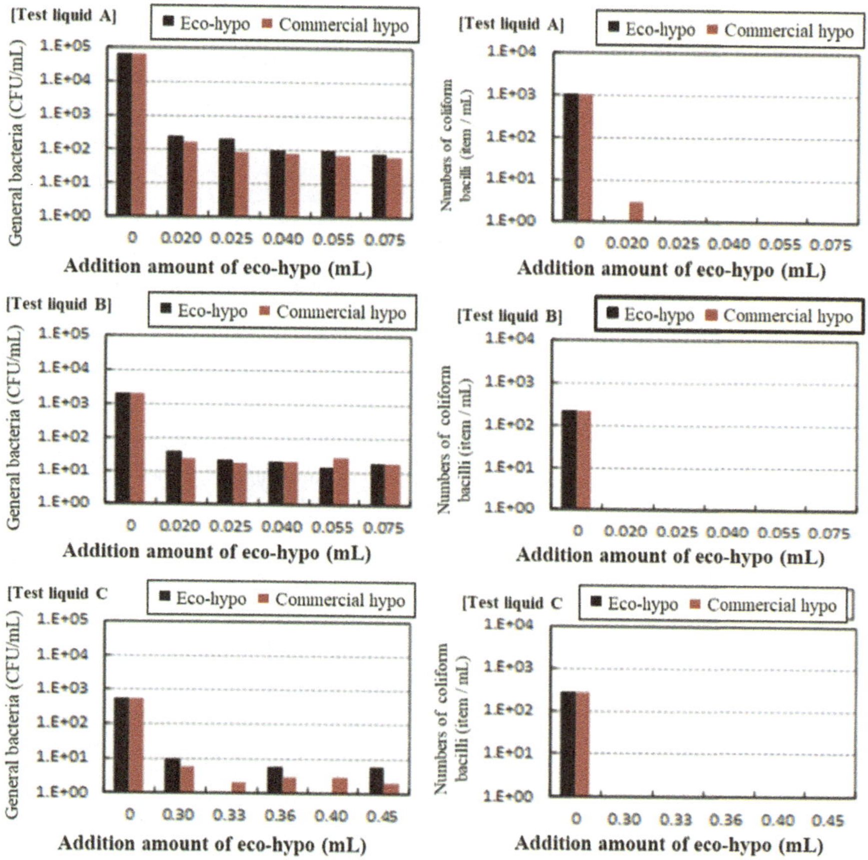

Fig. 2.26 Confirmation experiment results as a sterilizing agent of eco-hypo (Lei et al. 2014; Tsuboi et al. 2015)

Depending on the distance between the waste disposal facility of the by-product salt source and the sewage terminal treatment facility of the eco-hypo use, the installation site of eco-hypo production facility differs, so the by-product salt is also used as a concentrate as it is or transported as solid-type salt, and dissolved and used in the production equipment. Here, it was assumed that the concentrate was transported to a sewage terminal treatment plant with a tank truck to produce eco-hypo. Regarding economic efficiency by municipality size, municipalities with a population between 20,000 and 1 million are set as Cases I, II, III, IV, respectively, to 20,000 people, 100,000 people, 200,000 people, 1 million people. Table 2.13 shows the FS examination conditions.

The amount of by-product salt generated per ton of combustible waste was set to 6–7 kg, and the amount of combustible waste generated per person was set to 0.8 kg.

Table 2.12 Non-diaphragm electrolysis by-product salt standard proposal by SR (Study Group on Salt Recycling System 2004)

		Item	Standard value
Recommended quality	Salt content	Na, Cl, K	Above 85%
		Ca, Mg	Below 0.5%
		SO_4	Below 5.0%
	Heavy metals	Fe	Below 50 mg/kg
		Mn	Below 2 mg/kg
	Permissible limit of toxic substance in elution test	Cadmium and its compounds	Below 0.1 mg/l
		Cyanide compound	Below 1 mg/L
		Organic phosphorous compound	Below 1 mg/L
		Lead and its compounds	Below 0.1 mg/L
		Hexavalent chromium and its compounds	Below 0.5 mg/L
		Arsenic and its compounds	Below 0.2 mg/L
		Mercury and alkyl mercury and other mercury compounds	Below 0.005 mg/L
		Alkyl mercury	Not detected
		PCB	Below 0.003 mg/L
		Thiuram	Below 0.06 mg/L
		Simazine	Below 0.03 mg/L
		Thiobencarb	Below 0.2 mg/L
		Selenium and its compounds	Below 0.1 mg/L
Reference quality		Boron and its compounds	Below 10 mg/L
		Fluorine and its compounds	Below 8 mg/L
		Ammonia nitrogen * 0.4 + (nitrate nitrogen + nitrate nitrogen)	Below 100 mg/L
		COD	Below 160 mg/L
	Toxic substance	DXNs	Below 5 pg-TEQ/g
	Others	Moisture content	Below 5%
		Insoluble content	Below 1%

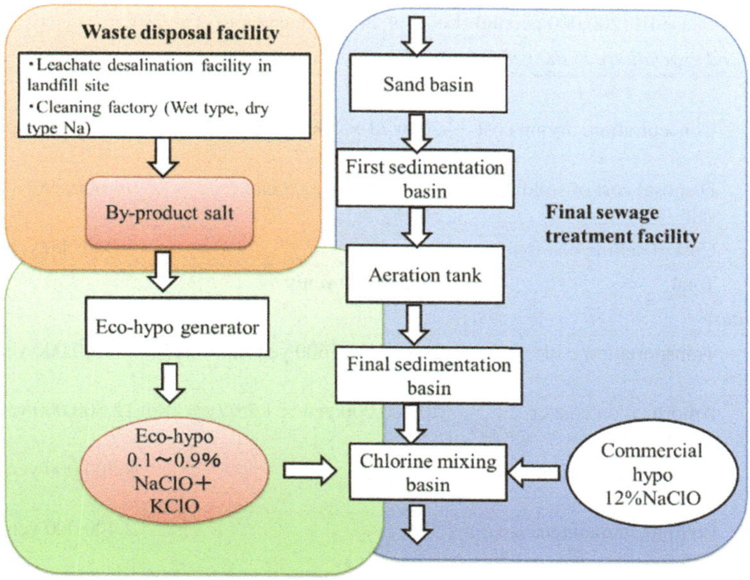

Fig. 2.27 Conceptual diagram of utilization of sterilizing agent of Eco-hypo

Table 2.13 Eco-hypo FS examination condition (Lei et al. 2014; Tsuboi et al. 2015)

Item	Unit	Value
Generation amount of by-product salt per ton of combustible waste	kg	6–7
Generation amount of combustible waste	kg/person/d	0.8
Electrolytic water concentration	%	5
Density of generated eco-hypo	%	0.3
Price of commercial hypo (10.5%)	%	40
Sale price of eco-hypo (hypo on the market × 0.6)	yen/kg	24
Disposal cost for drying solidified salt	yen/kg	40,000
Power expense	yen/kwh	17
Equipment redemption period	year	7

The concentration of desalted concentrate of leachate is set to 5%, the concentrated water is transported to a sewage treatment plant with a tank truck, and eco-hypo is generated and injected into sewage treatment water as sterilizing agents. The eco-hypo concentration was set at 0.3%. The sewer terminal treatment facility decided to collect the generated eco-hypo at a price of 60% of commercial hypo. Table 2.14 shows a calculation example for Case III (200,000 people). On the income side, it is possible to reduce the cost of dry solidification of leachate concentrated water at the concentration of 5% and eliminate the cost of disposing of industrial waste of

Table 2.14 Case III (200,000 people) (Lei et al. 2014; Matsumoto et al. 2013)

Income and expenditure in the generation side			
Income			
A	Concentration, drying cost	20 m^3/d × 5000 yen/m^3 × 365 d/y =	36,500,000 yen/y
B	Disposal cost of solidfied salt	1.0 t/d × 40,000 yen/t × 365d/y =	14,600,000 yen/y
C	Cost of selling eco-hypo		3,000,000 yen/y
	Total	54,100,000 yen/y	
Expenditure			
D	Transportation cost	20 m^3/d × 1600 yen/m^3 × 365 d/y =	11,700,000 yen/y
E	Amortization cost of equipment	70,000,000 yen × 1.25/7y =	12,500,000 yen/y
F	Power cost of facility operation		4,200,000 yen/y
G	Equipment maintenance cost		2,400,000 yen/y
	Total	30,800,000 yen/y	
Income and expenditure		Income–Expenditure	+23,300,000 yen/y
Sewerage income and expenditure			
Income			
H	Savings form the purchase of commercial hypo	60 kg/d/0.105 × 40 yen/kg × 365d/y × 36 =	+5,000,000 yen/y
Expenditure			
I	Purchase cost of eco-hypo	8,300,000 yen/y × 0.6 × 0.6 =	3,000,000 yen/y
Total of income and expenditure		Income–Expenditure	+2,000,000 yen/y
Overall income and expenditure			+25,300,000 yen/y

by-product salt. The collection fee of eco-hypo as income could be calculated. On the other hand, as expenditure, transportation cost of concentrated water, amortization cost of eco-hypo generation facility, power cost of facility operation and equipment maintenance cost is required. The examination result was shown in Table 2.15. In the case of a small scale of about 20,000 people, it is economically negative. It is clear that it is economically advantageous to generate and use eco-hypo at about 50,000 people scale or more. In addition, in the case of 1 million people, as more sewage terminal treatment plants are installed, the scale of each sewage treatment plant is expected to be 100,000–400,000 people. However, with regard to the economy, the actual system differs for each facility, so it is necessary to correct it according to the actual situation.

Table 2.15 Results of FS (Lei et al. 2014; Matsumoto et al. 2013)

Item	Case I	Case II	Case III	Case IV	Note
Municipality size	20,000 people	100,000 people	200,000 people	1 million people	
Generating quantity of by-product salt	0.1 t/d	0.5 t/d	1.0 t/d	5.0 t/d	
Volume of concentrated salt water	2 m^3/d	10 m^3/d	20 m^3/d	100 m^3/d	5% of solution
a. Cost of drying/concentration	5100	21,900	36,500	146,000	A
b. Disposal cost of solidfied salt	1500	7300	14,600	73,000	B
c. Transportation cost of concentrated salt water	1500	6600	11,700	51,100	D
d. Generation cost of eco-hypo	6400	12,300	19,100	55,600	E + F + G
e. Cost of selling eco-hypo	500	2500	5000	25,000	C
f. Savings from the purchase of eco-hypo	300	1700	3300	16,700	H
Income and expenditure in the generation side	−1000	+11,800	+23,300	+127,300	a + b + e − c − d
Sewerage income and expenditure	+200	+1000	+2000	+10,000	f
Overall income and expenditure	−800	+12,800	+25,300	+137,000	

Unit 1000 yen/y

2.2.6 *Generation and Use Example of Eco-hypo*

Matsuyama city, Ehime Prefecture, firstly in the world used the concentrated salt water produced by the leachate desalination treatment to generate eco-hypo by a non-diaphragm electrolytic method which was started to be used as a disinfectant at sewage terminal treatment facilities in the city. The brief introduction is as follows.

2.2.6.1 Overview

Yokotani Landfill Center of Matsuyama city is a landfill site with a capacity of 550,000 m^3 which began operation in April 2003 (Heisei 15), and accepts incombustible waste and incineration residues. Leachate was desalted by ED method. Concentrated salt was dried, solidified and subjected to external treatment. In order to expand the leachate treatment facility, it was decided to generate the eco-hypo unit and use it as a disinfectant at the sewer terminal treatment facility. Figure 2.28

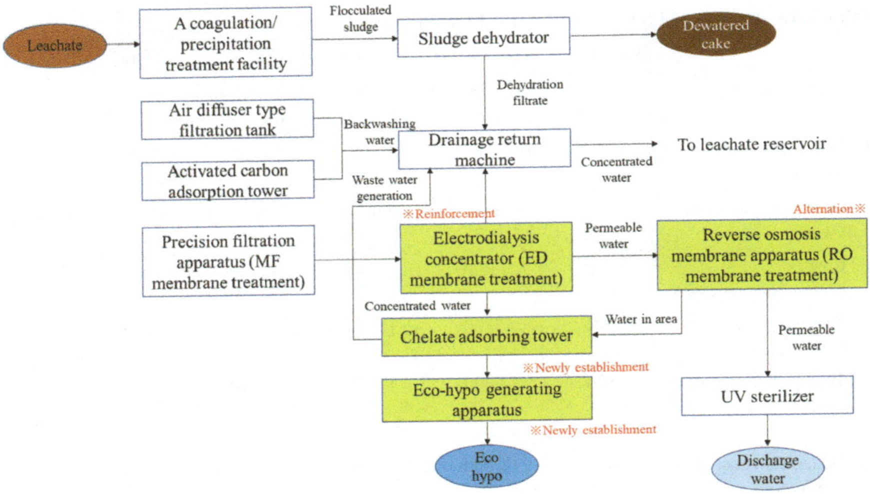

Fig. 2.28 Leachate treatment flow (Ito T 2017)

shows the leachate treatment flow. After conducting preliminary tests, actual machine tests, biological impact tests, etc., it started operation in April 2017. Conventionally, concentrated salt was dried and solidified in order to reduce cost, but eco-hypo was generated from concentrated water and dried, which reduced fuel cost of heavy oil, etc. Figure 2.29 shows the eco-hypo unit.

Fig. 2.29 Eco-hypo plant

2.2.6.2 Use of Eco-hypo

In the leachate treatment facility, when 115 m^3 per day of 13,000 mg/L chloride ion concentration is treated at the maximum operation, discharged water with chloride ion concentration of 200 mg/L is 95.7 m^3. When 108,000 mg/L of chloride ion concentration, 17.8 m^3 of eco-hypo with an effective chlorine concentration of 2000 mg/L can be generated.

After transporting this Eco-hypo to sewage treatment plants with tank lorries, it is stored in an eco-hypo storage tank and mixed with the existing commercial eco-hypo for use. While injecting a certain amount of eco-hypo, amount of existing commercial hypo is mixed against fluctuations in the flow rate of treated sewage water by control. The effective chlorine concentration of generated eco-hypo can be arbitrarily adjusted within the range of 1000–3000 mg/L by setting the current value.

2.2.6.3 Effect

With the introduction of eco-hypo generation unit, in addition to the reduction of the additional expansion and maintenance cost compared to the dry salt method, maintenance, and management expenses such as fuel costs and dry salt treatment costs used in the drying process is also reduced. Also, CO_2 emissions were also reduced by not using heavy oil in the drying process (Table 2.16). In addition, even in sewage terminal treatment plants, the amount of commercially available hypo purchase was reduced due to the use of eco-hypo.

Table 2.16 Eco-hypo economic effect (Matsuyama city) (ITO T 2017)

	Eco-hypo introducing apparatus	Conventional expansion	Reduction	Note
Maintenance cost	~900 million yen	~4 billion yen	~ 3.1 billion yen	※In addition to eco-hypo equipment, Maintenance cost includes the cost of leachate treatment facility expansion
Maintenance management cost	4600 yen/m^3	8700 yen/m^3	4100 yen/m^3	
Emission amount of CO_2	21 kg-CO_2/m^3	65 kg-CO_2/m^3	44 kg-CO_2/m^3	

2.2.7 *Eco-alkali and Eco-acid Recycling by Bipolar Membrane*

As a recycling method of by-product salt, there are achievements of using anti-freezing agent and disinfectant by eco-hypo generation, but their use is restricted to some areas with satisfying demand conditions. For this reason, a technique to generate acid and alkali from by-product salt and reuse them by using bipolar membrane electrodialysis (hereinafter referred to as "bipolar membrane") as an option of recycling will be introduced (Fig. 2.30).

2.2.7.1 What is Bipolar Membrane

The bipolar membrane is an ion exchange membrane having a structure in which an anion exchange layer and a cation exchange layer are laminated. If water was theoretically added a voltage of 0.83 V or more on both sides of the membrane, water in the membrane can be electrolyzed to acid (H^+) and alkali (OH^-) (Fig. 2.31).

For bipolar membrane electrodialysis, as shown in Fig. 2.32, by supplying inorganic salt (NaCl or the like) to a bipolar membrane, anion membrane and cation membrane. Ions migrate as shown by arrows. The H ion (H^+) electrolyzed with water and the anion (SO_4^{2-}) permeated through the anion exchange membrane were combined to produce an acid (HCl), while the OH ion (OH^-) electrolysis with water

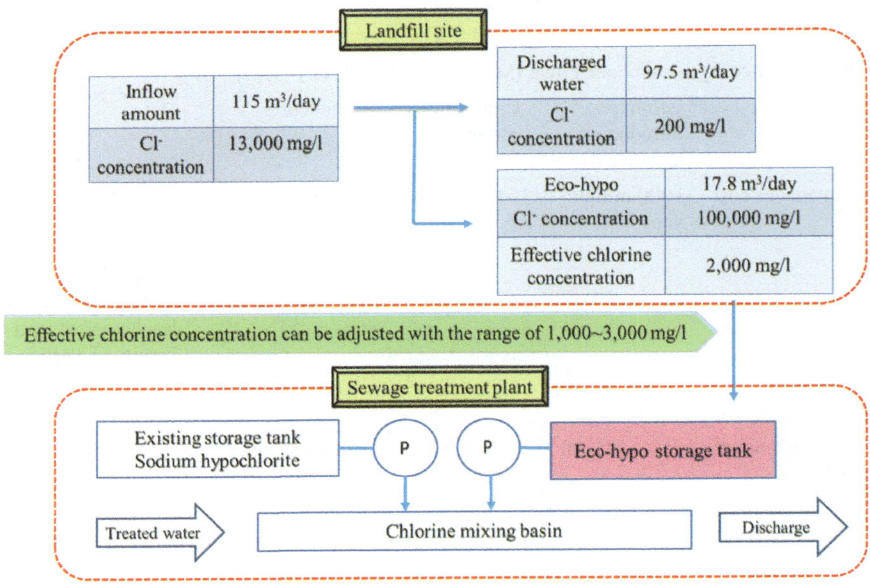

Fig. 2.30 Eco-hypo use flow (ITO T 2017)

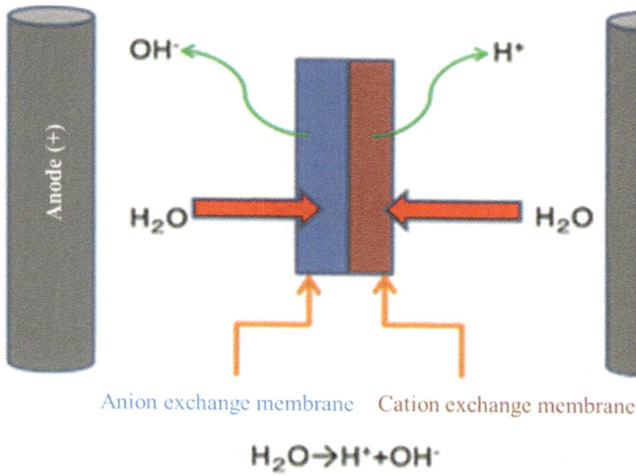

Fig. 2.31 Principle of bipolar membrane

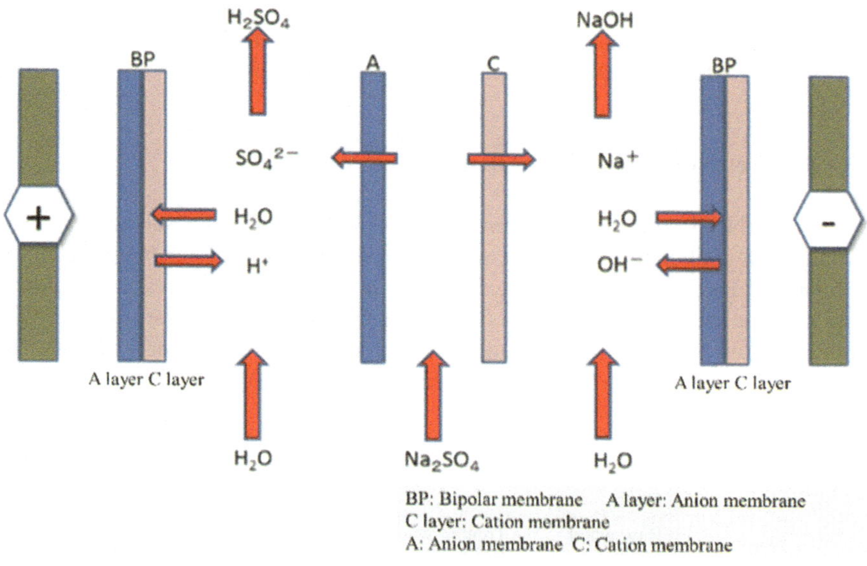

Fig. 2.32 Bipolar membrane electrodialysis

and the cation (Na^+) permeated through the cation exchange membrane are combined to generate alkali (NaOH) (Reverse reaction of neutralization proceeds).

The chemical formula is as follows.

$$NaCl + H_2O \rightarrow NaOH + HCl$$

Table 2.17 Ions of commercial salt and by-product salt (Takimoto et al. 2016)

Measuring ion	Commercial salt (wt%)	By-product salt (wt%)
Na^+	37.8	36.7
K^+	0.148	3.12
Ca^{2+}	0.036	0.066
Mg^{2+}	0.225	0.444
Cl^-	54.2	47.5
SO_4^{2-}	0	10.1

2.2.7.2 Experimental Subject By-Product Salt

After desalinating leachate from the municipal waste landfill site by electrodialysis membrane method, the concentrated salt is dried and solidified to get the by-product salt, which is used for the test (hereinafter referred to as "by-product salt"). In order to grasp the composition, the analysis was conducted on commercial salt and by-product salt. Commercial salt and by-product salt were dissolved each in 1 g of pure water to make 1L, and the concentrations of the respective ions were measured by anion and cation chromatography. The measured ions were Na^+, K^+, Ca^{2+}, Mg^{2+}, Cl^-, SO_4^{2-}. Table 2.17 shows the measured ions and the results.

2.2.7.3 Pretreatment (Calcium Removal)

A 4 mol/L NaCl solution (commercial salt (23.4%, byproduct salt 24.6%) was prepared for application to bipolar membrane devices. In bipolar membranes, as multivalent ions such as Ca^{2+}, Mg^{2+}, Fe^{2+} may interfere, so it is necessary to reduce these components to 1 mg/L or less. Therefore, Ca^{2+} and Mg^{2+} in the solution were removed by chelate adsorption method (Takimoto et al. 2016) (flowing with SV (1/h) = 10). The concentration in saline after removal is shown in Table 2.18.

Table 2.18 Result of Ca^{2+} removal test (Takimoto et al. 2016)

	Ca^{2+}		Mg^{2+}	
	Raw water (mg/L)	Treated water (mg/L)	Raw water (mg/L)	Treated water (mg/L)
Commercial salt	230	0.05	53	0.02
By-product salt	225	0.8	120	0.2

2.2.7.4 Acid and Alkali Generation by Bipolar Membrane

500 ml of saline water (4 mol/L as NaCl) which calcium ions, etc. were removed, 500 ml of ion-exchanged water (1 mol/L HCl and NaOH) and 500 ml of the electrode solution passed through the bipolar membrane and then the experiment was started. Figure 2.33 shows the apparatus used for the experiment (manufactured by Astom Co.).

The reason for putting acid and alkaline solution in advance is that electricity does not flow at the beginning of energization because conductivity is low when first starting the experiment with water. The reason for using the electrode solution is to use the same liquid as the solution used for the process in order to avoid contamination of impurities. When hydrochloric acid is used as an electrode solution, harmful Cl gas is generated. So, a solution containing Cl^- (NaCl, HCl) was not used. For the concentration of the electrode solution and the fluctuation of the balance amount, the concentration of the electrode solution does not fluctuate in the combination of the present membranes. However, there is a possibility that the amount of liquid or Na decreases little by little due to electrode reaction and volatilization of gas mist. Since part of the electrode liquid remains in the apparatus, the measured value is slightly lower than the concentration of the alkaline liquid added. The salt solution became acid at the start of water flow. However, there is a possibility that the amount of liquid

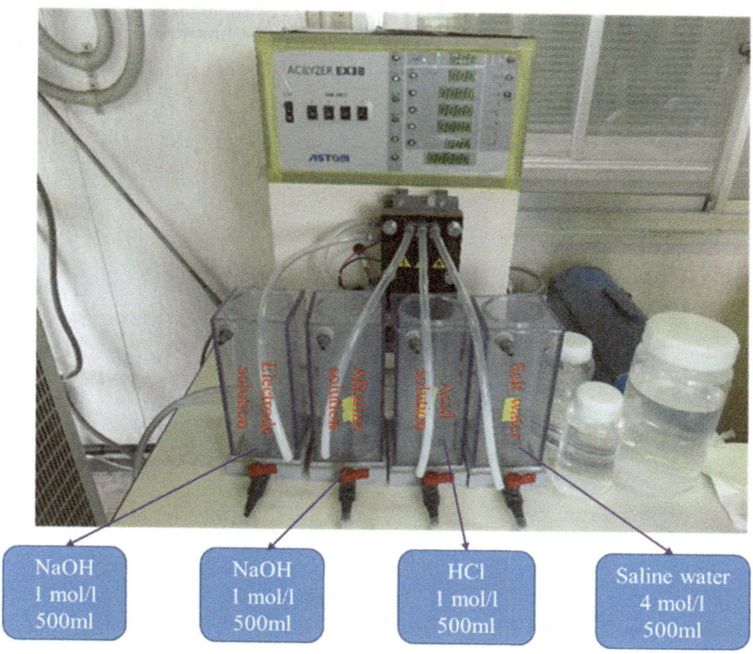

Fig. 2.33 Bipolar membrane device

or Na decreases little by little due to electrode reaction and volatilization of gas mist. Some of the electrode liquid remains in the apparatus. Therefore, the measurement value is slightly lower than the concentration of the alkaline solution added before. Relationships between electric conductivity of a salt solution, acid, alkaline solution and water passing time from the start of water flow are shown in Figs. 2.34 and 2.35.

In addition, the measurement results of the generated acid (HCl), alkali (NaOH) solution and desalted water concentration are shown in Table 2.19. By treating the by-product salt with the bipolar membrane, HCl in the recovered solution was 635 mL, 10.2 w/v% (From this result, 635–500 mL = 135 mL, 34.7% solution of HCl was produced from the by-produced salt). The amount of NaOH in the recovered solution was 675 mL, 6.7 w/v%. (From this result, 675–500 mL = 175 mL, 17.6% solution of NaOH was produced from the by-produced salt), the Cl concentration of salt water was desalted to 458 mg/L (99.6%) and the Na concentration to 208 mg/L (99.8%). So it was known that acid, alkali and desalted water equivalent to that of commercial salt can be produced. Regarding acid and alkali, they are not high-purity commercial products. For acid, H_2SO_4 is mixed in addition to HCl according to the composition of the by-product salt, and for alkali, KOH is mixed in addition to NaOH. That is why they are called eco acid and eco alkali. Normally, it is sufficient concentration as a substitute agent for hydrochloric acid (HCl: 36%) and caustic soda (NaOH: 20%) used as a neutralizing agent in leachate treatment facilities. Another advantage is that byproducts accompanying processing do not occur. In addition to eco-acid and eco-alkali, desalted water remains. As it has been highly desalinated to 500 mg/L or less, it is fully possible to use as reclaimed water or discharge. It can be an effective method as one of the by-product salt recycling methods.

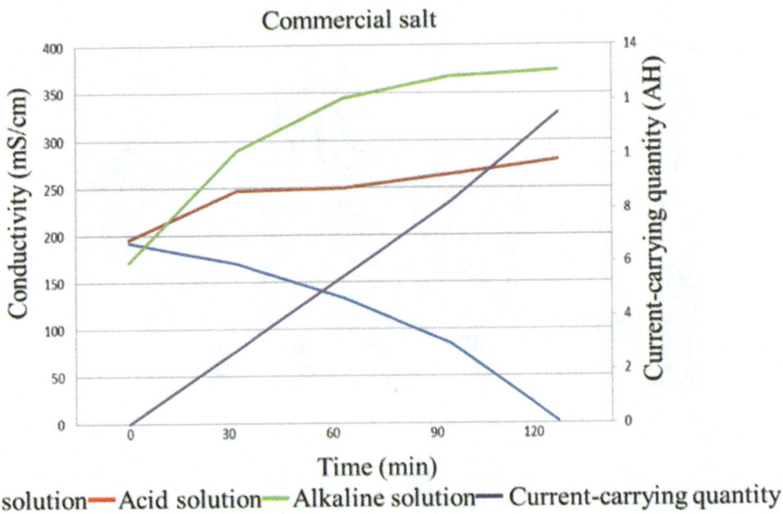

Fig. 2.34 Commercial salt (4 mol/L: NaCl) (Han et al. 2017)

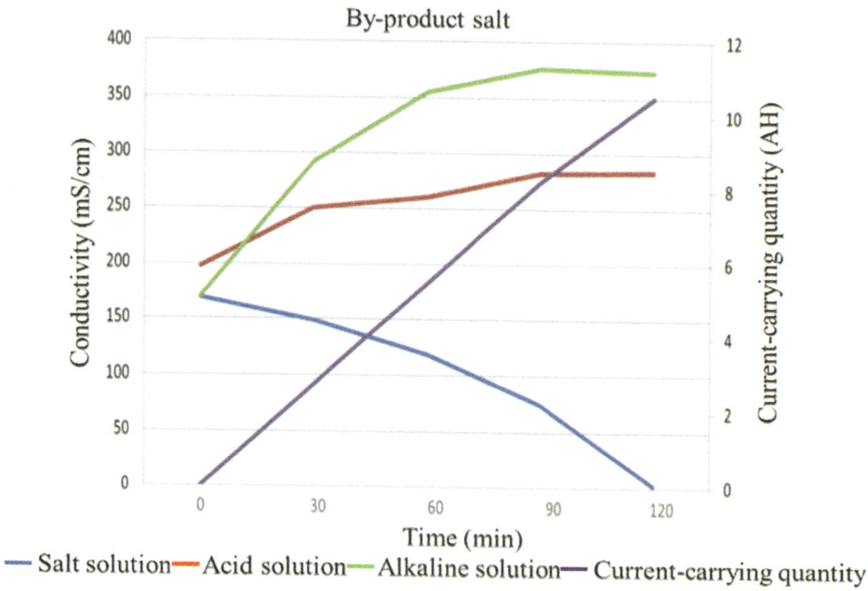

Fig. 2.35 By-product salt (4 mol/L: NaCl) (Han et al. 2017)

Table 2.19 Bipolar membrane experiment result (Han et al. 2017)

			Salt on the market	By-product salt
$(HCL + H_2SO_4)$		mL	658	635
		mol (%)	2.03 (11.0)	1.82 (10.2)
NaOH + KOH		ml	690	675
		mol (%)	2.06 (7.2)	1.84 (6.7)
Desalted water	Recovered solution	ml	152	190
	Cl^-	mg/l	339	458
	Na^+	mg/l	450	208

2.2.7.5 Appropriate Treatment of By-Product Salt, Towards Reuse

The root of by-product salt is NaCl, which was from seawater or rock salt. As the concept of recycling-based society spread, the amount of waste is decreasing. Also, due to the sophistication of exhaust gas treatment, the Cl emission to the atmosphere is also decreasing. Therefore, the amount of Cl in fly ash tends to increase. On the other hand, recently, sewerage facilities that enhance Cl- regulation for sewer acceptance are on the rise. For this reason, Cl lost its place to go but ends on the landfill sites. From now on, it is necessary to consider the landfill sites on the seaside or coastal as the promotion of byproduct salt recycling and incineration as the core waste treatment in Japan.

2.3 The Problem of Remaining Chelate

High concentration inorganic salts problem because of the increase in the ratio of incineration residue of landfill waste at landfill sites together with recent interference problems with the treatment of nitrogen and COD in leachate due to the influence of organic chelate used for stabilization treatment of fly ash became to come out. Here, the questionnaire survey revealed the actual situation of the chelate use of municipalities and the actual condition of the influence on leachate treatment. The organic chelating agent remaining in leaching water adversely affects the nitrogen treatment, so we introduce experimental verification examples and countermeasures.

2.3.1 Survey on the Actual Situation

2.3.1.1 Facilities to Be Surveyed

In the gasification melting furnace, the shaft furnace type accounted for 15 facilities (58%), and the fluidized bed type was 8 facilities (31%) (Figs. 2.36 and 2.37).

In order to collect data on the type of facility and use of chelating agent for municipal waste, based on the annual municipal waste disposal survey data (Ministry of the Environment 2011) conducted by the Ministry of the Environment annually, in the stabilization treatment of fly ash from 1309 incineration facilities nationwide, 151 places where chemical treatment or medicine treatment are being used are randomly

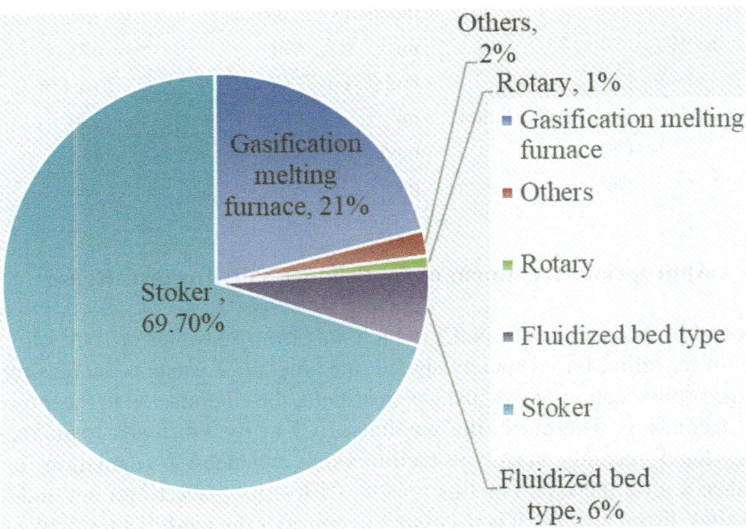

Fig. 2.36 Proportion of incineration facilities

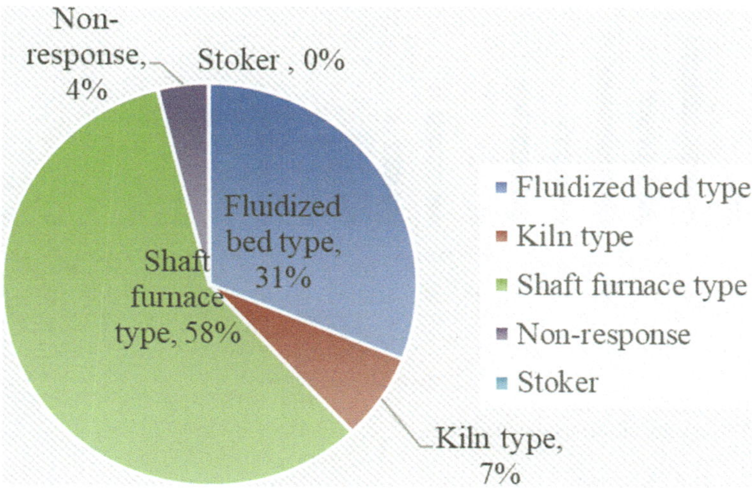

Fig. 2.37 Proportion of gasification melting furnace

extracted and a questionnaire survey was conducted. Also, in order to collect data on the amount of incineration residues to be transferred to the municipal waste landfill sites and leachate status, 107 sites performing denitrification in leachate treatment were extracted from 1857 landfill sites nationwide, based on the municipal waste disposal survey data (Ministry of the Environment 2011) and a questionnaire survey was conducted.

2.3.1.2 Survey Results (Higuchi S 2013)

① Incineration facility

Responses from 122 facilities to randomly extracted 151 facilities of municipal waste incineration facilities nationwide were got (response rate: 80.8%). The questionnaire results are shown below.

A. Types of incineration facilities

Stoker incinerators (69.7%) were used most frequently in incineration facilities and then gasification melting furnaces (21.0%) (Fig. 2.36). In the gasification melting furnace, the shaft furnace type occupied 15 facilities (58%), then the fluidized bed type, 8 facilities (31%) (Fig. 2.37).

B. The capacity of incineration facilities

On the processing capacity, the capacity of 50–100 t/day accounted for 18 facilities (15.0%), followed by the capacity of 150–200 t/day accounted for 16 facilities (13.3%). The result is shown in Fig. 2.38.

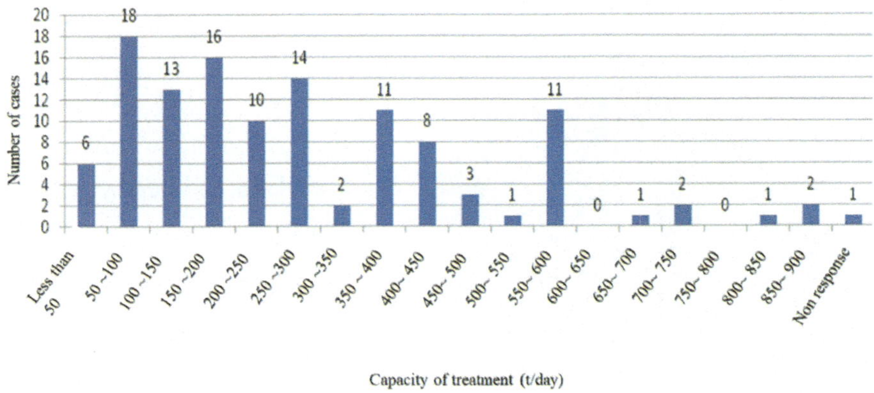

Fig. 2.38 Capacity of incineration facilities

C. Method of exhaust gas treatment

Dry type method accounted for 98 facilities (80%), and the wet type method was 16 facilities (13%). The survey result is shown in Fig. 2.39.

D. Type of chelating agent

The actual usage of chemicals was as follows. Organic dithiocarbamic acid type (hereafter DTC type) was 63 facilities (46%), piperazine type (hereafter PIP type) 41 facilities (30%), inorganic chemicals 7 facilities (5%) (Fig. 2.40).

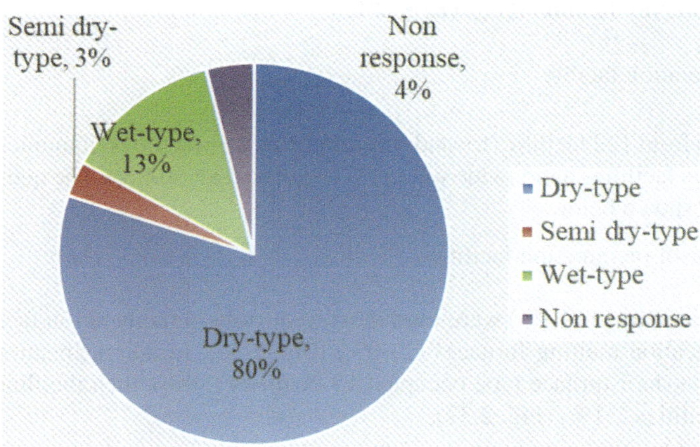

Fig. 2.39 Exhaust gas treatment system

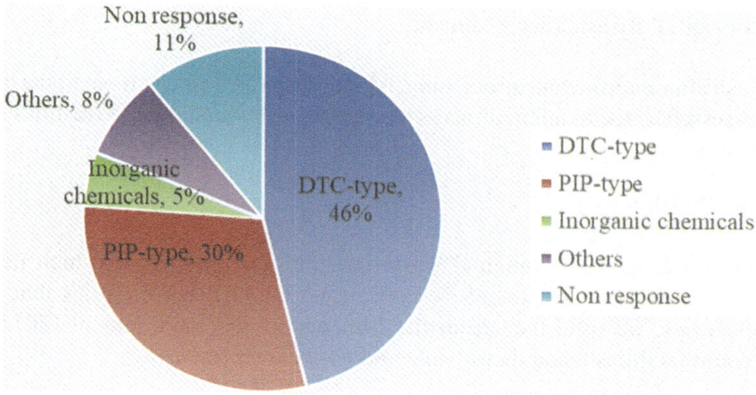

Fig. 2.40 Actual state of chelating agent usage

E. Chelate addition rate

91 facilities (74.6%) were treated fly ash with chemicals such as chelating agents. Facilities with the addition rate of chelating agent at 2.5–3.0% are the most, which is 26 (21.3%), followed by 10 (8.2%) at 3.5–4.0% addition rate. 3 facilities (2.5%) has the largest addition rate at 11.5–12.0% (Fig. 2.41).

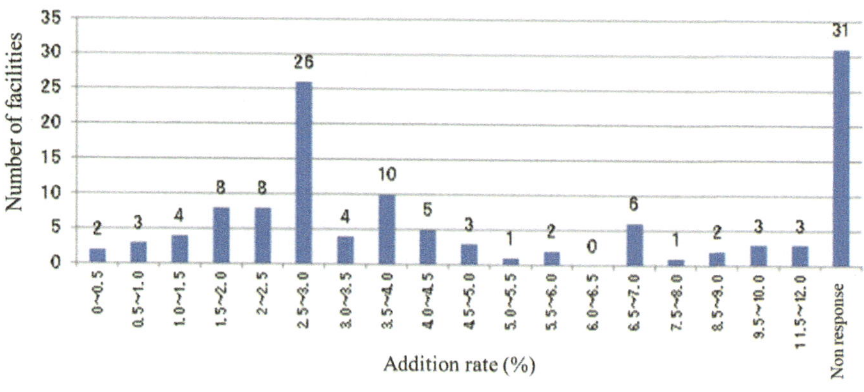

Fig. 2.41 Chelating agent addition rate

F. Disposal of fly ash after treatment

For fly ash after intermediate processing, 65 facilities (54%) carry it out to the landfill sites possessed by the municipalities as final disposal destinations, 38 facilities (30%) outsource it.

① Landfill sites

69 responses from 107 municipal waste landfill sites nationwide which perform leachate denitrification treatment (response rate: 64.5%). For facilities that didn't give responses, we used data from the Ministry of the Environment (2011) and didn't count facilities when their data could not be confirmed.

A. The situation of leachate treatment

The average quality of leachate treatment is shown in Table 2.20.

Regarding the situation of COD_{Mn}, 0–20 mg/L in leachate water is the most, accounts for 35.8% (29/81 facilities). Then, 20–50 mg/L is 27.2%, 50–100 mg/L is 19.8%, 100–200 mg/L shows 11.1%, and 500 mg/L or more is 4.9% (Fig. 2.42). In the water quality of treated water, 0–20 mg/L accounts for 78.6% (77/98 facilities), and 100 mg/L or more is 1.0% (Fig. 2.43). In processing efficiency, 70% or more accounts for 43.8% (35/80 facilities), 70% less is more than half (Fig. 2.44).

This suggests that COD components in leachate are likely to contain a large amount of COD_{Cr} which is considered difficult to remove by biological treatment. In the case of leachate water, 0–50 mg/L is the most of 54.9% (39/71 facilities), 21.1% at 50–100 mg/L, 18.8% at 100–200 mg/L, 4.2% at 300 mg/L or more Fig. 2.43). For treated water, 0–20 mg/L at 54.3% (51/94 facilities), 27.7% at 20–50 mg/L, 13.8% at 50–100 mg/L, and 4.3% at 100 mg/L or more (Fig. 2.45). Facilities with 50% or less treatment efficiency account for 60.9% (42/69 facilities) (Fig. 2.46). 62.5% of the facilities with processing efficiency less than 70% were caused by COD and 69.6% by T-N, which suggests that nitrification inhibition may be occurring.

Table 2.20 Actual condition of leachate treatment

Items of water quality	Unit	Leachate (average)	Treated water (average)	Removal rate (average %)
BOD	mg/L	177	11.9	69.3
COD	mg/L	110	13.8	58.2
T-N	mg/L	77.5	72.6	38.0
NH_4-N	mg/L	78.4	10.0	60.6
PH	–	7.9	7.3	–
EC	ms/m	962	971	–
Cl^-	mg/L	5959	4895	22.6
Ca^{2+}	mg/L	1026	359	17.9

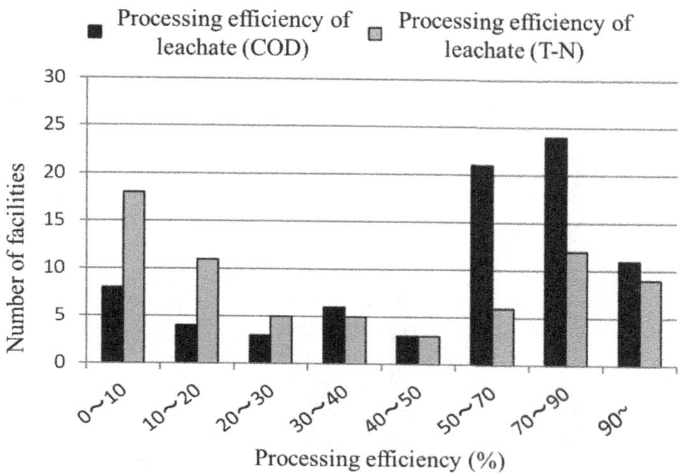

Fig. 2.42 Status of raw leachate

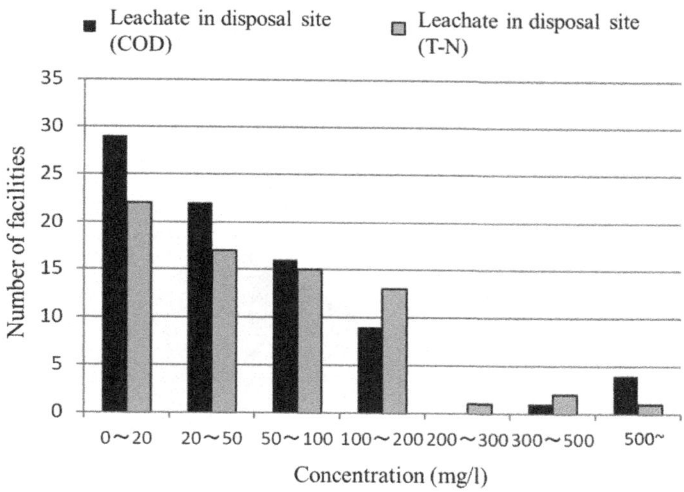

Fig. 2.43 Condition of treated water

2.3.2 Composition of the Chelating Agent

2.3.2.1 Composition of Commercial Chelating Agent

The composition of three kinds of commercial chelating agents as organic PIP type, DTC type, and inorganic phosphorus type, is shown in Table 2.21. It is known COD, S, T-N, Cl, and thiourea with high concentrations are contained in the organic chelating agent.

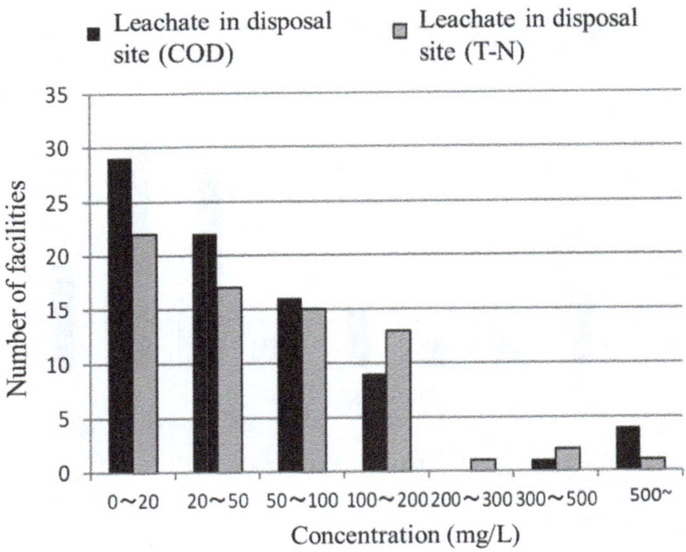

Fig. 2.44 Status of processing efficiency

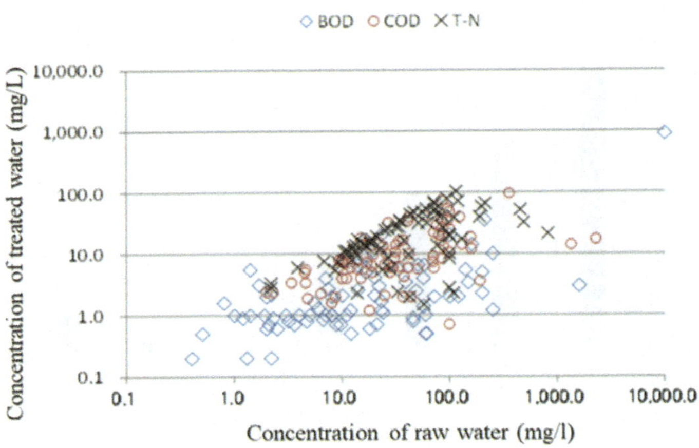

Fig. 2.45 Status of raw water quality and treated water concentration

2.3.2.2 Thiourea-Like Substance

Thiourea has a structure in which the oxygen atom of urea is replaced by a sulfur atom and is represented by CH_4N_2S, is mainly used as a raw material for urethane resin. Thiourea is used as a nitrification inhibitor mixed with fertilizer in order to maintain ammonia which is effective as nitrogen components of fertilizer. Also in the field of wastewater treatment, when ammoniacal nitrogen is oxidized to nitrous acid or nitric

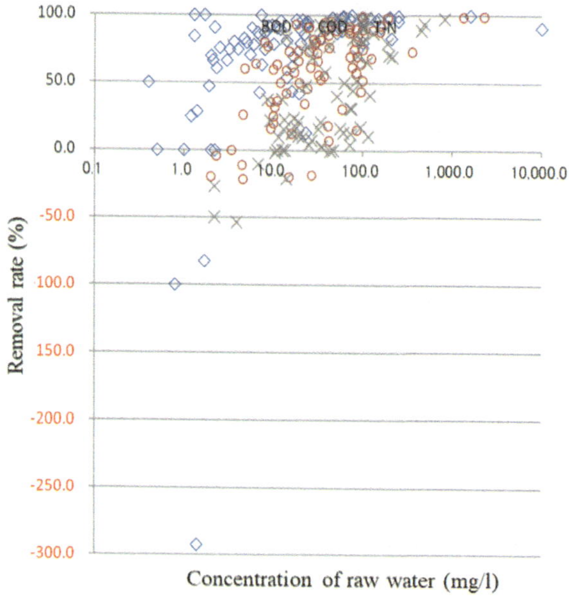

Fig. 2.46 Status of removal rate for raw water

Table 2.21 Composition example of commercial chelating agent

Item	Unit	Organic type		Inorganic type
		PIP type	DTC type	Phosphorus type
COD_{Cr}	mg/kg	321,000	399,000	6900
COD_{Mn}	mg/kg	195,000	314,000	4080
S	mg/kg	100,000	115,000	<300
T-N	mg/kg	32,200	40,300	80
T-P	mg/kg	<10	<10	<228,000
Cl^-	mg/kg	89,431	110,169	<40
pH	–	13.7	>14	<1
Specific gravity	–	1.235	1.2	1.58
Thiourea	mg/kg	1500	2000	–

DTC type: Dithiocarbamine
PIP type: Piperazine type

acid, the amount of oxygen consumed by the proliferation of nitrifying bacteria is added to the amount of oxygen consumed for decomposition of organic substance during BOD analysis for 5 days (N-BOD), which may be higher than the actual BOD (C-BOD). Therefore, in order to suppress nitrification, allylthiourea is added to exclude these effects to measure true BOD in some cases. Figure 2.47 shows the chemical structure of commercial industrial organic chelate and allylthiourea used

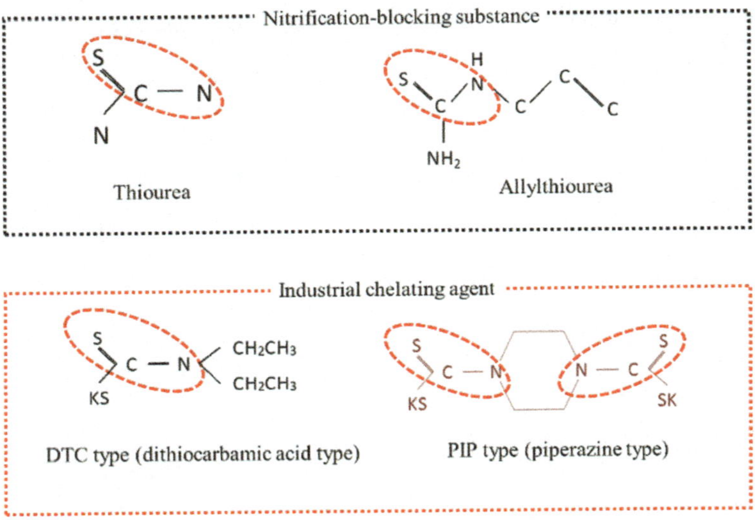

Fig. 2.47 Chemical bond between nitrification inhibitor and industrial chelate

for nitrification inhibitors. Based on the similarity between the chemical structures of thiourea or allyl thiourea known as nitrification inhibiting substances and the chemical structures of industrial organic chelating agents, industrial organic chelating agents are considered to be thiourea-like substance.

2.3.2.3 Nitrification Inhibition

The organic chelating agent contains 1500–2000 mg/kg of thiourea-like substance. It is said that concentrations of 37.4 μmol (about 3 mg/L) for thiourea, 10 μmol (about 2 mg/L) for sodium diethyldithiocarbamate (DDTC), 1.0 μmol (about 0.1 mg/L) for allylthiourea are used to inhibit ammonia monooxinase by 70% (Yamanaka 1999; Yoji 1996). The authors added a chelating agent to the leachate, conducted a nitrification experiment, and confirmed that nitrification inhibition was caused by the chelating agent (Uchida and Higuchi 2012). That is, the organic chelating agent itself is a nitrification inhibiting substance.

2.3.3 *Effect of Chelate Treatment*

Table 2.22 shows composition examples of general incineration residue (stoker furnace, HCl gas discharge value 50 ppm, high reaction slaked lime used).

It is set that the landfill ratio of this incineration residue to 20% fly ash and 80% bottom ash in terms of weight ratio and set that 3.0% by weight ratio of DTC type

Table 2.22 Composition example of incineration residue

Item	Unit	Dry Ca	
		Fly ash	Bttom ash
COD_{Mn}	mg/kg	3150	4000
COD_{Cr}	%	2.95	2.28
T-N	mg/kg	67	580
T-P	mg/kg	5700	2140
K	%	4.18	0.712
Ca	%	27.6	16.2
Cu	mg/kg	490	1890
Pb	mg/kg	2100	1420
Hg	mg/kg	14	0.06
Cl	%	20.1	0.932
S	%	2.54	0.484
pH	–	12.2	12.5

chelating agent with the highest usage ratio was added based on actual survey, T-N and COD content per 1 t of incineration residue landfilled from the chelate composition example in Table 2.21 are obtained as shown in Figs. 2.48 and 2.49. As shown in Fig. 2.48, when 3% of the DTC type chelating agent is added to the fly ash, the T-N source in the incineration residue is 64.5% from the bottom ash, 1.9% from fly ash and 33.5% from the chelating agent. Similarly, as shown in Fig. 2.49, for COD_{Mn}, the bottom ash is 70.8%, fly ash is 13.9%, and the chelating agent is 15.3%. It is known

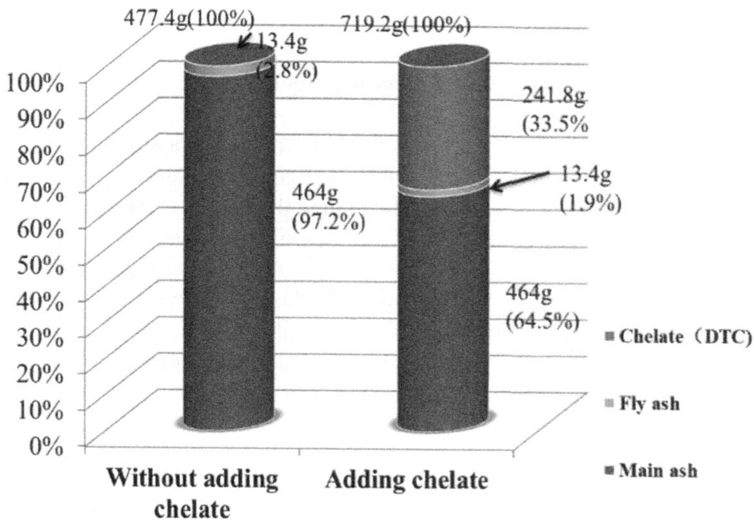

Fig. 2.48 Nitrogen source per ton of incineration residue

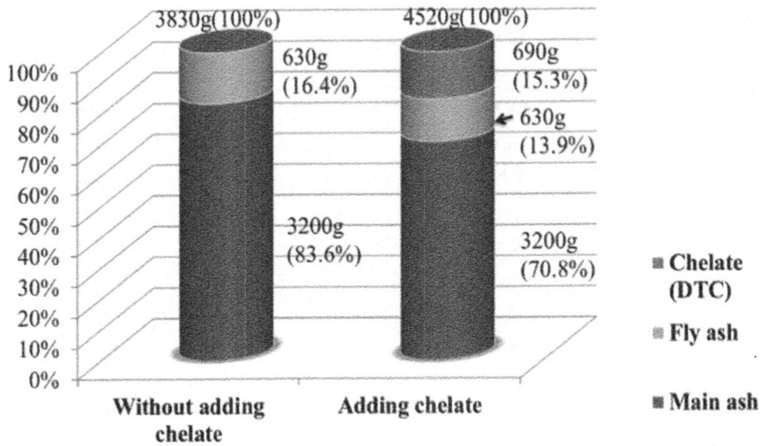

Fig. 2.49 COD_{Mn} supply source per incineration residue 1

that the proportion of the organic chelating agent is high, especially the nitrogen. On the other hand, T-N and COD_{Mn} source is extremely small for inorganic chelating agents. Furthermore, since the organic chelating agent also contains thiourea-like substance, if the incineration residue subjected to the stabilization treatment of the fly ash with the organic chelate is disposed of in landfill sites, nitrification inhibition and COD treatment efficiency decrease in leachate treatment would be caused.

2.3.4 Chelate Countermeasure

2.3.4.1 Change to the Inorganic Chelating Agent

When slaked lime is used in the exhaust gas treatment, fly ash is highly alkaline. On the other hand, since the inorganic chelating agent, for example, the phosphoric chelating agent is acidic, it is necessary to change the injection device or the like to acid-resistant material. The addition rate increases, nitrification inhibition of leachate water and COD treatment problem are solved. Although it is a fundamental solution, cooperation between incineration departments and landfill departments is necessary. Since many municipalities entrust maintenance and management, it is necessary to negotiate with the problems of the performance guarantee contract of incineration facilities of companies entrusted.

2.3.4.2 Add Residual Chelate Decomposition Step After Ozone Treatment or Accelerated Oxidation Method Etc. to the Existing Leachate Treatment

Processing efficiency is improved by adding processing equipment. However, the limit of quantification of residual chelate is 5 mg/L at present. On the other hand, since the residual chelate inhibits nitrification even at 1.0 mg/L in the case of the DTC type, it is necessary to improve the analysis accuracy of residual chelate.

2.3.4.3 Avoid Leachate Storage in the Landfill Site

Nitrogen and COD concentration increase in addition to residual chelate as the internal storage of leachate water. For this reason, it is possible to reduce leachate by adding leachate water adjustment facility or capping in the landfill completion zone, thereby eliminating elution and then reduce the impact.

2.3.4.4 Dilution Treatment of Leachate

In the initial stage of the landfill, if the residual chelate concentration increases, processability is secured by dilution.

2.3.4.5 Proper Management of Residual Chelate

It is said that concentrations of 37.4 μmol (about 3 mg/L) for thiourea, 10 μmol (about 2 mg/L) for sodium diethyldithiocarbamate are used to inhibit ammonia monooxinase by 70% (Yamanaka T 1999). In addition, from the results of nitrification inhibition experiments (Uchida and Higuchi 2012; Uchida et al. 2015), it is shown that the effect of organic chelate on bio-nitrification gives 10% inhibition at about 3 mg/L. Therefore, it is necessary to confirm 1.5 mg/L concentration of residual chelating agent of DTC type and PIP type in the incineration residue to be transported to landfill sites.

2.3.5 *Analysis of Residual Chelate*

When the fly ash with current organic chelate is disposed of in landfill sites, it will hinder the leachate management even at low concentration. Therefore, the analytical method with the lower limit of quantification of the residual chelating agent of 1 mg/L is shown.

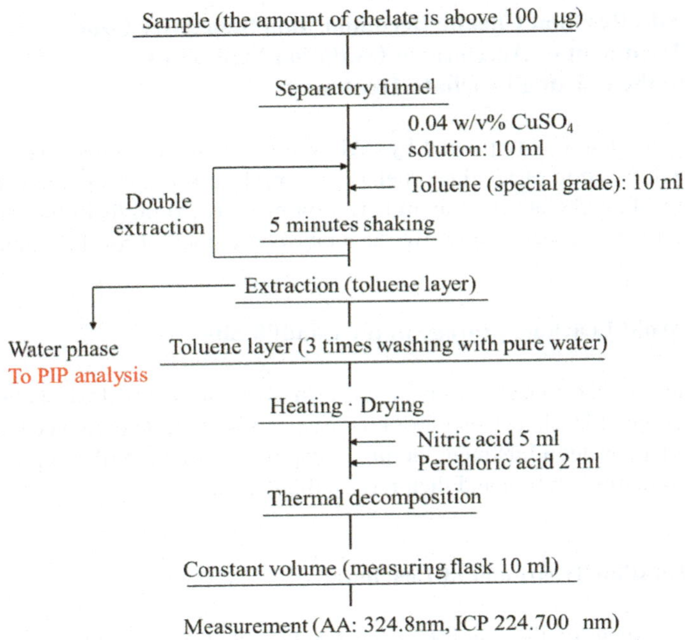

Fig. 2.50 DTC type chelating agent quantitative flow (Higuchi and Uchida 2018)

2.3.5.1 Determination of DTC Type Chelating Agent

Since the DTC-type chelating agent is soluble in a solvent, copper is added to a sample to which a DTC-type chelating agent is added to generate a copper-chelate complex and extracted with toluene using this characteristic, and then the extraction solvent is heated and dried. The quantity of copper is determined by using nitric acid-perchloric acid and then Inductively Coupled Plasma Emission Spectroscopic Analyzer (ICP-AES) or Inductively Coupled Plasma Mass Spectrometer (ICP-MS), Atomic Absorption Spectrometer (AA). Figure 2.50 shows the flow of DTC chelate analysis method.

2.3.5.2 Determination of PIP Type Chelating Agent

Since it is difficult to dissolve the PIP type chelating agent in a solvent, copper sulfate is added to the chelating agent to form a copper-chelate complex, and then it is collected on the filter paper using 0.45 μm membrane filter, and after acid treatment of each filter paper, measure the emission line near the wavelength of 224.700 nm using ICP-AES (ICP-MS, AA etc.) and quantify copper. For the calibration curve, a sample with a certain amount of chelating agent added is similarly treated, and the

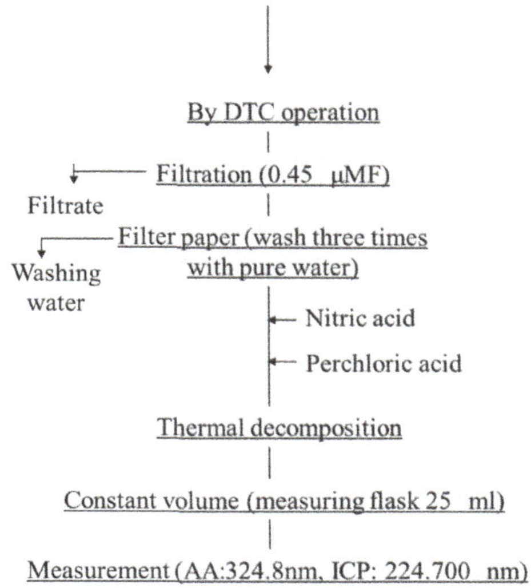

Fig. 2.51 PIP type chelating agent quantitative flow (Higuchi and Uchida 2018)

amount of chelate is calculated from the obtained copper concentration. Figure 2.51 shows the flow of the PIP-based chelating agent analysis method.

2.4 Covered Type Landfill Site

The covered type landfill site generally refers to a managed landfill site where roofs and walls of steel structure and membrane structure are installed. It is also called closed system landfill site. The closed system landfill site was proposed as an urban landfill site that utilizes the medium depth underground space as the landfill site by Prof. Masataka Hanashima of Fukuoka University in the 1980s. Figure 2.52 is an example of a conceptual diagram assuming that it will be installed with an

Fig. 2.52 Example of a conceptual diagram of covered type landfill site

incineration facility. Since the beginning of the 2000s, the landfill site with covering facilities based on the conventional open-type landfill site, was constructed as the first style facility in Yamagata village, Nagano prefecture, it started to spread rapidly. Currently, in addition to the municipal waste landfill sites, more than 60 facilities are constructed including the industrial waste landfill sites. As the background, it has a good landscape and not affected by a guerrilla rainstorm, etc., or it is possible to prevent the scattering of waste to the outside. On the other hand, there are many problems unsolved regarding the method of abolishing covered type landfill sites.

2.4.1 Definition of the Covered Type Landfill Site

The covered type landfill site is generally called a closed system landfill site as described above and refers to one having a roof structure. However, besides one with a roof structure, there is a landfill site that blocks rainwater and performs leachate treatment. For example, many foreign countries including Europe and the United States cap the surface of the landfill layer with sheets etc. after the end of landfill, to prevent inflow of rainwater and air so as to set an anaerobic state in the landfill layer, and generate methane gas, which is recovered and burned to generate power with gas turbine, and recover energy. That is, while landfilling, it is open type but after completion, it has a covering structure covering the surface with a sheet, etc. Although this is due to the difference in the waste management policy of full landfill, as the precipitation is high in Japan, capping when landfilling is used in some managed type industrial waste landfill sites. Figure 2.53 shows an outline of an example of Fig. 2.54. These cases are conducted in areas with high precipitation, areas other than landfill area are covered with sheets and surface water is collected. Rainwater is prevented from penetrating into the landfill layer, and leachate is treated. In order to move the sheet according to the situation of the landfilling work, it is made of PVC with the size of 20–30 m of one side and is held down by the sandbag so as not to be blown by the wind.

Although it takes time to maintain and manage the movement of the sheet in that way and to remove rainwater, the leachate management is the same as covered type landfill site. For this reason, the covered type landfill site is classified as covered type and capping type as shown in Table 2.23, and these are collectively defined as the covered type landfill site.

2.4.2 Advantages of the Covered Type Landfill Site

The following are the advantages of the covered type landfill site.

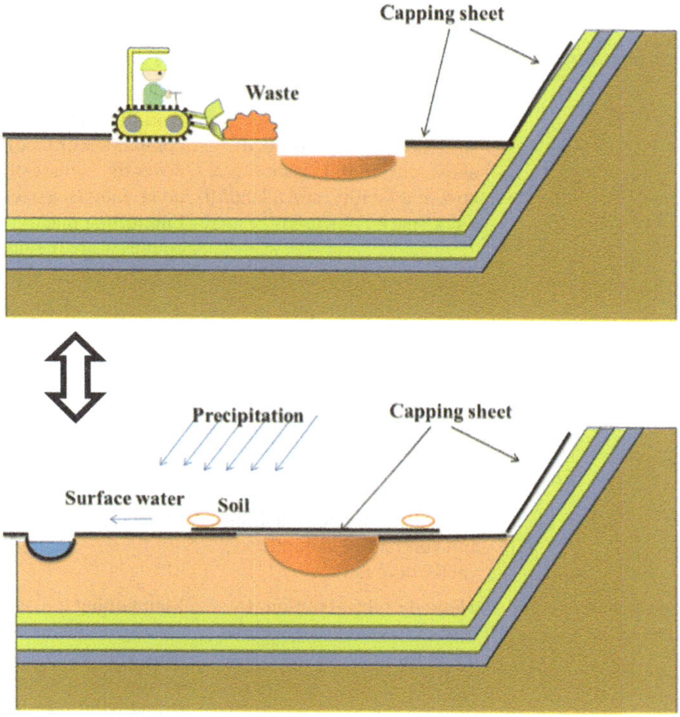

Fig. 2.53 Overview of landfill with capping

Fig. 2.54 Capping landfill case (ERC Takashiro)

Table 2.23 Definition of covered type landfill site

Classification of covered type	Roof type	Capping type
Outline	A disposal site where a steel structure, a membrane structure, etc. are installed above the landfill surface as a covering facility, and a landfill work space is secured	Move the capping sheet during landfilling. After the completion of landfilling, cover the surface of the landfill layer with a sheet etc. Landfilling methods in various countries are included
Landfill during precipitation	Landfill canbe done regardless of the weather, as precipitation can not enter it	As the opening is the smallest one where landfill is possible, it is possible in principle. Precipitation on sheet surface is excluded as surface water
Stabilization of landfill waste	Washing out and organic substance decomposition are performed by the improvement of the semi-aerobic function by artificial water sprinkling and ventilation (natural ventilation, fan)	It is premised to secure the semi-aerobic function. Water sprinkling equipment will be installed on the lower surface of the capping as required
Points of attention in operation and maintenance	Method of the abolition in case of treated water without discharging	Maintaining the opening and closing of the opening during heavy rains and the rainy season. Keep the sheet during strong wind. Abolishment method
Cause of cost increase	Cover and post, sprinkling equipment Desalting equipment and by-product salt disposal costs when using treated water as water for sprinkling	Capping sheet (during the landfilling, after the completion of the landfilling), transfer cost, the water spinkling equipemnt under capping
Cause of cost decrease	The capacity of leachate treatment facility and leachate adjustment facility becomes small	The capacity of leachate treatment facility and leachate adjustment facility becomes small
Townscape	Townscaping measures are possible as building design	There is almost no landfill waste. Sheet materials
Evaluation	It is easy to obtain agreement of the local residents. Because the cost is high, it is excellent in townscape and operation and maintenance	The construction cost is cheaper than that of the normal Open type. The landscape is slightly inferior to the roof type and takes a long time to maintain, but the function is almost the same as the roof type (economical covered type)

2.4.2.1 Landscape

Landfilling work is hardly visible, especially the roof type, as the entire facility is covered by the structure, waste will be shielded from the outside when the landfilling work starts. The landscape gets better and the image for the landfill site improves.

2.4.2.2 Landfill Management

There is concern about the influence of scattering by wind and malodorous gases such as hydrogen sulfide depending on the waste during landfill work. These can be prevented by coating facilities. Moreover, landfill work can be done in daytime and artificial sprinkling at night as no landfill work is done. And it is possible to do landfill work regardless of the weather.

2.4.2.3 Stabilization of Roof Type Landfill Layer

The final landfill site of the roof type performs landfill in the closed space. In this case, ventilation is done for the landfill work environment preservation. Ventilation is performed by forced or natural ventilation. For this reason, light negative pressure is generated in the interior. So the semi-aerobic function in the landfill layer is strengthened, and the pH lowering and organic substance decomposing effect are enhanced (Fig. 2.55). In other words, the open-type semi-aerobic function causes air to flow from the end of the water collecting pipe due to thermal convection arising from the temperature difference inside the landfill layer. In addition to this, the

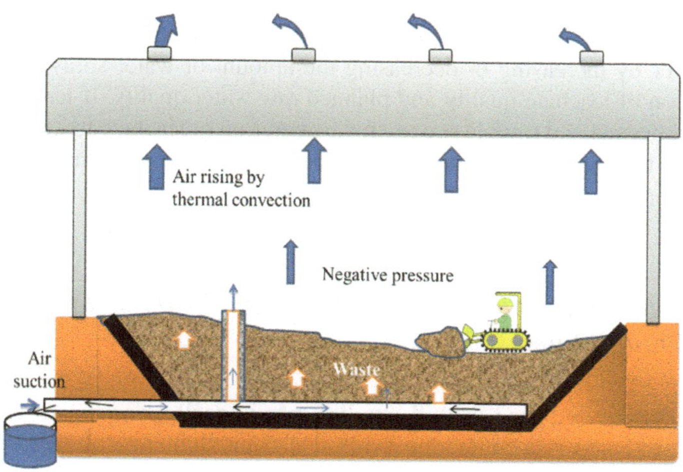

Fig. 2.55 Roof-type semi-aerobic function

roof type landfill site can generate a negative pressure inside the structure and heat convection of the air (in the case of natural ventilation, the upper air is warm, and the bottom air is low, thermal convection ventilation is performed), which can promote the semi-aerobic function. In addition, while the open-type landfill site is basically to wash out by natural precipitation and detoxification by leachate treatment, by artificially controlling the most effective wash-out water volume and the method of sprinkling, safe leachate processing and stabilization of the landfill layer become possible. That is, it has the possibility of early stabilization.

2.4.2.4 Leachate Management

It is possible to rationally control leachate which is one of the most characteristic features of the covered type landfill site. There is a high probability the open type landfill site will be forced to temporarily store the leachate inside the landfill layer due to insufficient leachate adjustment facility capacity since generated leachate exceeds planning capacity as the heavy rain in rainy and typhoon season. For this reason, leachate treatment facilities and leachate control facility scale are often provided with extra power in general. Nevertheless, due to the recent guerrilla heavy rain, the internal storage frequency of the leachate in the landfill layer is getting higher. When the internal storage time of leachate is prolonged, elution of inorganic salts accelerates, Cl^{-} and Ca^{2+} concentrations rise, organic concentration such as T-N, BOD, and COD also increases. As a result, the inflow load into the leachate treatment facility also increases, which will hinder leachate treatment. Meanwhile, since the artificial sprinkling water is used instead of natural precipitation at covered type landfill sites, leachate exceeding the planned treatment capacity is not generated by a constant amount of water spray, so leachate is not stored internally in the landfill layer, and its adjustment equipment can be set to the minimum scale. Since the amount of water spray can be controlled, it is possible to promptly stabilize the landfill layer by increasing or decreasing the amount of water spray according to the condition of leachate quality and planned raw water quality. If leachate quality becomes abnormal and trouble occurs in the operation of leachate treatment facility, it is possible to stop spraying during the cause investigation and countermeasure time so that rational leachate management can be done.

2.4.3 Issues and Countermeasures for the Covered Type Landfill Site

Although the covered type landfill site has many excellent aspects such as good landscape and heavy rain countermeasures, it also has challenges. Issues and current countermeasures will be discussed.

2.4.3.1 Economically

The roof type covered type landfill site is costly due to the coating facility, but on the other hand, it can make the leachate treatment facility and adjustment facility compact. One indication is that the cost can be increased by the coating facility and reduced due to the compactification of the leachate treatment facility, and the maintenance and management expenses reduction until the abolition can be balanced. The cost of covering facilities increases from landscape measures etc., so the overall cost tends to rise recently. On the other hand, the capping type has few cases, but the construction cost can be considerably reduced as compared with the roof type. Compared with the open type, the leachate treatment facility can be made compact, which is inexpensive. On the other hand, because of the movement of the capping sheet, replacement of the sheet due to deterioration of the ultraviolet rays, rainwater removal control, etc., maintenance and management costs increase. However, for the total cost, the capping sheet type is more economical than the roof type. There is an introduction of air dome which has many achievements in Korea as one of cost reduction measures of roof type. The air dome covers the entire disposal facility with a sheet and continues to supply air to keep the interior situation, and it is possible to withstand wind pressure by withdrawing air at typhoon time. It has merits such as excellent landscape, low cost, repeatable use, etc. However, since Japan has a high cost of electricity and if it is used for a long time, which remains a problem under the Building Standards Act. Therefore, there is no example of use in Japan yet. However, there is a possibility that facilities that conform to Japan's climate and regulations will become popular in the future. Figure 2.56 shows the example of Air Dome landfill site in Daegu, South Korea.

2.4.3.2 Sprinkling and Stabilization

The coating type landfill site is stabilized by washing out with artificial sprinkling, moisture supply and bioreactor function by ventilation.

Roof Type

Roof type artificial sprinkling water is done by sprinkler etc. The amount of water sprinkling is determined by the solid-liquid ratio (ratio of water volume to landfilled waste weight). The solid-liquid ratio is set to 1.0–3.0 as a standard based on the type of landfill waste and wastewater standard which aims to abolition (JWMS Conference 2010).

The general total solid-liquid ratio is set to 1.5–2.0 from landfilling to landfill closing, till the abolition of open-type landfill sites. Figure 2.57 shows the procedure for determining leachate treatment capacity and water sprinkling from liquid: solid ratio. The leachate treatment capacity is determined as follows: set the total leachate volume from the liquid: solid ratio to the waste amount to abolition period and set the

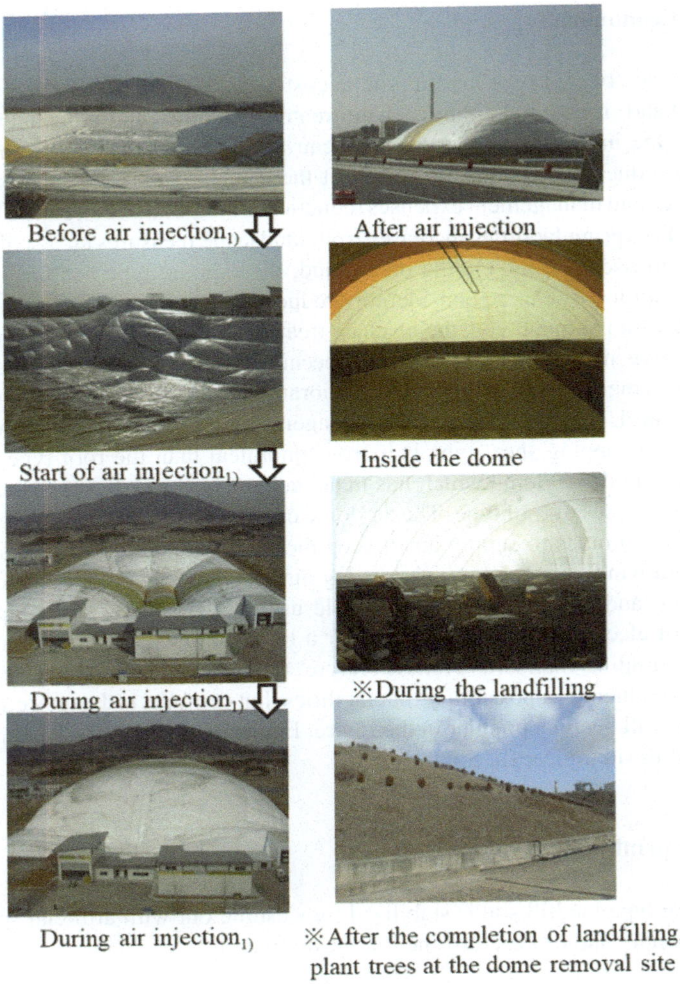

Fig. 2.56 Korean air dome landfill site (Provided by Professor Nam-Hoon Lee of Anyang)

leachate volume in different landfill period by considering the maintenance policy of landfill site, for example, pretreatment wash and leachate quality control by dilution water. The amount of water to be sprinkled is set taking into consideration the amount of evaporation from the amount of leachate.

Table 2.24 shows an example of the leachate volume calculated from the liquid: solid ratio. The waste landfill amount is 75,000 m^3. The covering earth amount is 22,500 m^3, and the landfill capacity is 97,500 m^3, the landfill area is 10,000 m^2. The landfill period is 15 years, after the landfill is over, the period until abolition is 15 years, so the total management period is 30 years. If the solid-liquid ratio is 3.0, the total leachate volume is 225,000 m^3. If it is operated 300 days per year, leachate

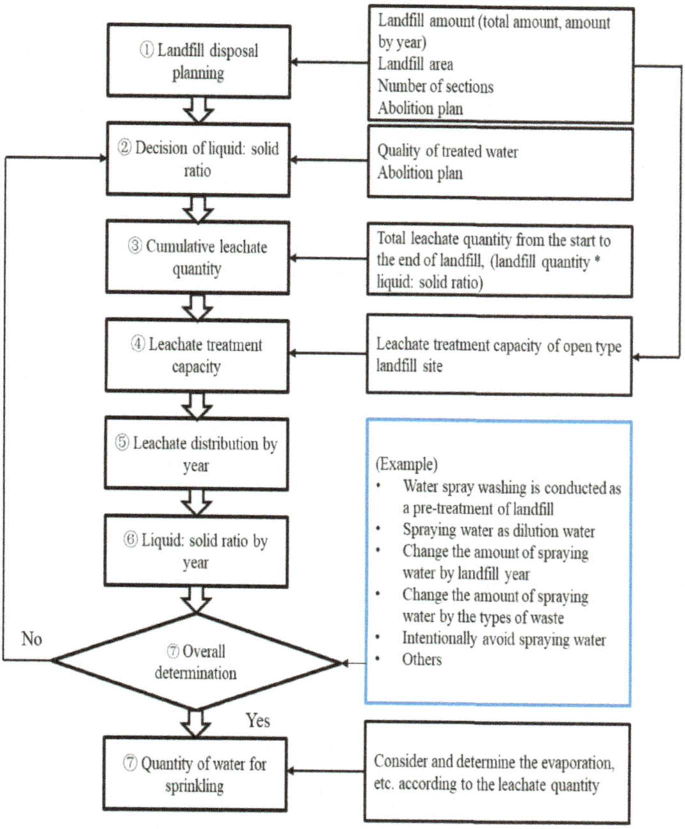

Fig. 2.57 Capacity of leachate treatment facility, flow for determining water spray in covered type landfill site

treatment capacity is 25 m^3/d. Considering the evaporation amount, if 60% of the amount of water sprinkled became leachate, the total amount of sprinkling water is 375,000 m^3. In addition, the rational formula is used to obtain the leachate amount in the case of open type, assuming two partitions. It is decided between 18 and 90 m^3/d when calculated by the following formula.

$$Q = \frac{1}{1000} * i * (C1 \cdot C2 * + C2 * C2)$$

Q: Volume of leachate (m^3/day)

i: The annual average daily precipitation is 4.0 mm/day, the daily conversion value of the maximum monthly precipitation is 20 mm/day.

C1: Seepage coefficient during the landfill is 0.5.

Table 2.24 Leachate water volume calculation example using liquid: solid ratio

Years	Landfill amount (m^3)	Cumulative landfill amount (m^3)	Volume of treated water by period (m^3)	Volume of cumulative treated water (m^3)	Liquid: solid ratio	Cumulative liquid: solid ratio
1–5	5000 × 5 = 25,000	25,000	5 × 300 × 25 m^3/day = 37,500	37,500	1.5	1.5
6–10	25,000	50,000	5 × 300 × 25 m^3/day = 37,500	75,000	0.75	1.5
11–15	25,000	75,000	5 × 300 × 25 m^3/day = 37,500	112,500	0.5	1.5
16–20	0	75,000	37,500	150,000	0.5	2.0
21–25	0	75,000	37,500	187,500	0.5	2.5
26– 30	0	75,000	37,500	225,000	0.5	3.0

A1: Landfill area of the section is 5000 m^3 during the landfill.

C2: Seepage coefficient is 0.3 in the section where landfill is completed.

A2: Landfill area of the section where landfill is completed is 5000 m^2.

In the case of the coating type, since it is 25 m^3/d, the leachate treatment facility capacity becomes compact. Furthermore, in the case of the open type, a large leachate control facility is needed, so the facility at the covered type landfill site is economical. In the case of municipal solid waste in which the incineration residue occupies the majority of the landfill waste, salt is washed out by sprinkling water, and due to the negative pressure effect caused by the room ventilation mentioned in (3), decomposition of low-concentration organic substance is carried out towards stabilization in the latter part of the landfill after pH decreases, Therefore, the stabilization of the landfill layer is considered to be equal or superior to the open type.

Capping Type

The capping type is generally thought to be anaerobic because air supply from the landfill surface is blocked by the capping sheet and stabilization is delayed. Figure 2.58 is a gas monitoring example of vertical degassing pipes installed in the sheet covered part of the managed landfill site where the capping sheet was laid and moved repeatedly.

During the period of capping of arrows to eliminate surface water, precipitation exceeding 300 mm is present, and surface water removal is carried out. During that time, anaerobic gases such as hydrogen sulfide and CH_4 gas are not generated. The

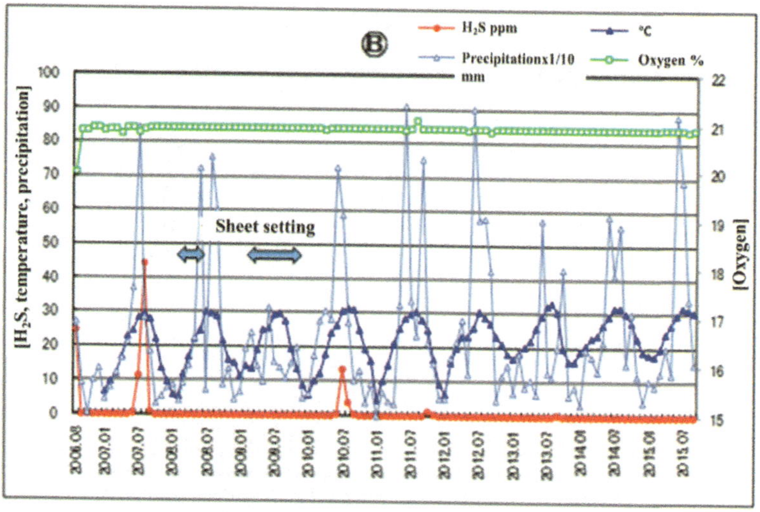

Fig. 2.58 Capping sheet and hydrogen (Higuchi et al. 2015), sulfide (Higuchi S 2011)

oxygen concentration is also 21%, which shows that a sufficient inflow of air is carried out. Although the temperature is initially affected by the outside air from 30 to 5 °C, the minimum temperature rises gradually and tends to rise from 18 to 33 °C. It is inferred that the interior is beginning to be aerobic by the airflow of collection pipe etc. On the other hand, hydrogen sulfide gas (hereinafter referred to as H_2S) rises to 45 ppm, 15 ppm, respectively, before capping. The precipitation at that time is from 750 to 800 mm. It is presumed to be a temporary anaerobic condition in large precipitation period, when the landfill surface is filled with rain and becomes saturated, causing aeration blocking. Although it was concerned that the sheet capped part hindered the breathability and became anaerobic, the landfill surface water was eliminated by capping and it is confirmed that air supply of 25 m^3/min from the water collecting pipe is able to maintain aerobic state.(Nagano et al. 2016). For this reason, sheet capping does not inhibit semi-aerobic function but functions rather effectively. In Japan, a country with high precipitation, it is stabilized basically by washing out the landfill layer by precipitation and then treat the leachate. For this reason, the capacity of the leachate adjustment facility and leachate treatment facility are large, and the cost burden is large. And there is no provision for sheet capping in Japan's landfill site maintenance standard. However, in the area where the annual precipitation exceeds 2000 mm/year, the scale of the facility increases, and the cases of covered type landfill site construction also increase due to the risk of outflow accident of untreated leachate caused by recent guerilla heavy rain. Under such circumstances, the landfill method using sheet capping together in a high precipitation region is capable of safe leachate water management at a low cost. In addition, except for the rainwater removal function, sheet capping can be said to be one of means for exerting a semi-aerobic landfill function, since breathability of the landfill is maintained (Fig. 2.59).

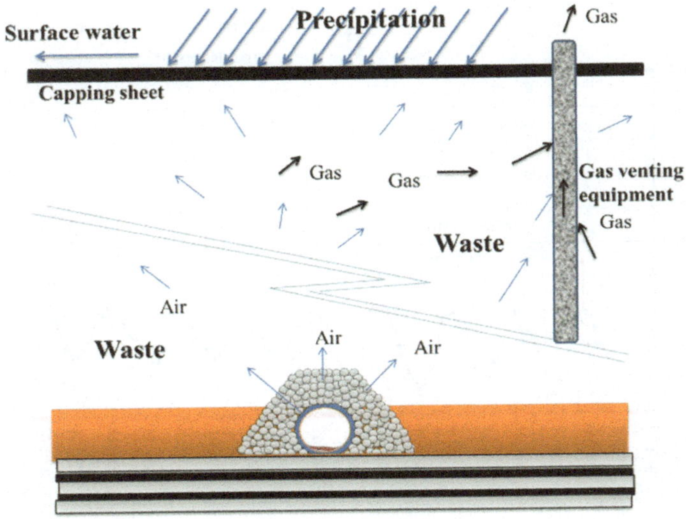

Fig. 2.59 Concept diagram of capping type (Landfill period)

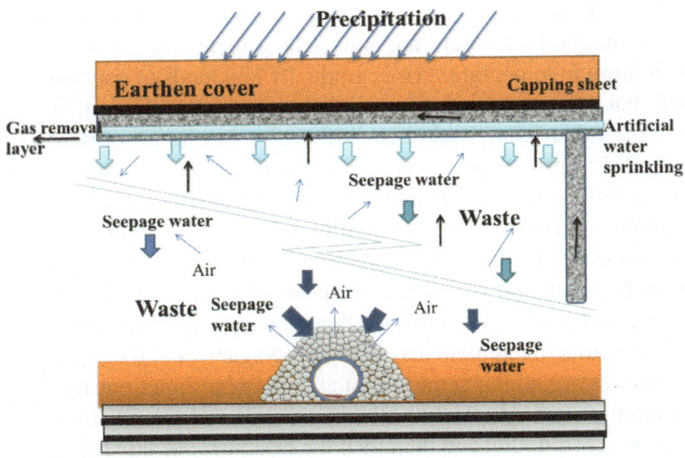

Fig. 2.60 Conceptual diagram of capping type (landfill completed)

The capping type does not interfere with the semi-aerobic function, but in regions with low precipitation, it may be thought that the organic decomposition stops due to lack of water supply and may inhibit stabilization. For this reason, it is necessary to move the opening portion as necessary and supply moisture. Since the entire surface of the landfill layer is capped after the landfill is finished, it is necessary to install an artificial water supply device under the capping sheet to stabilize it (Fig. 2.60).

2.4.3.3 Release of Treated Water

Most of the covered type landfill sites do not discharge leachate treated water or sewage discharge. The reason is that river discharge of leachate treated water is often not possible due to the opposition of river water rights holders and residents. In the case of non-discharge, leachate treated water is recycled as sprinkling water, so desalination treatment is carried out mostly because the landfill waste is mainly incineration residue. When desalination treatment is performed, by-product salt (concentrated salt) is generated, so problems of storage and proper disposal of by-product salt also occur. When leachate or its treated water is discharged to the sewer system, it will be based on the criteria for accepting wastewater to the sewage system. After the completion of the landfill, the leachate falls below the wastewater standard, or even it is time to abolish, if the river discharge cannot be done, it is still necessary to continue sewage discharge semi-permanently.

2.4.3.4 Abolition of the Covered Type Landfill Site

Fifteen years have passed since the covered type landfill site emerged, so there are landfill sites where the abolition period will come from landfill completion. For repositories where leachate treated water can be discharged, it is possible to abolish based on the abolishment criteria. However, when leachate treated water is not discharged, or when sewer discharge is carried out, the abolition method is not clear. Generally, it is considered to abolish by remaining the roof or capping after removing the roof as measures to prevent rainwater from entering. Meanwhile, Article 1(5) D of the Technical Standard for Waste Disposal Law stipulates that "a landfill site is a place that has solid and durable structures of pipe duets and other collection facilities that effectively collect retained water, etc. and can promptly discharge it ~omission~. In the case of landfill sites where the necessary measures are taken to prevent rainwater from entering, facilities mentioned above are not required because the landfilled municipal waste will not decompose or generate retained water. (Excluding water surface landfill sites). Installation of leachate collecting pipe and leachate treatment facility is excluded (Japan Environmental Sanitation Center Japan Environmental Sanitation Center 2014). Depending on the performance guidelines, the final landfill site of general waste is obliged to install a water collecting pipe even in a covered repository, but its function will also be stopped if stopping watering. As a result, if Article 1 (5) D is applied, "landfill where the necessary measures are taken to prevent rainwater from entering" is applicable to the landfill site, but for provision "landfill municipal waste which does not decompose and generates retained water", incineration residue, molten slag, etc. are considered to be applicable although there are no detailed provisions. It is considered possible to stop sprinkling water and to abolish it after taking measures to prevent rainwater from entering at the time when the landfill layer became stabilized and will not decompose by artificial sprinkling through landfill period and after the landfill. In this case, technical standard for maintenance of the landfill site of municipal waste pursuant to the provisions of Article 8-3, paragraph 1 of the Act, (17) "Regarding the landfill prescribed in the proviso, it is necessary to cover the surface of a nonwoven fabric laid in a waterproofing layer provided with any of the requirements of 1 (a)–(c) of the same item and then covered the surface with earth and sand, or something which is equal to or more than the effectiveness of the waterproofing, light shielding, the strength and the durability."24)In other words, if it falls under the "landfill site specified in the proviso," a waterproofing layer equivalent to the bottom waterproofing will be installed on the surface and final earth covering will be carried out, which is a surface layer specification similar to the criteria of EU. In the case of industrial waste, "landfill site specified in the proviso" is not applicable. For these reasons, the prefectural governors judge individually on the abolition of the covered type landfill sites.

2.5 Sea Surface Landfill Site

There are 26 Sea surface landfill sites of municipal waste in Japan in 2004 (Heisei 26) (14 other private sites including the Osaka bay regional offshore environment improvement center etc.). It is 1.5% for 1698 facilities in total but accounts for 27.2% in landfill capacity ratio (Ministry of the Environment, Waste of Japan 2016). Although the number of sites is smaller, it can be seen the reliance on sea surface landfill site is high from the viewpoint of securing disposal capacity. Although there are cases of constructing landfill sites in the sea overseas, it is common to landfill the sea with dredged earth and sand, then build a land once and then construct the landfill site at its top. It is seen that the case of the Semakau sea surface landfill site of Singapore where incineration residues are disposed is similar to the landfill method of Japan, which waste is directly dumped to the sea surface. Japan has constructed landfill sites on the sea before incineration spreads. Similar to the inland landfill site, the landfill waste differs depending on the landfill timing, and the quality of the leachate (residual water) and stabilization conditions also differ.

2.5.1 General Structure of Sea Surface Landfill Site

Basically, the structure of the sea surface landfill site is the same as the landfill site constructed on land, but it is sometimes impossible when applying structural standards, maintenance standards and abolishment criteria, etc. as the waste is dumped into the water. For example, regard to retained water collecting and drainage facilities, "For water surface landfill site, it is necessary to build solid and durable structures (retained water collecting and drainage facilities) that can effectively discharge retained water and spill water". In general, spill water (water intake facility in the inland landfill site) collecting and drainage facilities can be used as retained water collecting and drainage facilities, so they are not installed. That is, the functions of

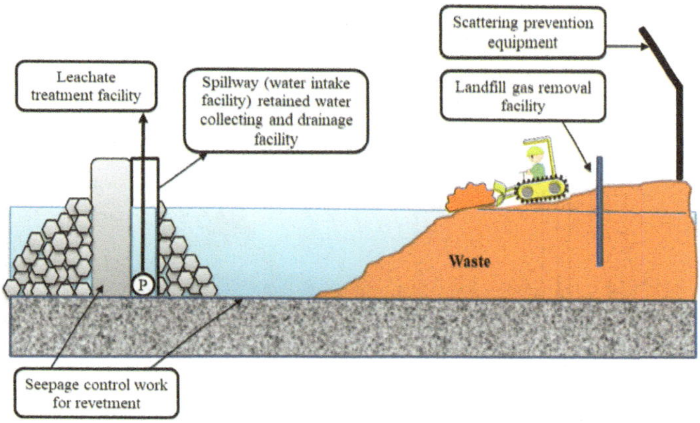

Fig. 2.61 General structure of sea surface disposal site

the inland landfill site and the sea surface landfill site are different. The inland landfill site has both storage and purification functions, in particular, it does not cause leachate to stay in the landfill layer and could be cleaned by precipitation, and has a purification function to lower pH by air circulation, while the sea surface landfill site mainly has a storage function, and the purification function is cleaning by precipitation and spillage water treatment (leachate treatment). Figure 2.61 shows the general structure of the sea surface landfill site.

2.5.2 *Water Balance of Sea Surface Landfill Site*

Sea surface landfill site begins with the dumping of waste, progresses in the procedure of land formation, pound (residual water surface) and water disappearance by waste dumping.

① Early stage of landfill

The early stage of landfill starts with the input of waste to the water. Leachate volume (Q) is the amount obtained by subtracting water surface evaporation amount (E1) from input waste amount (W) and precipitation amount (R). When the land is formed by the progress of landfill, further subtract the soil surface evaporation amount (E2). That is

$$Q = (R - E1 + E2) + W$$

to express (Fig. 2.62).

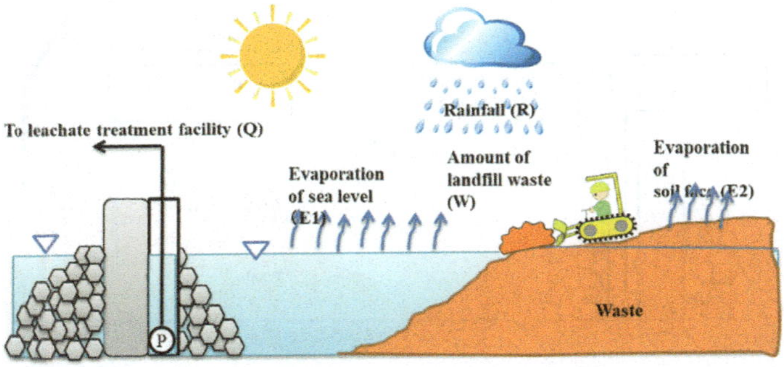

Fig. 2.62 Early stage of landfill (Water surface landfill)

② Middle stage of landfill

When landfilling, the water surface gradually disappears, and in generally remained water is called pond. The pond is used to eliminate surface water from the landfilled part and to withdraw leachate (Fig. 2.63). The water balance is basically the same as ① but with the decrease of surface water, the evaporation decreases and the soil surface evaporation from the land area increases.

③ Late stage of landfill

As landfill progresses further, it becomes land only, leachate volume (Q) is obtained by subtracting soil surface evaporation amount (E1) and surface water amount (S) of the completed surface from precipitation amount (R), which is the same as an inland landfill site.

$$That\ is\ Q = (R - E2) - S$$

to express (Fig. 2.64).

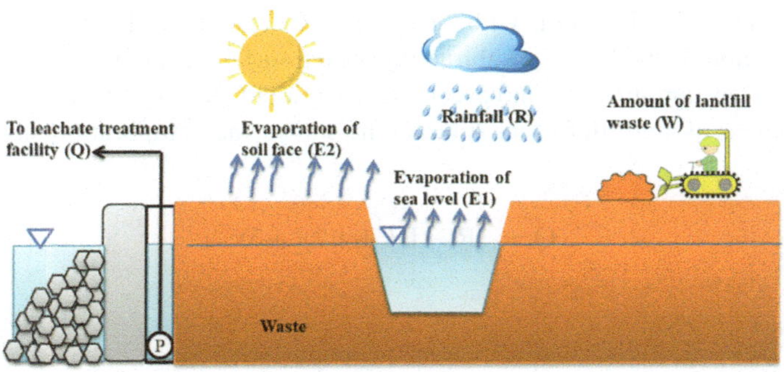

Fig. 2.63 Middle stage of landfill

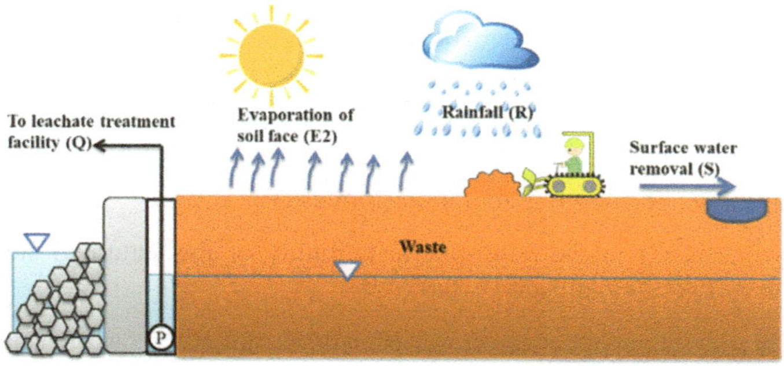

Fig. 2.64 Late stage of landfill

④ Water Balance of Sea Surface Landfill site

Thus, the generation mechanism of leachate volume changes depending on the landfill stage for the sea surface (water surface) landfill site.

Here, the water surface evaporation height is calculated by the Penman equation (Nakagawa S 1985). Since the water surface evaporation amount is measured at local meteorological stations, the data is referred to in some cases.

$$Q = R - E1 - E2 + W - S$$

Q: Leachate quantity Q (m^3/day)

R: Amount of precipitation (m^3/day) $R = 1/1000 \times A \times r$

A: Landfill area (m^2) = ($A = A1 + A2$)

r: Height of precitipation (mm/day)

E1: Amount of water surface evaporation (m^3/day) $E1 = 1/1000 \times A1 \times e1$

A1: Area of water surface (m^2)

e1 = Heigh of water surface evaporation (mm/day)

E2: Amount of water surface evaporation (m^3/day) $E2 = 1/1000 \times A2 \times e2$

A2: Area of water surface (m^2)

e2 = Heigh of water surface evaporation (mm/day)

W: Landfill amount of waste (m^3/day)

S: Rainwater rejection amount (m^3/day) $S = 1/1000 \times A2 \times C$

C: Rainwater rejection rate (0.3–0.9)

Penman equation

$$E1 = \frac{\Delta}{\Delta + \gamma} Rno + \frac{\gamma}{\Delta + \gamma} E_a$$

E1: Amount of water surface evaporation (mm/day)

Δ: Gardient (mb/°C) in the atmospheric humidity of the saturated steam vapor pressure curve T_a (°C) γ: albedo (= 0.05)

Rno: Evaporatiuon equivalent of net radiation of virtual water surface (mm/day)

E_a: Drying capacity of atmosphere $E_a = (e^{\ast}_{a} - e^{a}) f(u)$ $e^{\ast}_{a}$: Saturated steam vapor pressure on the evaporation surface (mb)

e^a: Atmpspheric water vapor pressure (mb)

f(u): Wind velocity function (mm/day/mb)

Regarding the soil surface evaporation amount, the possible evaporation amount is calculated by the Thornswait method and the Blaney Criddle method (Higuchi S 1995), and the actual evaporation amount is set to 60–70% of the possible evaporation amount from the construction of the soil covering and the vegetation management situation on the landfill completion side in many cases.

Balney Criddle equation

$$E_r = 25.4 \cdot K \cdot C_j \cdot t_j$$

$$C_j = \frac{d_j}{\sum d_j} \times 100(\%)$$

E_r: Amounbt of evaporation that may be genrated monthly (mm)

D_j: Hours of sunlight in the month (h)

C_j: Ration of monthly sunshine hours to annual sunshine hours (%)

t_j: Monthly average temperature (Fahrenheit) F = 1.8C + 32

K: Coefficient is based on the plants. It is 0.6–0.8 in the case of irrigation reservoir and forest reservoir

E2: Actual evaporation of soil surface = E_r × (0.6–0.7)

2.5.3 Issues of Sea Surface Landfill Site

The merit of the sea surface landfill site is its large disposal space, which can be secured with economical construction cost and it also has merit for landfill waste containing high concentration salt such as incineration residue. On the other hand, decomposition of organic substance is delayed due to anaerobic conditions because the landfill layer is immersed in seawater in the early stage of the landfill, especially in the case that landfill waste is mainly incineration residue, the decomposition of COD and T-N derived from residual chelate is delayed as described in Sect. 2.3. Furthermore, as mentioned in Sect. 2.2, most of the incineration facilities use calcium-based chemicals for exhaust gas treatment, so pH in leachate water is high and reduction is slow.

Table 2.25 shows the structure of the inland landfill site and the sea surface landfill site and comparisons of maintenance and management.

Table 2.25 Comparison between inland and sea surface landfill site

		Inland landfill site	Sea surface landfill site
Structure	Seepage control	Storage structure + Double seepage control sheet, etc.	Lower the internal water level than the revetment + bottom seepage control work + external water surface
	Water collection	Collecting pipe in the bottom	Water intake pit, etc. (No collecting pipe)
	Leachate treatment	Ca removal + Biological treatment + physicochemical treatment (flocculation sedimentation, filtration, activated carbon) + desalting	Biological treatment + physicochemical treatment (flocculation sedimentation, filtration, activated carbon)
Maintenance and management	Internal water level	As a rule, do not store water inside (internal storage is only carried out during heavy rains)	Almost always internal storage condition
	Landfill structure	Semi-aerobic	Anaerobic
	Internal water movement	Precipitation → Penetration → Water collection	Precipitation + internal sea water → residue (mostly no movement)
	Intermediate earthen cover	It is unnecessary when the ignition loss is below 15% (in many cases, soil is covered to prevent scattering)	None (same as inland when the water surface disappears)

As shown in Table 2.25, in the case of the sea surface landfill site, because there is almost no water transfer in the landfill layer and almost no air into the landfill layer as it is in immersed state, the purification function and pH decrease in the landfill layer can hardly be expected compared with the inland landfill site. Therefore, it is difficult to expect stabilization and abolition with the same standards as inland landfill sites.

2.5.4 Countermeasures

Considering the structure of the surface landfill site, the following points can be considered as the sea surface landfill site constructed and operated so far and the sea surface landfill site to be constructed in the future.

① Existing sea surface landfill site

As for existing sea surface landfill sites, carrying out long-term maintenance and management out towards abolition, or applying abolition criteria only to the sea surface landfill site should be considered (For example, it is judged not with percolating water but with residual water (semi-surface water)). As mentioned above, there is no air circulation for the landfill site of mainly incineration residue, so it takes a considerable period to satisfy the abolishment criterion, especially for pH. As a measure against pH, there are methods of pond aeration or air circulation to the landfill layer to neutralize, or oxidizer (hydrogen peroxide or ozone) injection, but considering the size of sea surface landfill site, considerably large scale and investment are required. Figure 2.65 shows the oxidizer injection image illustration, and Fig. 2.66 shows the image illustration of leachate circulation.

② The newly constructed Sea surface landfill site

For the sea surface landfill site to be constructed, it is necessary to create a facility that considers abolition. Abolition inhibiting factors of the sea surface landfill site are

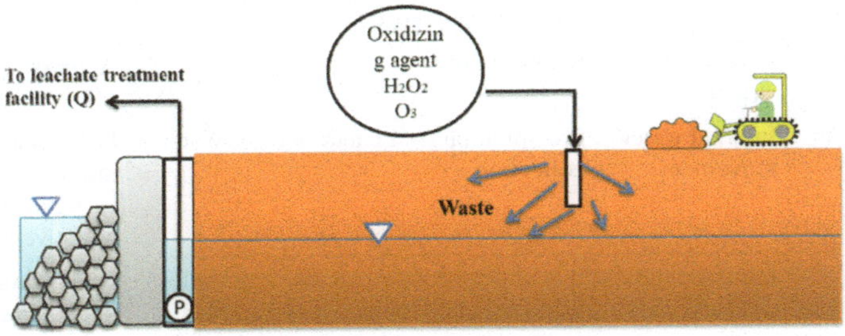

Fig. 2.65 Oxidant injection into the landfill layer

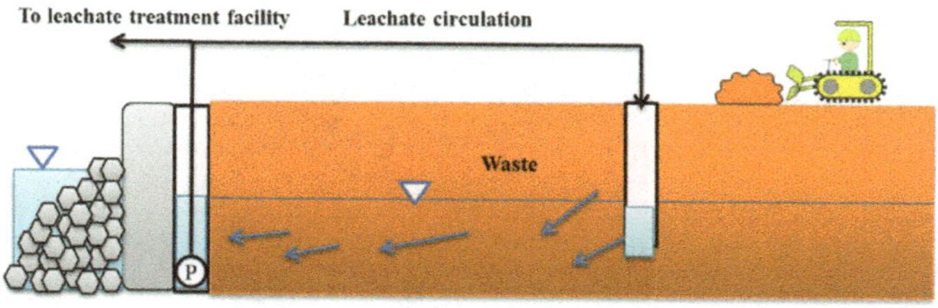

Fig. 2.66 Leachate circulation image

T-N, COD (from organic chelate used to stabilize fly ash because of slow biological decomposition as the landfill layer is immersed) and pH (influenced by lime used for incineration exhaust gas treatment). Since these causes are clear, the following can be considered as countermeasures in the case of new construction.

a. Restrictions of landfill waste

Landfill waste to the sea surface landfill site shall be limited to biologically inactive waste such as incineration residue, crushed incombustible residue, inorganic sludge, etc. Particularly for incineration residue which uses $Ca(OH)_2$ for the exhaust gas treatment agent or uses organic chelate for fly ash stabilization treatment cannot be accepted. Or to say, make acceptance restrictions.

b. Revetment structure

Make the revetment structure watertight and set the collecting conduit. The water intake is from the bottom, and when the residual seawater disappears, the internal water level is lowered, and it is made a semi-aerobic state structure like the inland landfill site.

2.5.5 Advanced Use of Ultimate Land (Higuchi S 2015)

Ultimate land use of final landfill site is often at the surface, recently, cases have been highly utilized at sea surface landfill sites. In the ultimate land of sea surface landfill site, liquefied natural gas (LNG) base with LNG tanks of 180,000 Kl (diameter = 83 m, height = 54 m) × 2 and the world's largest LNG tanker (loading capacity 125,000–217,000 KL) pier facilities, etc. was constructed and completed in November, Heisei 26. The landfill site is a managed-type site where other companies landfilled sodium carbonate sludge, and its area is 25 ha, with the capacity of 450,000 m^3 (Fig. 2.67). Since the LNG is supplied from tanker to the tank by pipelines, it was necessary to penetrate the steel plate vertical wall of the revetment. Therefore, the penetrating part was enclosed with temporary steel plate water shielding. At that time the opportunity was obtained to observe the steel plate part

Fig. 2.67 Example of advanced use of ultimate land of sea surface landfill site (Tsuboi T et al. 2015)

Fig. 2.68 Steel plate water shield penetration part

in the sea for a long time (Fig. 2.68). Corrosion of the surface layer did not proceed and the water-blocking function was maintained.

① Landfill environment at the time of ultimate land use

When using the land of final landfill sites, after boring investigation in the landfill layer and out of the landfill site, survey of landfill waste materials, and quality analysis of percolating water and groundwater were conducted, and when no hindrance to living environment conservation was confirmed by the third party committee, the construction work started.

② Considerations when using ultimate land

During the construction of the LNG station, about 400 foundation piles per tank were placed, so groundwater survey outside the landfill site and percolating water survey inside the landfill site were conducted once a month. As a result, although the percolating water temporarily exceeded the control standard value due to the construction work, there was no abnormality in the sea area outside the landfill site and also the groundwater. In the case of using this site, pile foundation was possible because the bottom of the land was rock and the permeability coefficient is small.

③ Current situation

The case was a construction of a large-scale LNG base at the ultimate land of the managed sea surface landfill site and may enjoy the benefit of rock bed. But large LNG tanks were constructed on the land with over 800 pile foundations, which was probably the first case in the world. It draws attention as a precedent example of large-scale ultimate land utilization of the future sea surface landfill site.

2.6 Semi-aerobic Landfill

The semi-aerobic landfill, which is the basis of our landfill technology, was researched and developed by Fukuoka University's Professor Hanashima in the 1960s as a bioreactor type landfill technology and spread throughout the country. Recently it has been transferred to other countries as a global eco-friendly landfill technology that suppresses the occurrence of methane gas. On the other hand, in Japan, efforts are made to recycle and reduce waste because it is difficult to secure landfill sites as it's a narrow country, after further reduction, detoxification, stabilization by intermediate processing such as incineration, landfilling the residue becomes the basis of waste management. For this reason, the landfill waste is mainly incineration residue or incombustible crushed residue, evolving independently to salt elution suppression function in high salt or high alkali environment, neutralization function by aeration, and to low load type bioreactor function.

2.6.1 Concept

The basic idea of the semi-aerobic landfill is "to utilize the purification function of nature and its decomposition mechanism." Natural purification function is the decomposition of organic substance by aerobic microorganisms and fungi and it is promoted by the gas release by thermal convection using decomposition heat, air supply and rinsing by precipitation and water supply function to microorganisms.

2.6.1.1 Decomposition of Organic Substance in Daily Life

Although it is rare to actually see how organic decomposition takes place in our daily life, returning dead bodies of small animals to soil and composting kitchen waste at home are some examples. Paradoxically it is full of cases preventing organic substance from decomposing in everyday life. It is food preservation. The easiest example to understand is refrigeration and freezing by refrigerators. And also the same as vacuum packed food. These are technologies that preventing decay of organic substance by cutting off the temperature and air outside, when microbial and fungal activities can not be performed. It is predecessors' wisdom of life that we should not forget. Salted foods such as pickles, karashi mentaiko, dry foods such as dried shiitake mushrooms, dried fruits, salted and dried fish and dried squid are also the same. These are technologies born of microorganisms and bacteria that dislike salt. Also, preservation of special rice cakes in water is the closest to the structure of the landfill site, so the semi-aerobic landfill type will be explained by this case. As recently there are more generations who do not know special rice cakes, so we will explain it firstly (Fig. 2.69).

Aerobic decomposition

$$C_xH_yO_z + \left[\frac{4X+Y}{4} - \frac{Z}{2}\right]O_2 \rightarrow XCO_2 + \frac{Y}{2}H_2O$$

Anaerobic decomposition

$$CH_aO_bN_c + \frac{4-a-2b+3c}{4}H_2O$$
$$\rightarrow \frac{4-a+2b+3c}{8}CO_2 + \frac{4+a-2b-3C}{8}CH_4 + cNH_3$$

In the days of the Showa era in the past, we made rice cakes in homes and towns at the end of the year, and use them to make special rice cakes for decorating the New Year, or cooked with a soup with special rice cakes) to prepare for the New Year. But when the New Year was over and mold started to grow on it after a while. So the special rice cake was soaked in water to prevent the occurrence of mold, and taken out from the water, boiled or baked and then ate until early spring. Because special rice cake mold is aerobic bacteria, it works actively when there is plenty of air and decomposes special rice cake by eating it. It is inconvenient for humans, so immersing it in water to block it from the air is one of the food preservation methods that suppress the activity of mold. Modern kagami mochi is a vacuum pack that pulls out air to prevent the occurrence of mold, so many young people have never seen molds themselves.

Let's replace the special rice cake with organic waste such as food residue, and replace container to add mizumochi with the landfill site. When there is water in the landfill site, waste is blocked from air and decomposition does not proceed. It does not mean no proceed. Anaerobic bacteria act in the absence of air, and anaerobic decomposition begins after the acid generation phase. However, without water, aerobic

Fig. 2.69 Predecessor's wisdom

bacteria that prefer oxygen in the air are activated and aerobic degradation proceeds. When comparing anaerobic and aerobic decomposition, the aerobic decomposition has a higher rate of decomposition and the fermentation heat is also 60–70 °C, higher than the 36 °C of anaerobic decomposition. The organic decomposition formulas of aerobic decomposition and anaerobic decomposition are shown below.

According to it, aerobic decomposition decomposes into carbon dioxide and water, but in anaerobic decomposition carbon dioxide, methane and ammonia are produced. In aerobic decomposition, water and oxygen (air) are necessary to activate the aerobic

Fig. 2.70 Food preservation methods

microorganisms, and the temperature is also necessary. Some of the food preservation methods born from predecessors' wisdom of life like special rice cakes in water, are "salting" typified by umeboshi and pickles, "drying" typified by dried fish and dried fruits, or "cold storage " underground. In modern daily life, there are "vacuum pack" which blocks air and "refrigerated" (Fig. 2.70).

2.6.1.2 What's Semi-aerobic Landfill

The semi-aerobic landfill is a landfill technology to promote decomposition of organic waste by aerobic microorganisms and fungi. In order to make waste an aerobic condition, air must be sent by blowers or compressors. However, as shown in Fig. 2.71, air can be supplied by natural aeration from the water collecting pipe installed at the bottom of the landfill to the landfill layer without using energy from blowers, etc.

Heat is generated by aerobic decomposition. Gas is heated and discharged into the atmosphere from degassing pipes by thermal convection. As the gas is discharged into the atmosphere, the inside of the landfill layer becomes negative pressure. At this time, as the bottom collecting pipe is open to the atmosphere, air flows into the landfill layer from the water collecting pipe, and the inside of the landfill layer becomes an aerobic condition, and aerobic decomposition is promoted. That is, leachate flows at the bottom of the collecting pipe, and air flows in the opposite direction to the leachate at the upper part. When leachate stagnates in the landfill layer, the water level exceeds the top of the collecting pipe and the water collection pipe is buried

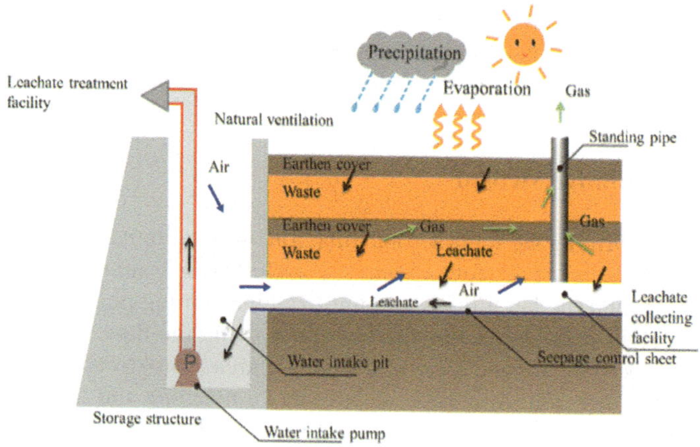

Fig. 2.71 Conceptual diagram of semi-aerobic landfill

in the leachate, the inflow of air is blocked and the inside of the layer becomes anaerobic.

2.6.2 *Function Check Method of Semi-aerobic Landfill*

Since the semi-aerobic landfill became the basic structure of the landfill site in Japan, most domestic landfill sites claim they are semi-aerobic landfill. However, in reality mostly they only have the name of semi-aerobic, and many facilities do not have semi-aerobic functions. The reason is often due to clogging of the water collecting pipe and poor drainage in the landfill layer. In addition, due to the spread of incineration facilities, the raw wastewater quality of BOD, which is an inorganic indicator in leachate, is declining as landfill waste is changed to inorganic substances mainly. Therefore, confirmation methods of whether or not the semi-aerobic landfill is functioning will be stated.

2.6.2.1 Qualitative Method

In every day's maintenance and management, the color of leachate can be used as an indicator to qualitatively confirm When air flows into the landfill layer, the iron in the covering soil, etc. is oxidized, and the color of leachate is brown or reddish brown. On the other hand, when the inside of the landfill layer is in an anaerobic state, iron is reduced to show a black color. In particular, when the leachate stays in the landfill layer, and is in the internal storage state, the inflow of air through the water collecting pipe is blocked, and leachate becomes black. Unavoidably, it is necessary to eliminate internal storage state before the leachate in the storage structure becomes black. If leachate turns black, a large amount of air is required for oxidation in the leachate treatment process, and the amount of flocculant added increases.

2.6.2.2 Quantitative Method

Methods for confirming the semi-aerobic function by quantitative methods include ① measuring the amount of air flowing into the landfill layer from the end of the water collecting pipe, and the landfill gas amount from the venting pipe and gas analysis. The amount of air and gas are obtained by multiplying the flow velocity measured by the flow velocity meter and the cross-sectional area. Smoke can be injected from the end of the water collecting pipe with a smoke machine and it is also possible to visually check. Also, there is a method of periodically inserting a self-propelled camera in the water collecting pipe, photographing, checking the sediment deposition situation inside the pipe, clogging of the water collecting pipe perforated section, and cleaning as necessary. Particularly since the proportion of incineration residues has increased recently, Ca in the incineration residue elute into leachate and be oxidized to generate Ca scale, which may cause clogging, so regular cleaning is desired. It can be cleaned relatively easily with a combination of a camera used for the maintenance and management of the sewer pipe duct and high-pressure washing water. However, it is necessary to consider the arrangement and the diameter of the water collection pipe at the time of designing and construction so that the camera can be inserted. Figure 2.72 shows the example of measuring the amount of air flowing from the water collecting pipe. Figure 2.73 is a case of capturing in the collection pipe. Earth and land deposited in the pipe and scale are observed. Figure 2.74 is a photograph after washing with high-pressure washing water.

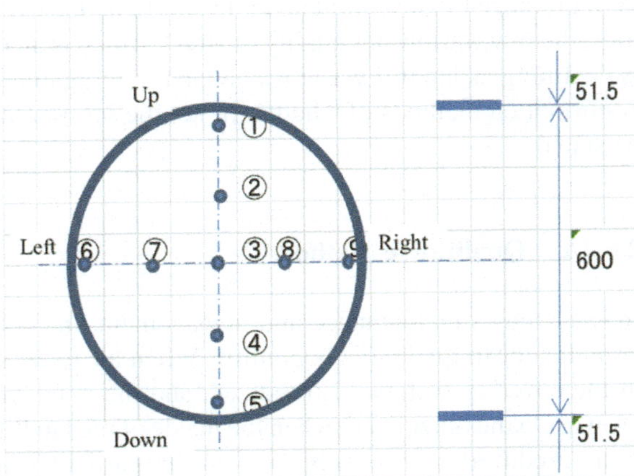

Fig. 2.72 Measurement example of air flow rate of collecting pipe (Higuchi et al. 2015)

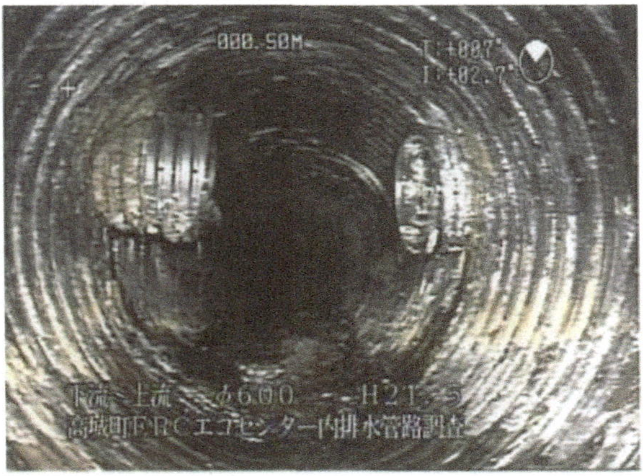

Fig. 2.73 Capturing photo in water collection pipe (Higuchi et al. 2015)

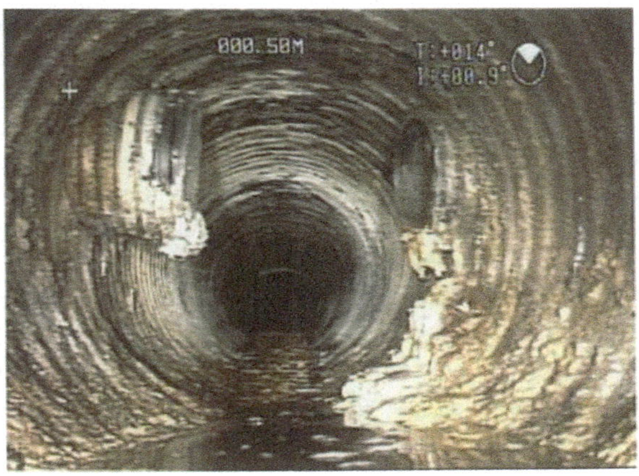

Fig. 2.74 Water collecting pipe after washing (Higuchi et al. 2015)

2.6.3 Inhibitors of Semi-aerobic Landfill

The semi-aerobic landfill functions by supplying the air flowing from the water collecting pipe into the landfill layer. The obstructing factors are cases caused by the landfilling work as well as the storage of leachate in the landfill as described above, and cases due to landfill waste materials and climate.

2.6.3.1 Landfill Work

When rain falls on the protective soil of the unfilled land in the early stage of the landfill, the surface of the protective soil is disturbed. The sediment of large particle size precipitates first, and the small grain size silt precipitates at last according to

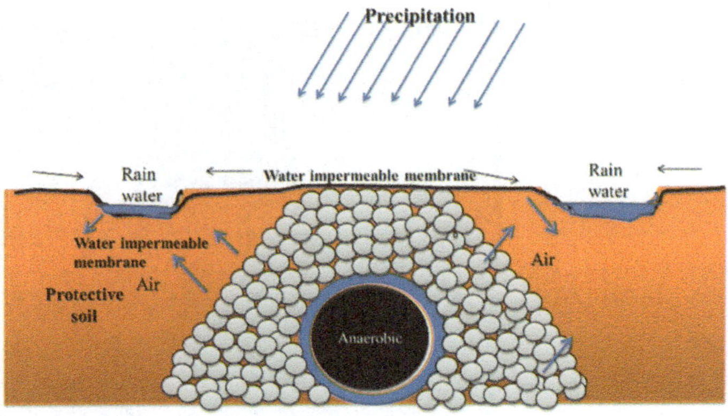

Fig. 2.75 Example of blocking water collecting pipe with protective soil (before the landfill)

the Stokes' law. Since the silt portion is impermeable, impermeable membrane is formed. Rainwater is pooled because of the membrane and cannot penetrate into the protective soil (Fig. 2.75).

As a result, the water collection efficiency decreases and the internal water level tends to rise, and the anaerobic state tends to occur. Moreover, if the protective soil is constructed in the same way as the top end of the collecting pipe in order to facilitate the landfilling work, the soil particles disturbed when raining will block the top end of the collecting pipe, impairing the water collecting and ventilation function, and the inside of the landfill layer becomes anaerobic (Fig. 2.76).

For this reason, it is recommended that the top end of the water collecting pipe and the protective soil end be separated by 50 cm or more on the trunk line as described in the planning design procedure (Fig. 2.77). So the water collection pipe performs

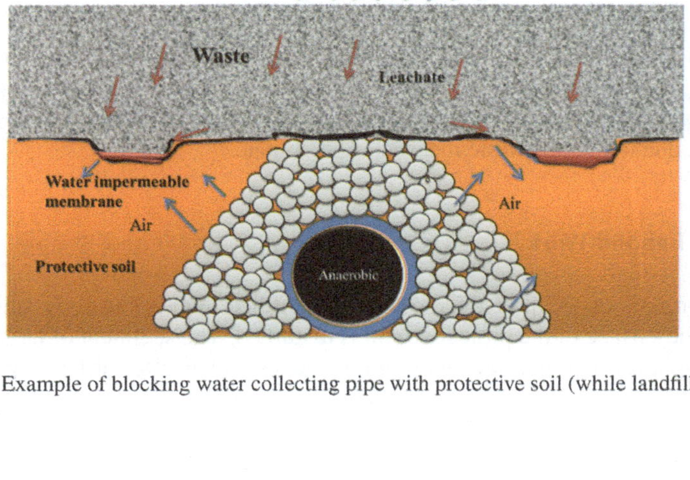

Fig. 2.76 Example of blocking water collecting pipe with protective soil (while landfilling)

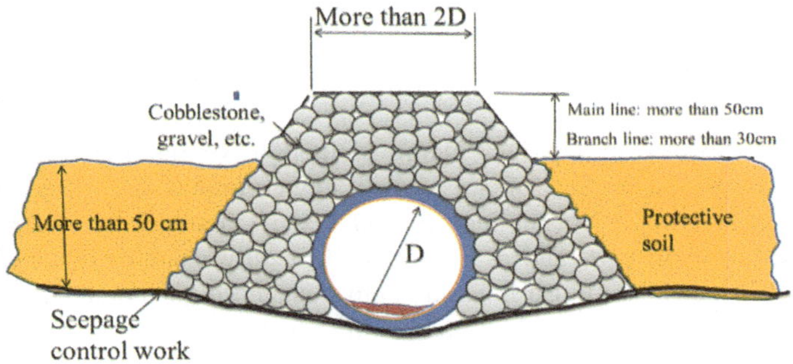

Fig. 2.77 Water collecting pipe and protective soil

air supply and leachate drainage function to maintain the semi-aerobic landfill and it is the most important facility as the landfill gas removal facility. Recently, cases of using grain crushed stone for protective soil are increasing to raise water collection and ventilation efficiency.

2.6.3.2 Landfill Waste Materials

The food preservation method is shown in Fig. 2.69 is contrary to the concept of semi-aerobic landfill aiming at organic substance decomposition. In the 1960s when the concept of semi-aerobic landfill was established in Japan, the incineration rate was <50%, and there was no exhaust gas regulation. So it was consistent with the conditions to demonstrate bioreactor function. Since 1990, after the incineration rate of combustible waste became nearly 100% and the exhaust gas regulation became strict, quality of landfill waste changed to high salt, high alkali and turned to be an anti-bio environment. From this, it can be considered at the landfill site mainly of incineration residue, the bioreactor function does not work in the early stage. The semi-aerobic function works in the late-stage of the landfill when the chloride is lowered and neutralized by rain.

2.6.3.3 Climate

Japan is located in the Asian monsoon area and has a temperate climate with high precipitation. It is not an exaggeration to say that semi-aerobic landfill was born because of the background. Therefore, it is thought that semi-aerobic landfill is difficult to be used in dry or low-temperature cold areas.

2.6.4 Intermediate Processing Residue Landfill and Semi-aerobic Landfill

Recent landfill waste in Japan is mainly composed of intermediate processing residues and the proportion of waste with high organic substance content is decreasing. For general waste, incineration residue and crushed incombustible residue are the main part, and as described in quality changes of landfill waste in Sect. 2.1, especially fly ash contains high concentration inorganic salts as Cl, Ca, K. For this reason, the inside of the landfill layer becomes a high alkaline high salt environment, and it seems that the bioreactor function, which is the maximum function of the semi-aerobic landfill, hardly works. However, as shown below, under any harsh environments bacterial flora is formed according to the environment due to contamination of bacteria from the outside and decomposition proceeds only slightly though.

In Fig. 2.78, the simulated landfill tank with a diameter of 300 mm and a perforated height of 1000 mm is filled with incineration residue (Desalting agent: highly reactive slaked lime, fly ash stabilizer: PIP, bottom ash 70%, fly ash 30% by weight ratio), and leachate monitoring is performed by artificial sprinkling. The bacterial flora was examined every six months from immediately after the start of the experiment. After the start of the experiment, the common soil bacterium Ralstonia was main, but gradually the environment became wet, Acinetobacter genus, widely distributed in the natural environment, increased and is going to stabilize. The pH of the leachate

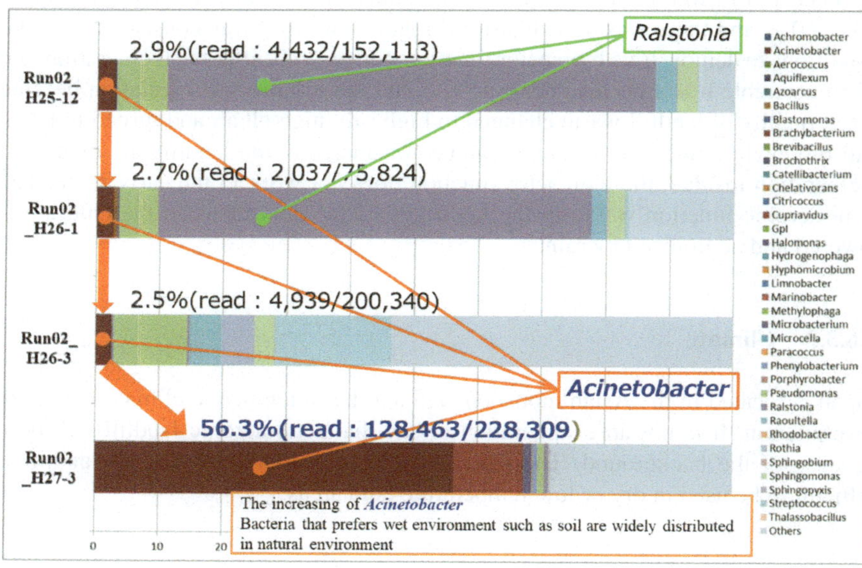

Fig. 2.78 Bacterial flora in incineration residue

at the beginning of the experiment was 12–13, and the Cl^- covers 15%, and after 2 years pH was about 12 and the Cl^- was lowered to 300 mg/L or less.

Because there are many cases where slaked lime is used as an exhaust gas treatment agent in incineration residue, the fly ash contains a large amount of inorganic salts such as Cl and Ca. In addition, when organic chelate is used for the stabilization treatment of fly ash, it contains much T-N and COD because of chelate. In the case of landfill disposal, inorganic salts are leached at high concentration because of their high solubility in water. As described above, the biological activity condition is low in high salt and high pH environment, but the salt decreases relatively fast. Figures 2.79 to 2.80 shows the change of pH and residual chelate over time in the leachate, when simulated landfill tank filled with incineration residue and is maintained in semi-aerobic and anaerobic (internal storage by leachate) state.

From the figure, it is found that pH and residual chelate are also higher when in the anaerobic (immersed) state. This is because the contact time with water is longer and the elution amount is also larger. Particularly when it is exhibiting a high pH due to the effect of slaked lime, an exhaust gas treating agent, it is necessary to neutralize it by contact with air and lower the pH. So, it is important not to let the water stay in the landfill layer in order to suppress elution of pollutants contained in the waste and manage appropriate leachate treatment even in the landfill site mainly of the incineration residue. On the other hand, decomposition of organic substances, which is a feature of the semi-aerobic landfill, is considered to be hardly progressed because of high pH. It is thought that during the high pH, as it is washed out by precipitation, it flows out as leachate.

It is considered that the decomposition by microorganisms begins gradually with the decrease of pH and that the semi-aerobic function performs in low load condition

Fig. 2.79 pH due to different landfill structure

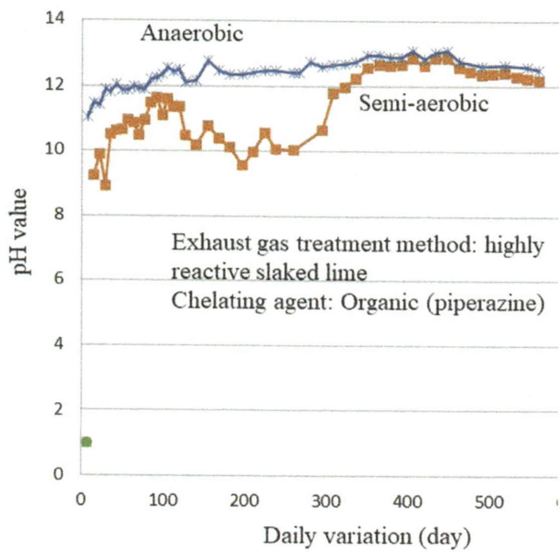

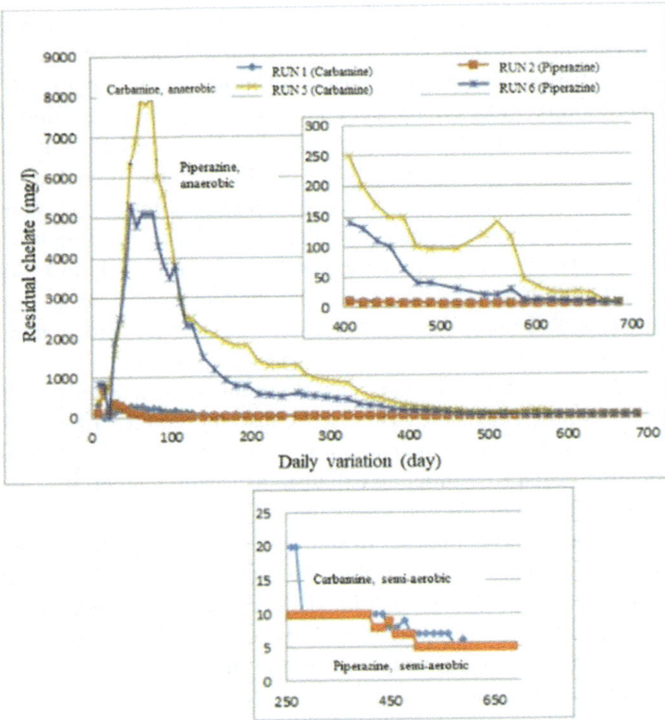

Fig. 2.80 Residual chelate due to difference in landfill structure

of organic substance in the late stage of landfill. That is, even in the landfill site of main incineration residue, it is believed that semi-aerobic landfill is effective as it suppresses the high concentration of inorganic salts, reduces the pH at the early to the middle stage of the landfill and bioreactor functions in the low-load state in the late stage of the landfill. However, since T-N and COD derived from residual chelate are hardly biodegradable substances, semi-aerobic bioreactor does not function. For this reason, it is important to avoid storage of leachate in the landfill site of intermediate processing residues, especially incineration residue, and stabilize fly ash by minimizing residual chelate at incineration facilities.

2.6.5 *Semi-aerobic Landfill at a Covered Type Landfill Site*

Semi-aerobic behavior functions because of the landfill waste materials and air inflowing into the landfill tank at the covered type-type landfill site. Rather, due to the ventilation facility for maintaining the work environment indoors, the indoor air is forcibly excluded, and negative pressure is generated in the landfill layer. The

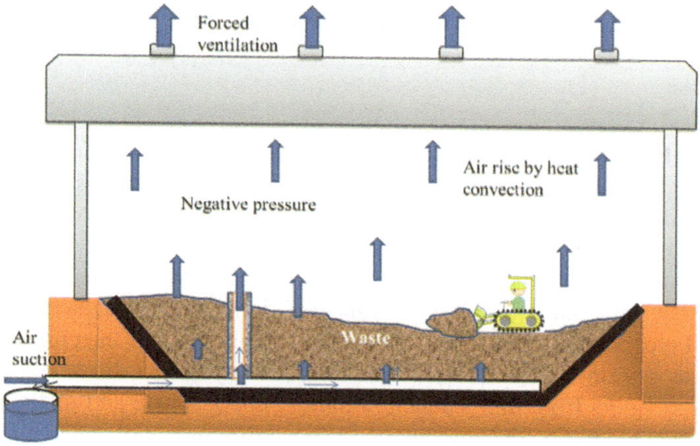

Fig. 2.81 Air flow in a covered type-type landfill site

air circulation effect is much improved than the open-type landfill site. Figure 2.81 schematically shows that negative pressure is applied to the interior of the room by exhaust from the operation of the roof ventilator, and the inflow amount of air increases due to suction by negative pressure to the air flowing from the water collecting pipe. In addition, in the covered type-type landfill site, there is no internal storage of leachate in the landfill layer by artificial sprinkling to stabilize landfill waste and leachate control system. So, residual chelate which interferes with leachate treatment and rapid increase in salt concentration is suppressed, good maintenance and management become possible.

2.6.6 Semi-aerobic Landfill at the Full Landfill Site

As stated in Chap. 1, in 1960, the incineration rate was already approaching 50%, and it seems that the full landfill site disappeared around 1990. For this reason, data showing the effect of the semi-aerobic landfill at the actual full landfill site is extremely small. Among them, a case of a large simulation experiment conducted in Beijing by the Urban Construction Institute of China and Fukuoka University will be introduced.

2.6.6.1 Demonstration Experiment in China

a. Purpose

As the landfill site in China has a high proportion of full landfill that does not perform the incineration, high concentration of organic leachate (COD_{Cr}: 10^3–10^4 mg/L) occurs at most landfill sites, which exacerbates the local environment. Strict wastewater regulation (COD_{Cr}: 100 mg/L, etc.) was carried out with the aim of 2011. So, it is required to develop an economical leachate treatment system which clears wastewater regulation for high concentration organic leachate. In addition, the landfill gas power generation is stopped at the time when the concentration of the CH_4 gas becomes 30% or less, and the efficiency is low. However, even after the stoppage, odors from high concentration organic leachate and methane gas are generated, and there is a concern that the local environment will deteriorate. In view of these backgrounds, environmental impact reduction type landfill system which can stabilize newly at an early stage by suppressing the load on leachate treatment facilities and CH_4 gas generation is required.

b. Experiment outline

Seven steel lysimeters (1000 mm in diameter × 3000 mm in height) were used and filled with waste. Assuming anaerobic landfill, semi-aerobic landfill and aerobic landfill, incorporate geonet for promotion of drainage to see the effect of early stabilization, add-on landfill to see the effect of landfill layer thickness and a chemical oxidation method to reduce COD after the end of the landfill were incorporated. Experimental conditions are shown in Table 2.26, and simulated landfill tank outline is shown in Fig. 2.82.

The experimental conditions are that aerobic landfill 4 layers, semi-aerobic landfill 2 layers, anaerobic landfill 1 layer, and in the aerobic layer 1 and the semi-aerobic layer 6, in order to promote the aeration and water flowing effect, geonet was installed in the waste layer. The aerobic layers 1 and 2 were filled with 100% of waste, the aerobic layer 3 33% of waste and the aerobic layer 4 was filled with 66% of waste.

The packing density was 0.5–0.6 t/m^3, the sprinkling water was artificial sprinkling and sprinkled amount was 7L/week · m^3 referring to annual average precipitation of 465 mm/year in Beijing. The aeration tank was aerated at 12 l/min · m^3.

As shown in Fig. 2.82 the diameter of the lysimeter is 1 m, the height is 3 m, and the capacity is 2.355 m^3. The inner wall of the Lysimeter was subjected to antiseptic treatment, and also heat retaining in winter as Beijing's climate. To prevent the diffusion of generated gas, a lid was attached to 31 the top. A gravel layer of 30 cm was installed at the bottom, and a perforated pipe of 5 cm in diameter was installed on the gravel.

An air piping capable of forced ventilation was installed in advance at the bottom part, and a vent pipe was installed on the upper part of the perforated pipe. There, the waste was landfilled by 270 cm inside the tank. At the upper 30 cm of the landfill

Table 2.26 Experiment condition of simulated landfill tank (Gao H 2010)

Landfill simulated tank	1	2	3	4	5	6	7
Landfill structure	Aerobic	Aerobic	Aerobic	Aerobic	Semi-aerobic	Semi-aerobic	Aerobic
Amount of filling waste	900 kg	900 kg	300 kg	600 kg	900 kg	900 kg	900 kg
Volume of water for spraying	7 l/week	7 l/week	7 l/week	7 l/week	7 l/week	7 l/week	7 l/week
Ventilation volume	22.6 l/min	22.6 l/min	7.53 l/min	15.1 l/min	–	–	–
Note	Waste filling rate 100% Install geoknitted for ventilation and water flow	Waste filling rate 100%	Waste filling rate 33%	Waste filling rate 66%	Waste filling rate 100%	Waste filling rate 100% Install geoknitted for ventilation and water flow	Waste filling rate 100%

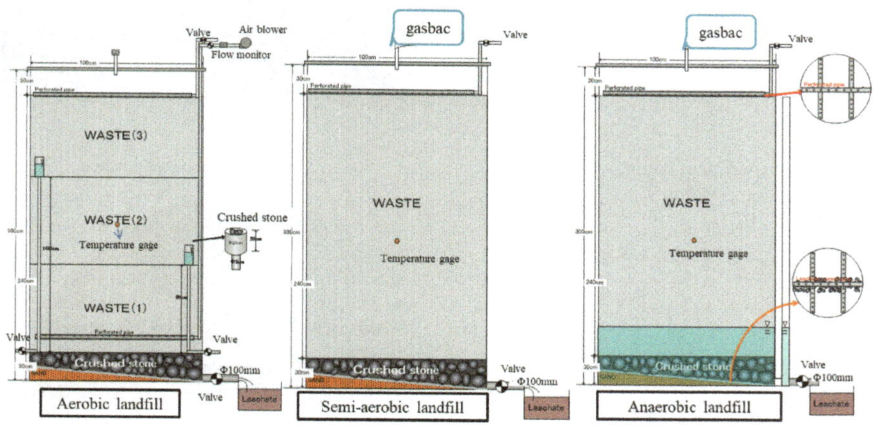

Fig. 2.82 Outline of simulated landfill tank (Gao H 2010)

surface, a water sprinkler was installed, and finally, the lysimeter was closed with a lid (Figs. 2.83, 2.84). Figure 2.85 shows the composition of filled waste.

The waste used in the experiment was living waste from waste collection center of Shijingshan District, Beijing. The composition ratio was 43.9% for kitchen waste, 9.9% for plastics, 7.8% for paper, 5.1% for earth and sand, and 27.7% for water and others.

Fig. 2.83 Simulated landfill tank (Xu W 2010)

Fig. 2.84 Simulated landfill tank (winter) (Xu W 2010) season

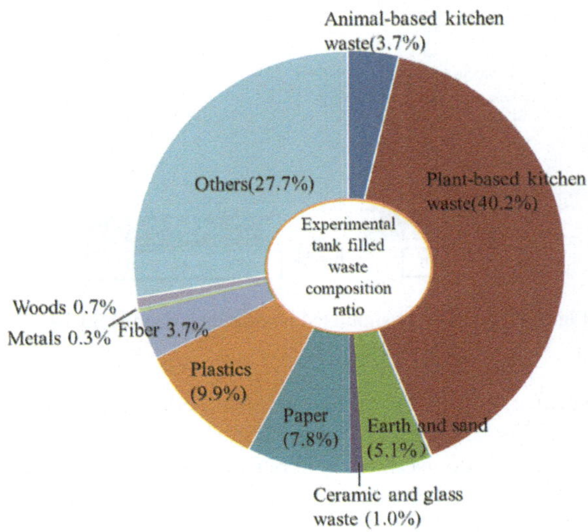

Fig. 2.85 Experimental tank filled waste composition ratio (Xu W 2010)

c. Experimental results

After packing in the lysimeter, the aerobic tank forcedly started aeration. After packing in the lysimeter, the aerobic tank forcedly started aeration. The gas measurement was performed once a month. The amount of water spraying was calculated based on the average precipitation in Beijing over the last 10 years and set to 7 l of tap water per week. The experiment started from May 2008, despite heat retaining in January and February 2009, the simulated landfill tank froze due to the low temperature, so leachate sampling became impossible, which made the experiment stop after 60 days.

- Temperature in tank

Changes of tank temperature with time are shown in Fig. 2.86. In the early stage of the landfill, although the outside temperature and the anaerobic tank are almost interlocking, the temperature in the anaerobic tank has been shifting by about 5 °C higher than the outside temperature. The semi-aerobic tank is 5 °C higher than the outside temperature in the early stage. After 300 days, it has been 10 °C higher. For the aerobic tank, the temperature was 15–20 °C higher than the outside temperature and reached 50 °C at the lapse of 50 days. However, after 130 days, the difference with the outside temperature gradually disappeared and it remained within the range of 5–10 °C. The reason is considered to be that the activity of aerobic bacteria initially increased, and the temperature rose, and in the later period, the aerobic bacteria were inactivated due to the reduction of organic substance.

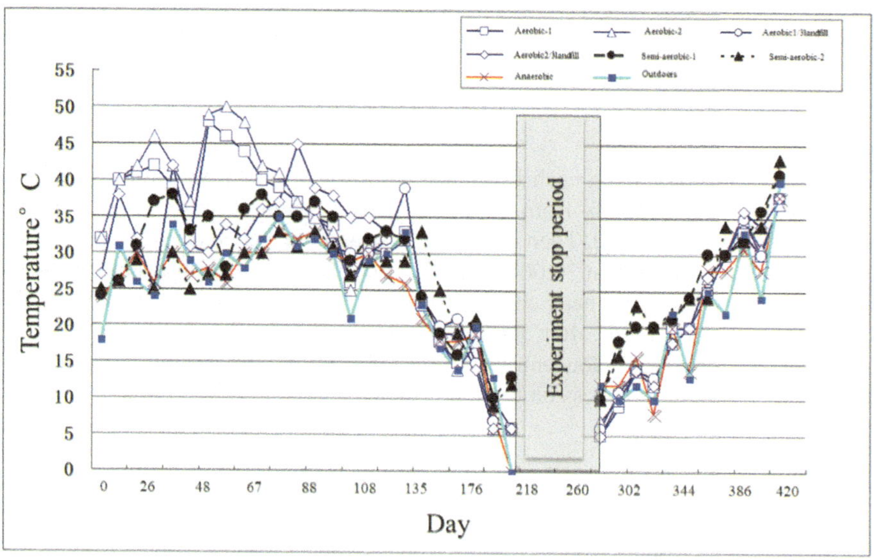

Fig. 2.86 Changes over time of temperature (Gao H 2010)

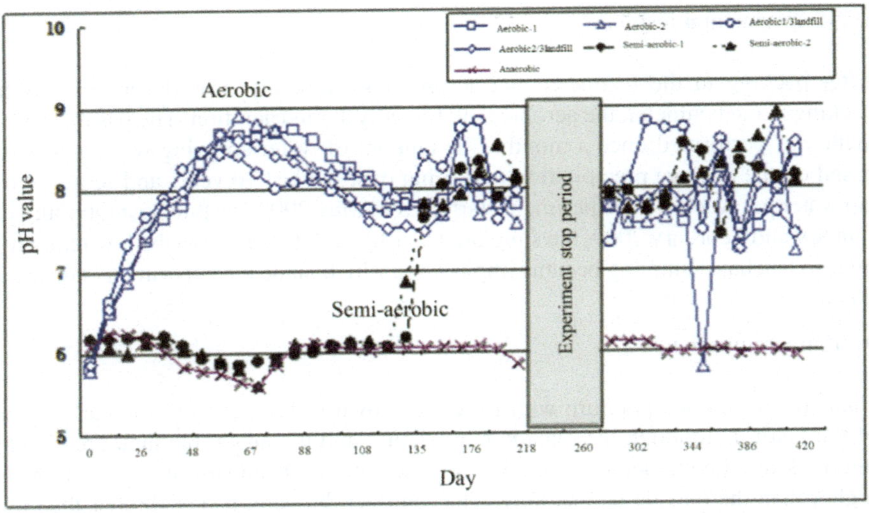

Fig. 2.87 Changes over time of pH value (Gao H 2010)

- PH

Changes of pH with time are shown in Fig. 2.87. In all the tanks, the initial pH was weakly acidic around 6.0, but only the anaerobic tank remained continuously around pH 6.0. The semi-aerobic tank sharply increased after 120 days and kept at pH 8–9. The pH of the aerobic tank increased immediately after waste was filled and reached 8–9 after 40 days. The reason for the low pH of the anaerobic tank is thought to be the influence of organic acid.

- BOD

The BOD of leachate from each lysimeter is shown in Fig. 2.88. In the anaerobic tank, it decreased gradually from the initial peak concentration of 35,000–10,000 mg/L after 200 days passed. Then it began to rise and reached a maximum of 50,000 mg/L, and remained high at 30,000–40,000 mg/L. The semi-aerobic tank gradually decreased from the initial peak concentration of 30,000–1000 mg/L or less on the 176th day, 400–800 mg/L after 300 days and 100–200 mg/L on 390th day. After reaching the initial peak concentration of 15,000 mg/L, the aerobic tank sharply decreased to 100 mg/L or less on 88th day, then changed at 10–100 mg/L.

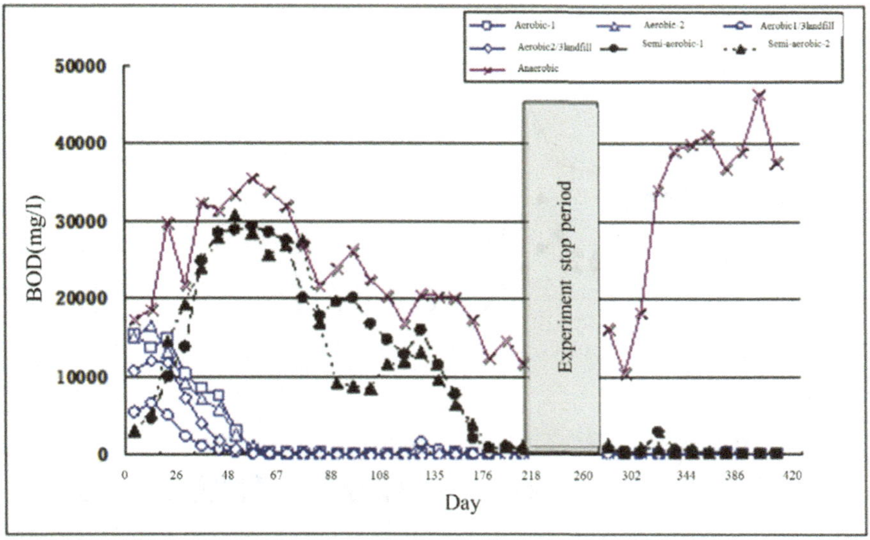

Fig. 2.88 Changes over time of BOD (Gao H 2010)

- CODcr

Changes of CODcr with time are shown in Fig. 2.89. Figure 2.90 is an enlarged representation of the CODcr concentration of 3000 mg/L or less.

In the anaerobic tank, it decreased gradually from the initial peak concentration of 81,000–45,000 mg/L after 200 days passed. After the experiment stopped, it began to rise and reached a maximum of 66,000 mg/L. Then it gradually began to decline and remained at a high concentration of around 50,000 mg/L. In the semi-aerobic tank, it gradually decreased from the initial peak concentration of 61,000 mg/L and became <1000 mg/L on the 176th day. And it slightly increased to 4500 mg/L after the experiment stopped. Since then, it has been around 900 mg/L from the 390th day. In the aerobic tank, the initial COD concentration was 38,000 mg/L, which showed a sharp decrease trend thereafter. It decreased to 1000 mg/L or less from the 150th day onwards and then changed to around 500 mg/L.

- NH_3-N

Changes of NH_3-N with time are shown in Fig. 2.91. Figure 2.92 is an enlarged representation of the low concentration range of NH_3-N concentration of 500 mg/L or less. For the anaerobic tank, it showed a sharp upward trend from the initial concentration of 650 mg/L and reached 1000 mg/L when the experiment stopped. After that, it reached 1400 mg/L on the 274th day and gradually increased continuously. For the semi-aerobic tanks, the tank 1 and tank 2 showed 1370 and 1500 mg/L respectively

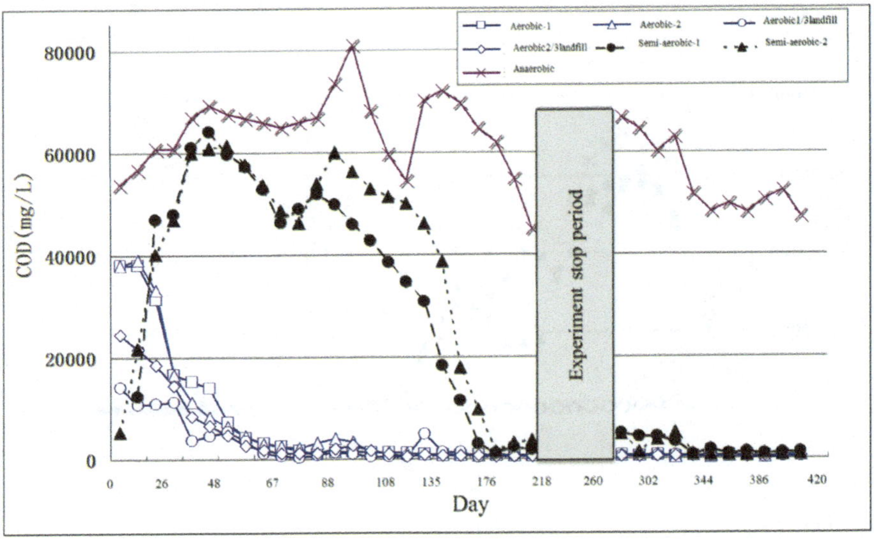

Fig. 2.89 Changes over time of CODcr

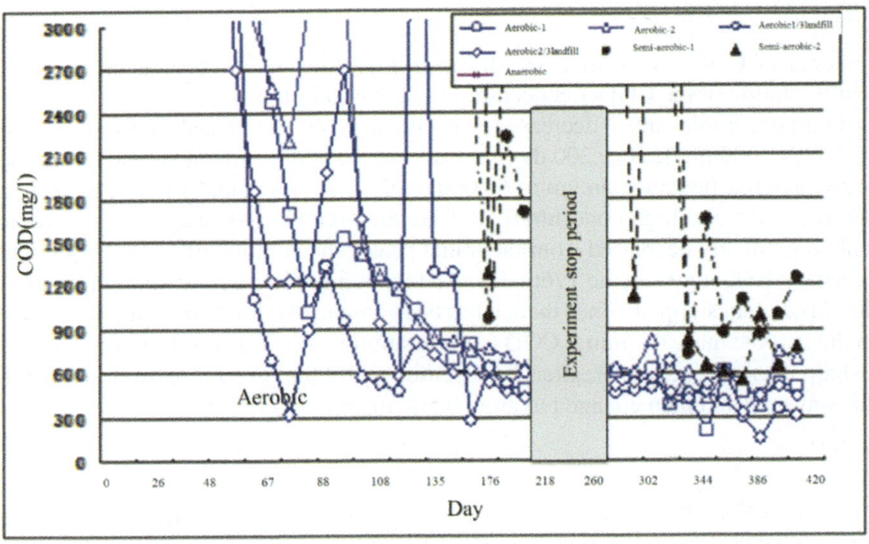

Fig. 2.90 Changes over time of COD_{cr} (Low concentration range) (Gao H 2010)

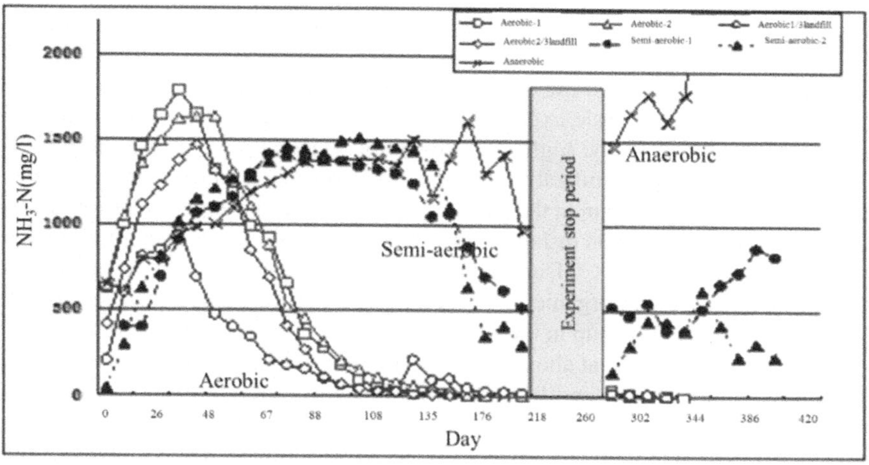

Fig. 2.91 Changes over time of NH_3-N (Gao H 2010)

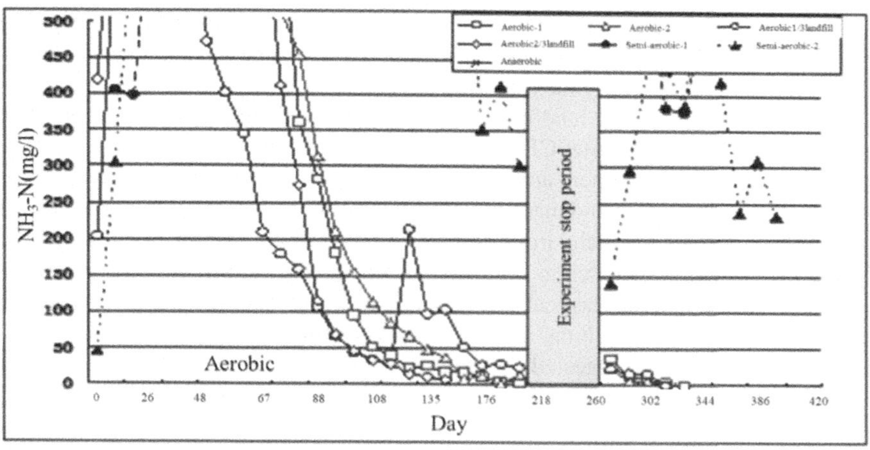

Fig. 2.92 Changes over time of NH_3-N (Low concentration range) (Gao H 2010)

on around 90th day and showed a gradual decrease trend thereafter. From the 218th day, the tank 1 became 500–800 mg/L and the tank 2 was 200–400 mg/L.

For aerobic tanks, both the tank 1 and the tank 2 had an initial peak concentration of 1790 mg/L, the tank 3 (1/3 filled) was 1000 mg/L, while the tank 4 (2/3 filled) was 1500 mg/L. After that, it showed a sharp declining trend and decreased to <50 mg/L in all aerobic tanks on day 150 (after the 316th day, the NH_3-N experiment ended).

2.6.6.2 Effect of Semi-aerobic Landfill on Full Landfill

As a result of simulation for the full landfill or the landfill in the urban area of China, it was the same as the basic experiment in Japan. The early stabilization speed of the aerobic landfill was the highest, followed by the semi-aerobic landfill and the anaerobic landfill. The stabilization speed of semi-aerobic landfill is slower than the experimental result in Japan in the past is that the scale of the experiment was scaled up and that waste materials in Japan and China, especially water content was large. However, both BOD and CODcr stabilized in almost one year after completion of landfill under high-load organic environment, and they were expected to stabilize early even at further scale up in the field. In climate change as well, the stabilization rate does not change even at about 30% (600 mm/year) in precipitation compared to Japan, even under severe conditions such as low-temperature freezing in winter, the bioreactor function recovers promptly as the temperature rises, the climate coverage of the semi-aerobic landfill is judged to be wide. In addition, it was suggested that the landfill height of aerobic landfill is almost proportional to the initial leachate concentration, and concentration control by accumulated landfill method is possible. Regarding the aerobic tank, it stabilized in about one year in the experiment. For this reason, when the aerobic landfill was carried out, the capacity of the blower capacity was set as landfill waste for 1 year, suggesting that reasonable maintenance can be carried out by appropriately moving the blowing point.

Currently, the anaerobic landfill is implemented in full landfill in China and other areas, and at some landfill sites CH_4 gas recovery power generations are set. However, maintenance and management are necessary even after recovering CH_4 gas, and it is said that the maintenance and management period after the completion of the landfill is over 30 years. Under such circumstances, it is possible to stabilize air supply by setting the structure to switch to semi-aerobic or aerobic landfill after gas recovery. As a result, LCC can be drastically reduced as well. Incineration facilities began to spread in some areas of China, and the is approaching Japan's state in the 1980s. From these facts, it is suggested that the introduction of semi-aerobic and aerobic landfill can significantly reduce the generation of greenhouse gases such as CH_4 and there are sufficient economic merits.

2.7 Abolition of the Landfill Site

In 1998 the abolition criteria for the landfill sites were established. Generally, the landfill site is often built with 15 years of landfill. For this reason, the landfill site constructed after the abolish criteria has been established will reach the end of the landfill in 15 years in around 2013 and will be abolished based on the criteria. In general, there are many sites continuing the landfill for more than 15 years due to the extension of life to reduce the amount of waste. Meanwhile, it is difficult to set strict wastewater standards such as Cl^- and in the above-mentioned covered type landfill site, maintenance and management are necessary for a considerably long period after the completion of the landfill, and it takes time, so the existence of landfill sites become a concern. Also, with regard to the landfill site constructed before 1998, since the current abolition criteria are not applied, most of the landfill sites abolished now were constructed before 1998.

2.7.1 Closure and Abolition of the Landfill Site

Figure 2.93 shows the outline from the start to the abolition of the landfill site.

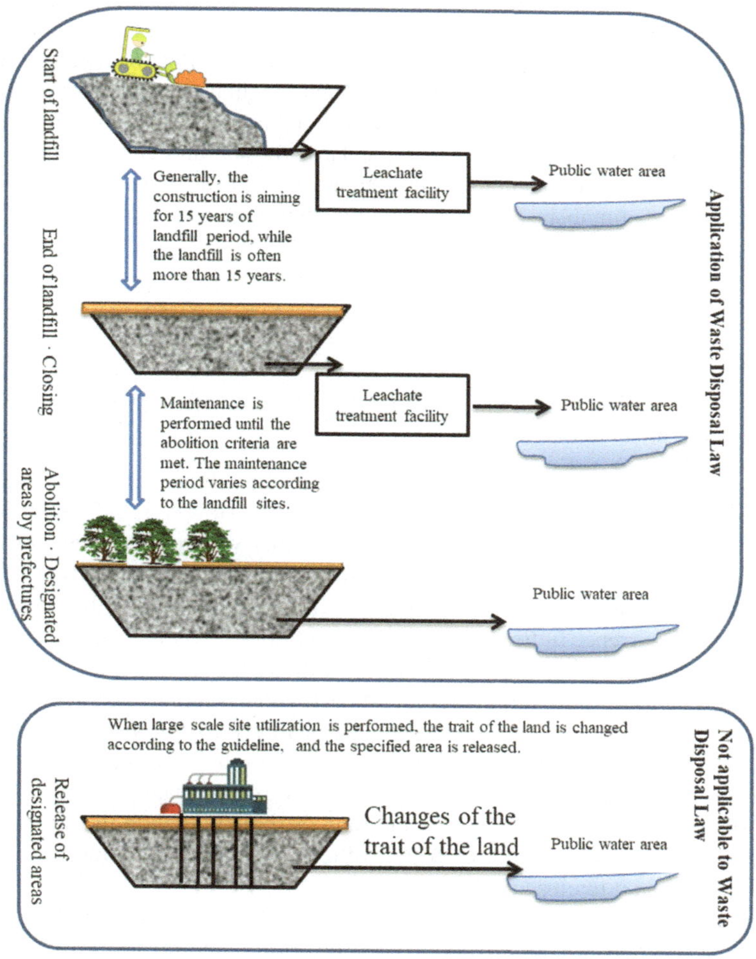

Fig. 2.93 Schematic drawing of abolition of landfill site

The landfill site is generally built with a landfill period of around 15 years. After completion, measures are taken to close down, and maintenance will be done until the landfill layer is stabilized. After that, abolition procedures are carried out after satisfying the abolish criteria. The period from closure to abolition varies depending on the maintenance plan of the landfill site and landfill waste materials, generally, it takes more than ten years. That is, landfill period and the period till the abolition will take several decades of maintenance management. In the period of maintenance management, standards according to standard ministerial ordinance apply. After the abolition criteria are satisfied and the abolition has been carried out, the landfill site will no longer be applied to the maintenance standard but based on Article 15–17 of the Waste Disposal Law, prefectures designate waste as a designated area as land underground, and the waste disposal law continues to be applied. Thereafter, when changing the trait of the land by large scale site utilization, the specified area is released when satisfying "criteria on the method of land traits change" in Article 12–40 of the Waste Disposal Law Enforcement Regulation and will be utilized. In other words, the waste disposal law will not apply at the time of the release of the designated area, and environmental standards for soil and groundwater will be applied. However, environmental quality standards for soil are not applicable in cases where it can be distinguished from the general environment even after the release of the designated area.

2.7.1.1 Closure

According to the Ministerial Ordinance, the following is stipulated on the closure of landfill where the landfill was completed.

① When the landfill is completed, earth and sand with a thickness of approximately 50 cm or more should be used to cover the opening (However, for landfills where the necessary measures are taken to prevent rainwater from entering, it shall be closed with a covering which is equal to or better than the waterproof work). To prevent scattering and outflow of waste from the opening of the landfill site, and the occurrence of fire, odor and the penetration of rainwater, it is necessary to close the opening of the landfill site with earth and sand of a thickness of about 50 cm or more after rolling compact. Other similar coverings are those having the same strength and water permeability as 50 cm thick earth and sand. Regarding covered type landfill sites, from the viewpoint of preventing penetration of rainwater, lying a nonwoven fabric on the waterproofing layer, and furthermore, covering with a soil and sand or something with the effectiveness of waterproofing and light blocking, sufficient strength and durability equal to or more than it to close.

② For closed landfill sites, necessary measures to prevent damage to the covering should be taken. For closed landfill sites, necessary measures to prevent damage to the covering should be taken.

③ Make records of inspections, checks and other measures carried out according to types and quantities of landfill waste and maintenance and management of landfill sites, and to preserve records until abolition.
④ If there are internal partition facilities at the landfill site, the types and quantities of landfilled waste shall be recorded for each section surrounded by the partition.
⑤ In addition, when inspections of the retaining wall and discharged water, repair of waterproof work, etc. are carried out, the results shall be recorded and preserved until the abolition.

2.7.1.2 Abolition

Table 2.27 shows a summary of the abolition criteria for the landfill site.

Regarding the qualitative expression of the abolition criteria and concrete survey methods, in March 2002, the Waste Society of Japan (now, Waste Resources Circulation Society) Waste Landfill Disposal Research Subcommittee issued a survey on the waste landfill site waste disposal criteria Evaluation method (Waste Landfill Research Group of Waste Society of Japan 2002) has made various interpretations and suggestions. The outline is as follows.

① Individual Criteria 6. "Generation of gas is hardly observed or increase in gas generation amount is not recognized for 2 years or more." The amount of landfill gas is defined as the sum of the amount of methane gas and the amount of carbon dioxide gas. For more than two years, it has been measured at least once every three months in all vent pipes, including data for summer (July or August) When the gas generation amount is less than the measurement lower limit (about 1 l/min), it is said that the abolition criterion is satisfied. Also, if necessary, it is desirable to add methane gas composition judgment conditions, and it is preferable to measure the concentration of methane gas under covering at least one place in 1000–1500 m^2 except for the vent pipe, When the total methane gas composition is 5% or less, satisfy the abolishment criteria. is described.
② Individual criteria 7."The inside of the landfill is not abnormally high temperature with respect to the surrounding underground temperature." "The difference between the temperature inside the landfill site and the surrounding ground is <20 °C." is described. Regarding the method of temperature measuring, "appropriate part is selected from ventilators, and the depth is judged not to be affected by the temperature of the earth surface measured at 1 m interval in the vertical direction from the ground surface using thermocouple such as thermocouple type. It will be compared with the underground temperature of the same depth in the surrounding land. In addition, refer to the "Stabilization Monitoring Manual" for the selection of temperature measurement points inside the landfill site." is described.
③ Individual Criteria 12. (1) "It meets the standard for drainage standard for 2 years or more"

Table 2.27 Outline of the abolition criteria (Environment Agency 1998)

	Stable type landfill site	Controlled landfill site (municipal waste disposal site)	Strictly controlled disposal site
Common standard	1. Measures to prevent odor emission have been taken outside the landfill site		
	2. Measures to prevent fire have been take		
	3. Measures to prevent the generation of insects, etc. have benn taken		
	4. There is no risk that the groundwater quality meets the standard or does not meet the standard		
	5. There are no problems in terms of preserve the living environment		
Individual standard	6. It is admitted that there is almost no gas generation or not increased for 2 years		–
	7. The temperature inside the landfill site does not show an abnormally high temperature than the surrounding ground temperature		
	8. The opening is closed by a cover of more than 50 cm		
	9. Landslide, settlement prevention work, rainwater drainage facilities comply with structural standard	11. It meets the structural standards except for enclosures, labels, adjustment reservoir and leachate treatment facilities	14. Landslide, settlement prevention work, partitioning facilites conform the structural standard
	10. Leachate meets the following standards. (1) Conforms the groundwater standard. (2) BOD is below 20 mg/L.	12. The following items can be met for more than 2 years. (1) Drainage standard: one or more times in 6 months. (2) BOD, COD, SS: one or more times in 3 months	15. Closed by a cover that has the same effect as the perimeter partitioning facilities
		13. No sinks, cracks or other deformations were observed in the cover of the landfill site that only fills waste that does not contain rainwater and without rotting and produce retained water	16. Measures have been taken regarding waste and perimeter partitioning facilities

Fig. 2.94 Measurement example with a gas measuring instrument

Since the wastewater standards are standards prescribed in the maintenance and management plan, when voluntary wastewater standard values including unregulated water quality items are decided in the maintenance plan, it is the abolition criterion. For example, when removal of chlorine ion or calcium ion is specified as wastewater standard of a maintenance plan when BOD wastewater standard value is set to 10 or 5 mg/L by living environment influence survey, and it becomes the abolition target value. In other words, the more advanced the leachate treatment facility at the time of construction of the landfill site, the stricter the abolishment requirement becomes.

④ Individual Criteria 13. "A landfill site where rainwater does not enter and landfill only waste that does not decompose or generate retained water" and "a structure that rainwater can't enter" include a covered-type landfill site as roof-type or capping. "waste that does not decompose, or generate retained water" is not specifically described, but it is considered to correspond to melts, incineration residues, etc.

Figure 2.94 shows the measurement example with a gas measuring instrument.

2.7.1.3 Designated Area and Its Release

The landfill site deviates from the application of waste treatment law. However, in the future, regarding the site where there is a fear that obstacles to the preservation of the living environment may arise if the trait of the land is changed due to the use of the landfill site, the prefectural governor will, in accordance with Article 15–17 of the Waste Disposal Law, designate the land area as the designated area, the waste disposal law will continue to apply. The obstacles to the preservation of the living environment include contamination of public water areas and groundwater due to scattering and outflow of waste and leachate when excretion of land at the time of the ultimate land use. Emanation of gas, occurrence of insect pests, etc. odor, noise ad vibration occurrence to the surrounding environment may also be obstacles to preservation of the living environment. The landfill sites could be the designated areas are those constructed after 1970 when the Waste Disposal Law was enacted.

In the case of using the land (ultimate site) designated in the designated area, land can be used by the release of the designated area. The release of the designated area is possible when the total amount of waste is removed or when the waste is decomposed and stabilized so that the gas and leachate are not greatly different from ordinary land and appearance cannot be distinguished from the soil. If the stabilized waste remains after the designated area is released, the soil environmental standards will be applied, but if the condition continues to be distinguishable from the general environment after the release of the designated area, the soil environmental standard is not applicable.

2.7.2 *Issues of Abolition*

After the abolition criteria were established, ten years have passed. Landfill sites applying abolition criteria also began to appear. Several issues appear among them.

2.7.2.1 Abolition of Covered Landfill Site

Recently, the abolition of covered-type landfill sites where construction cases increased was described in Sect. 2.4, and it is shown below again.

Generally, it is considered to abolish covered landfill sites by remaining the roof or capping after removing the roof as measures to prevent rainwater from entering.

Meanwhile, Article 1 (5) D of the Technical Standard for Waste Disposal Law stipulates that "a landfill site is a place that has solid and durable structures of pipe ducts and other collection facilities that effectively collect retained water, etc. and can promptly discharge it.... In the case of landfill sites where the necessary measures are taken to prevent rainwater from entering, facilities mentioned above are not required because the landfilled municipal waste will not decompose or generate retained water. (Excluding water surface landfill sites). Installation of water collecting pipe and leachate treatment facility is excluded (Japan Environmental Sanitation Center 2014). Depending on the performance guidelines, the landfill site of municipal waste is obliged to install a water collecting pipe even in a covered-type, but its function will also be stopped if stopping sprinkling. As a result, if Article 1 (5) D is applied, "landfill where the necessary measures are taken to prevent rainwater from entering" is applicable to the landfill site, but for provision "landfill municipal waste which does not decompose and generates retained water", incineration residue, molten slag, etc. are considered to be applicable although there are no detailed provisions. It is considered possible to stop sprinkling water and to abolish it after taking measures to prevent rainwater from entering at the time when the landfill layer became stabilized and will not decompose by artificial sprinkling through landfill period and after the landfill. In this case, technical standard for maintenance of the landfill site of municipal waste pursuant to the provisions of Article 8-3, paragraph 1 of the Act, (17) "Regarding the landfill prescribed in the proviso, it is necessary to cover the surface of a nonwoven fabric laid in a waterproofing layer provided with any of the requirements of 1 (a)–(c) of the same item and then covered the surface with earth and sand, or something which is equal to or more than the effectiveness of the seepage control, light shielding, the strength and the durability" (Japan Environmental Sanitation Center 2014a). In other words, if it falls under the "landfill site specified in the proviso," a seepage control layer equivalent to the bottom waterproofing will be installed on the surface and final earth covering will be carried out, which is a surface layer specification similar to the criteria of EU. In the case of industrial waste, "landfill site specified in the proviso" is not applicable. For these reasons, the prefectural governors judge individually on the abolition of the covered landfill sites.

2.7.2.2 Abolition of a Landfill Site that No Leachate or Treated Water is Released by Sewage

In cases of discharging by sewage, even if the leachate concentration is below the planned treated water concentration and becomes lower than the discharge water quality of the sewerage terminal treatment facility due to the stabilization of the landfill layer, if it cannot be discharged to the public water area around the landfill site, sewer discharge will be continued and sewer expenses must be paid even after abolition. In this case, the following can be considered.

① Conduct a discussion on discharge to public water areas.
② Since there is no water quality load to the sewer terminal treatment facility, exemption or reduction of the contribution money can be consulted with the sewage system administrators.
③ The abolition procedure applying the individual criteria 13 is applied. It is assumed that there is no discharge.

2.7.2.3 In the Case Where It is Installed with Incineration Facilities, etc., Leachate or Treated Water is Reused as Cooling Water and Not Discharged

If the operation of incineration facilities etc. is continued until the abolition of landfill sites, there is no problem, but if the incineration facilities, etc. are abolished and dismantled before the abolition, it is necessary to treat leachate and secure discharge destination.

① To suppress the generation of leachate by capping
② Leachate water is stored and transported to another water treatment facility by lorries, etc.
③ For abolition, procedures applying individual criteria 13 (capping, etc.) are carried out. It is assumed that there is no discharge. Regarding the abolition of either case, the prefectural governor will judge individually.

2.7.3 Abolition and Stabilization

Abolition can be explained by that the waste at the landfill site is stabilized and will not hinder the preservation of the living environment. If abolition measures are taken based on the abolition standards prescribed by the standard ministerial ordinance, landfill sites are judged to be stabilized at the time. However, considering the leachate which is one of the requirements for abolition, after rainwater has penetrated the landfill layer, it is expected to form water channels in the landfill layer, and it will run to leachate collection facility through the channels. In other words, until the precipitation is collected as gravity water in the leachate collection facility, it flows down through the channels formed, cannot contact uniformly with the landfill waste layer, so the pollutants in the landfill layer are impossible to be completely washed out (Fig. 2.95).

For this reason, it is conceivable to stabilize the entire landfill layer and set the time when it becomes the same as the general soil. That is, it can be considered that the time when the quality of the landfill waste reaches a certain level is one of the indicators of stabilization. Dr. Tanaka defines the final stabilization of the landfill site as "a (soil reduction) state where no change in the environment occurs when excavate waste and expose it to the atmosphere or rain" (Tanaka N 1999).

In other words, inactivation of landfill waste is considered to be stabilization of landfill sites. From these facts, stabilization based on the abolition criteria can be said to be "apparent stabilization" from the viewpoint of the quality of landfill waste. On the other hand, stabilization of landfill waste is the "true stabilization" we sought.

2.7.4 Abolition Promotion Technologies

Waste disposal systems have changed in response to changes in social conditions. With that, waste disposal technologies have also advanced. With these changes, the quality of waste in landfill sites has also changed. For this reason, even after

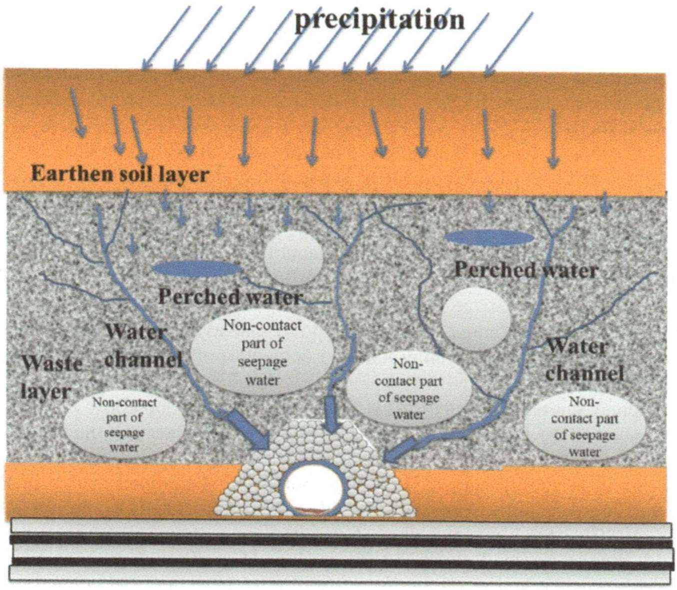

Fig. 2.95 Formation of water channels

the completion of the landfill, high concentration of leachate and gas continue to be generated and landfill sites where stabilization is delayed became negative legacies in the areas. By restoring or regenerating these landfill sites, it is possible to eliminate the negative legacies and solve the problem of insufficient landfill capacity by prolonging the life of them. For this reason, the development of countermeasure technologies and early stabilization technologies capable of responding to such situations is desired. These technologies are described in Chap. 4.

References

Environment Agency (1998) Regarding partial revision of the ordinance that sets technical standards related to the landfill site of municipal and industrial waste. The Environment Agency Press Release Presentation

Gao H (2010) Research on the development of landfill system with environmental burden reduction. Masters Thesis, Recycling and eco-technology specialty, Graduate School of Engineering, Fukuoka University

Han et al (2017) Study on recycling of by-product salt accompanying keachate desalination treatment. Brochure of JWMA Reseach and Case Presentation in 2017

Higuch S (1995) Planning and construction of landfill sites. Daily report, pp 156–157

Higuchi S (2011) Study on construction of recycling system for by-product salt discharged with waste treatment disposal. Reserch report of grant-in-aid for promotion of recycling-based society

Higuchi S (2013) Study on the effects of incineration flue gas treatment chemicals and fly ash chelate on landfill management (3k133007) Research report of grant-in aid for promotion of recycling-based society

Higuchi S (2015) Advanced use of the abolished landfill sate. Life Environ 60(2):29–33

Higuchi S, Uchida M (2018) Development of separation and determination method of organic chelating agent remaining in leaching water. Paper collections of 36th JWMA reseach and case presentation, pp 133–135

Higuchi S, Nagano S, Muroi O, Toyohisauroi S, Nakaguma D, Yagi Y (2015) Functional inspection and evaluation of semi-aerobic landfill sites. The 26th annual conference of LSMCWM, pp 465–466

Ito T (2017) Recycling of by-product salt to eco-hypo by matsuyama method. Fukuoka University Reseach Institute for Resource Recycling and Environmental pollution Control, 2017 achievement presentation lecture collection

Japan Environmental Sanitation Center (2014) Waste Disposal Law, pp 706

JWMS Conference (2010) Plan, design, and management guidelines for the landfill site, 2010 revised edn., pp 432

Lei et al (2014) Study on landfill characteristics of incombustible shererdded residue. In: Proceedings of the annual conference of the Japan Waste Management Association 35, pp 209–211

Matsumoto et al (2013) Study on recycling of desalted concentrated water into disinfectant by electrodialysis membrane. In: 36th JWMA Research and Case Presentation, pp 127–129

Matsumoto T et al (2013) Studies on the use of sterilizing agent. In: 24th annual conference of JSMCWM, pp 269–270

Ministry of the Environment (2011) Actual survey results of municipal waste disposal and facility maintenance status (Incineration facility data and landfill site data)

Ministry of the Environment, Waste of Japan (2016) Waste/Recycling Department of Minister's Secretariat. Heisei 26 edn., p 56

Nagano S, Higuchi S, Toyohisa S, Muroi, Tagi Y, Nakaguna D (2016) Functional inspection and evaluation of semi-aerobic landfill sites (part2). In: 37th JWMA Research and Case Presentation, pp 264–266

Nakagawa S (1985) Measurement method of evapotranspiration (vii) Penman Method, Agricultural Meteorology (J Agr Met) 41(3):281–282

Study Group on Salt Recycling System (2004) Research report on by-product salt recycling by waste treatment SR

Takimoto et al (2016) Study on generation of eco hypochlorite by diaphragm electrolytic method. Brochure of JWMA Research and Case Presentation in 2016

Tanaka N (1999) Environmentally safe waste landfill disposal technology. J Waste Res Soc 10(2):118–127

Tsuboi et al (2015) Study on recycling of desalted concentrated water into disinfectant by electrodialysis membrane. In: 36th Japan Waste Management Association (JWMA) Reseach and Case Presentation, pp 127–129.

Tsuno Y (1996) Study on identification of nitrification inhibitor of wastewater in sluge drying process research papers on environmental engineering, 33

Uchida M, Higuchi S (2012) Influence of chelate-treated fly ash on landfill management (part 3). In: The 23rd Annual Conference of JSMCWM D2-2

Uchida M, Higuchi S, Tameda K (2015) Study on behavior of chelate components in landfill site. Pub Interest Incorp Assoc, Japan waste management Association, 68(323):81–88

Ushikoshi et al (2013) Studies on eco hypochlorite formation using by-product salt generated from the waste treatment process III. In: 24th annual conference of JSMCWM pp 269–270

Waste Landfill Research Group of Waste Society of Japan (2002) Survey and evaluation methods of abolition criteria of landfill sites

Xu W (2010) Sturdy on the bioreactor landfill of municipal solid waste. Fukuoka University doctoral dissertation

Yamanaka T (1999) Biochemistry of autotrophic bacteria. IPC, p 75

Yoji T (1996) Study on Identification of nitrification inhibitor of wastewater in sludge drying process. Res Papers Environ Eng 33

Chapter 3
A Relationship Between Landfill Materials and Landfill Technologies

As mentioned above, landfill technology varies across landfill waste qualities. Japan started with whole amount landfill technology, and using a landfill technology for incineration residues, mainly composing inorganic substance at present. On the other hand, the developing countries began with sanitary landfill technology, and the technology transfer of Japan's "semi-aerobic landfill" is underway.

However, when the whole amount landfill was carrying out, Japan had not entered the period of rapid economic growth yet. It is no exaggeration to say that there is nearly neither the technology development nor the accumulation of technology existed at that time. The incineration facilities were introduced as soon as possible with the rapid economic growth, and the garbage issue attracted people's attention. When the "semi-aerobic landfill" was developed in the late 1960s, the ratio of incineration in our country was nearly 50% in statistics. Kitchen refuse was used in basic experiments and demonstration experiments of "semi-aerobic landfill". Since the waste in the filed were the mixture of kitchen refuse, incombustible waste and incineration residues, we still have less experience in whole amount landfill technology.

Besides, as waste quality also varies by lifestyles, cultural traditions, and climates, it is hard to state out the whole amount landfill technology clearly, and it is developing independently in both Western countries and Asian Countries. With the development of landfill technologies, more technology options can be provided, and different technologies for seepage control and leachate treatment can be introduced to developing countries according to their different economic conditions. The landfill technology after the introduction process of intermediate processing is mentioned in this chapter.

S. Higuchi, *Municipal Solid Waste Landfill Technology in Japan*, Environmental Science and Engineering, https://doi.org/10.1007/978-981-16-2734-7_3

3.1 Leachate Management in the Introduction Process of Intermediate Processing

Japan aimed to become the first country in the world to introduce the incineration facilities. The ratio of incineration residues in landfill has increased from the 1980s and exceeded 50% in the 1990s while the landfill of kitchen refuse was disappeared. Figure 3.1 shows the changes in landfill waste materials. We can find that it was not the whole amount landfill but the landfill mainly composing of organic substance before the 1970s in Japan. The incineration facilities were introduced in the 1980s. The period from that time to 1990 was called the period of the introduction process of intermediate processing.

The guidelines of MSW landfill sites in Japan was based on the explanation of the guidelines of waste treatment facility structure from Japan Waste Management Association in 1978. Since then, with the spread of incineration facilities, the landfill ratio of incineration residues had increased. The guiding principle of MSW landfill sites was revised in 1989. Then, it was revised again in 2000 and 2010.

3.1.1 Guidelines of MSW Landfill Sites (1978)

At that time, direct landfill of combustible waste was still carrying out. The decomposition of landfill layer was a bioreactor type using living organisms. Depending on

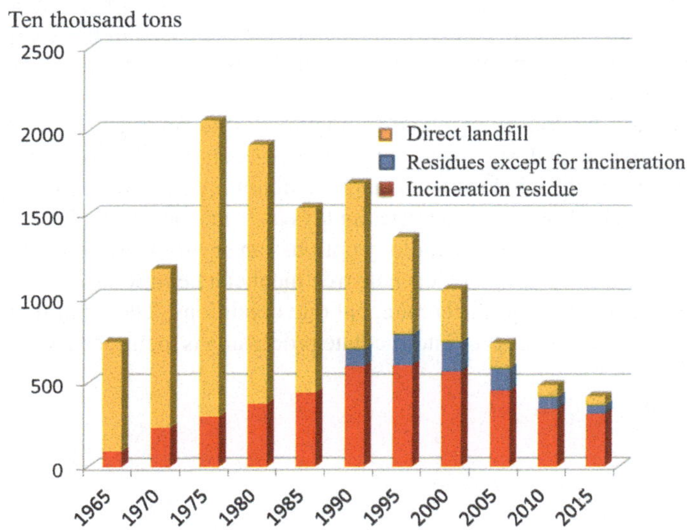

Fig. 3.1 The changes in landfill waste materials

Table 3.1 Examples of leachate quality

Landfill site	A	B	C	D
Chromaticity (°)	500–200		500–4000	
PH	6.0–7.0	5.2–6.1	7.4–8.3	7.2–7.8
BOD (mg/L)	100–6000	2000–7000	200–2000	50–1000
COD_{MN} (mg/L)	100–1000	200–500	100–1200	150–1000
SS (mg/L)	50–200	40–240	20–300	40–100
NH_4-N (mg/L)	150–200	5–300	50–400	100–250

the presence of air in landfill layer, the landfill structure is divided into "aerobic landfill structure", "semi-aerobic landfill", "anaerobic sanitary landfill" and "anaerobic landfill". The "semi-aerobic landfill" is recommended. It was not obligated to set seepage control work. According to the case studies, the leachate quality is shown in Table 3.1. "The leachate treatment method shall be determined in consideration of the amount or the quality of leachate and its variation, the conditions of water utilization in the discharge destination, the locational conditions of the treatment facility and its maintenance conditions" (Explanation of Guidelines for the Structure of Waste Treatment Facilities 1978). "Though the removal object and removal level determine the leachate treatment method, the biological treatment method is effective because the component of leachate is mainly organic substance. Besides, in the case of advanced treatment, physicochemical treatment is also used" (Explanation of Guidelines for the Structure of Waste Treatment Facilities 1978). As examples of treatment methods, aeration lagoon method, activated sludge method, and sprinkling filtration method are biological treatment methods. Coagulating sedimentation method, sand method, and activated carbon adsorption, etc. were introduced as physicochemical treatment methods. All these design values have not been shown.

3.1.2 Explanation of the Guidelines on MSW Landfill Sites (1989)

The incineration facilities began to spread in 1989s. As a result, both the ratio of incineration residue in the amount of landfill and the concentrations of inorganic salt in leachate has increased. The scale is attached to calcium collecting pipe and pumps for leachate treatment facilities. Besides, the Cl^- concentration increases, the corrosion of metal pipes and pumps become severe. When the agricultural water was used in the discharge destination, rice damage appears due to the increase of chloride ion concentration. Landfill sites where calcium removal equipment was added in the leachate treatment process and began to perform a desalting treatment began to appear. On the other hand, metals containing in fly ash have also been detected in leachate. Due to these inorganic salt problems, the number of MSW landfill sites which using seepage control sheet began to increase. Under these circumstances,

"The Guidelines on MSW Landfill Sites" was revised in 1989 and changed to "Explanation of the Guidelines on MSW Landfill Sites". In the case of raw water quality, "as the design wastewater influent quality is basically determined by landfill waste materials, it is necessary to fully understand the garbage disposal basic plan, the collection, and transportation plan, the intermediate processing plan before setting" (Explanation of Guidelines for the Structure of Waste Treatment Facilities 1978). The necessity to consider landfill waste materials in intermediate processing was required, the effects of Cl^- and Ca^{2+} were also mentioned. A specific criterion of design wastewater influent quality is introduced in Table 3.2.

Table 3.2 A specific criterion of raw water quality (Explanation of guidelines for MSW landfill sites 1989)

Item	Mainly combustible waste	Incombustible waste and incineration residue	Notes
BOD (mg/L)	1200	250	(1) Landfill structure: Semi-aerobic landfill
SS (mg/L)	300	300	(2) Landfill period: 5 years
COD (mg/L)	480	100	(3) Landfill thickness: 4 m
NH_4-N(T-N) (mg/L)	480	100	(4) The ignition loss of incineration ash: 8%
PH	It presents an acid reaction when the amount of putrefactive organic substance increases	It presents an alkaline reaction when the ignition loss of incineration ash decreases	(5) For TDS, the presence of HCl removal apparatus in the incinerator and disposal method of dust are noticed
TDS (mg/L)	It is an order on 103–104 mg/L		(6) When the dust is landfilled together, not only TDS but also Ca^{2+} and heavy metals are attracted attention
Number of colitis germ legions	Exceeds 3000		
Fe^{2+}	10 mg/L as usual		
Mn^{2+}	Trace as usual		
Other heavy metals	Undetected as usual		
Chromaticity	Dark brown–Light yellow		

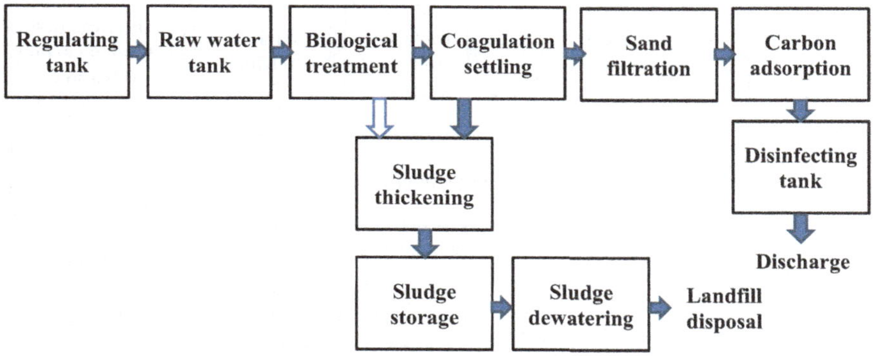

	pH	BOD (mg/l)	COD (mg/l)	SS (mg/l)	T-N
Raw water	6~9	250	150	300	100
Treated water	5.8~8.6	20	20	10	10

Fig. 3.2 Example of processing flow of incombustible waste (Explanation of guidelines for MSW landfill sites 1989)

On the other hand, "Although the leachate quality is determined by landfill waste materials, it is also influenced by landfill structure, landfill operation, the capacity of the landfill, and landfill duration. Therefore, it is desirable to make decisions regarding a measured result of the water quality of other MSW landfill sites where landfill waste materials and landfill structure are relatively similar."

As leachate treatment methods, contact aeration process is applied to conventional biological treatment. Alkaline coagulation sedimentation treatment and weak acidic coagulation sedimentation treatment is added calcium removal for physicochemical treatment. Figures 3.2 and 3.3, design specification such as overflow rate, residence time, filtration rate has been described.

Examples of processing flow by various kinds of landfill waste are introduced. Besides, seepage control work is also presented.

3.2 Landfill Technology for Intermediate Processing Residue

Depending on the policy of waste management, the intermediate processing methods are different. Japan is focusing on incineration and crushing sorting. On the other hands, some other international countries are centering on MBT method other than incineration. Since the intermediate processing methods vary depending on country conditions, we used the Japanese intermediate processing method to figure out the features of the final disposal technique in the case of landfill waste containing incineration residues and incombustible crushing residues and made discussed on it.

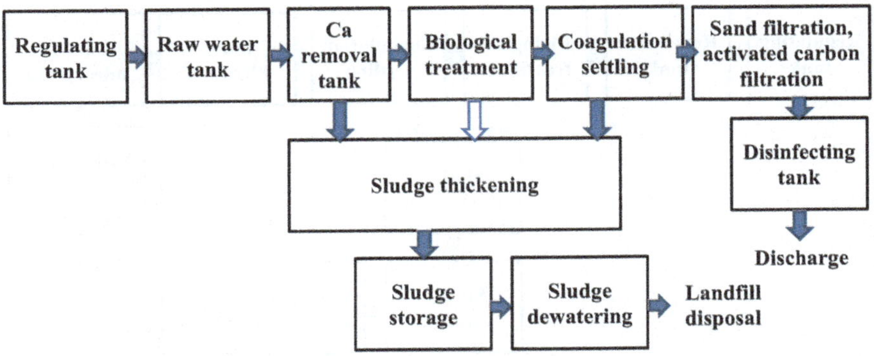

	pH	Ca (mg/l)	BOD (mg/l)	COD (mg/l)	SS (mg/l)	T-N
Raw water	5.8~8.0	1,500~2,500	250	100	300	100
Treated water	5.8~8.6	100	20	20	10	10

Fig. 3.3 Example of processing flow of incinerationresidue

The features were discussed in Chap. 2, "Issues and Measures for Landfill Technologies." However, the measure against a high concentration of inorganic salt and the measure against high pH value, besides, the measure against nitrogen caused by remaining organic chelating agents need to pay more attention. And the facilities, such as

- Leachate collecting facilities
- Landfill gas treatment facilities
- Leachate treatment facilities

need to be noticed.

For landfill operation,

- Landfill method
- Leachate management

are very important.

3.2.1 Leachate Collection/Drainage Facility

The ratio of incineration residues has exceeded 80% of landfill. With the increase of the ratio of Ca and Cl in incineration residues, the generation of Ca capacity is blocked and the water collection function is impaired. Meanwhile, the air circulation function through the water collecting pipe is also impaired. As a result, the air supply to the landfill layer is shut off, the neutralization in the high alkaline condition is delayed and the high pH value is continued for a long time. In other words, the "semi-aerobic landfill" will not work anymore. Essentially, the "semi-aerobic landfill" aims to activate aerobic microorganism and decompose organic substance. However, in the landfill sites which mainly composing incineration residues, the microbial activity is strongly inhibited because of its high pH value and high concentration of inorganic salt environment. For these reasons, the function of the "semi-aerobic landfill" from the beginning period of the landfill is required to decrease the pH value by air supply and to relax the high salt environment by the method of washout by precipitation. The function of collecting pipe is needed. The Ca content of the fly ash in the incineration residues is 20–25% (Noma and Kida 1996), it is dissolved and then ionized in the leachate, and the calcium capacity ($CaCO_3$) can be produced if it is oxidized.

The calcium scale adheres to the landfill layer, protective crushed stones for collecting pipe, collecting water pipe wall and the leachate intake pump, causes water collecting inhibition and device inhibition. Especially the calcium scale adheres to the landfill layer, Calcium scaleprotective crushed stones for collecting pipe and collecting water pipe wall, causes ventilation inhibition, and result in a substantial delay in pH reduction by washing out the salt. Therefore, it is desirable to the maintenance of collecting pipes such like using a self- traveling camera to monitor whether the perforated part of the collecting pipe is blocked and do a high-pressure cleaning in order to preserve the function of collecting pipes. Furthermore, the capacity prevention measure by landfill management which will be described later is also an effective measure.

3.2.2 *Landfill Gas Treatment Facility*

It is no exaggeration to say that there is almost no generation of landfill gas in the landfill where mainly composing incineration residues. In this type of landfill sites, landfill gas treatment facilities have a large proportion of complementary facilities for leachate collecting facilities. Since the calcium scale in landfill layer is impermeable, the osmotic water stagnates and does not move to the bottom of the landfill layer, which causes the washing function and ventilation inhibition. Therefore, it is one of the effective measures to frequently use landfill gas treatment facilities as complementary facilities for leachate collecting facilities (Fig. 3.4).

3.2.3 *Leachate Treatment Facility*

The leachate treatment from the landfill sites mainly composing incineration residues is mentioned in Sect. 3.2.2, "Issues of high concentration inorganic salt", the desalination treatment is carried out according to the removal of calcium scale and the condition of treated water discharge destination, etc. As byproduct salt is produced when the desalination treatment is carried out, it is necessary to use appropriate

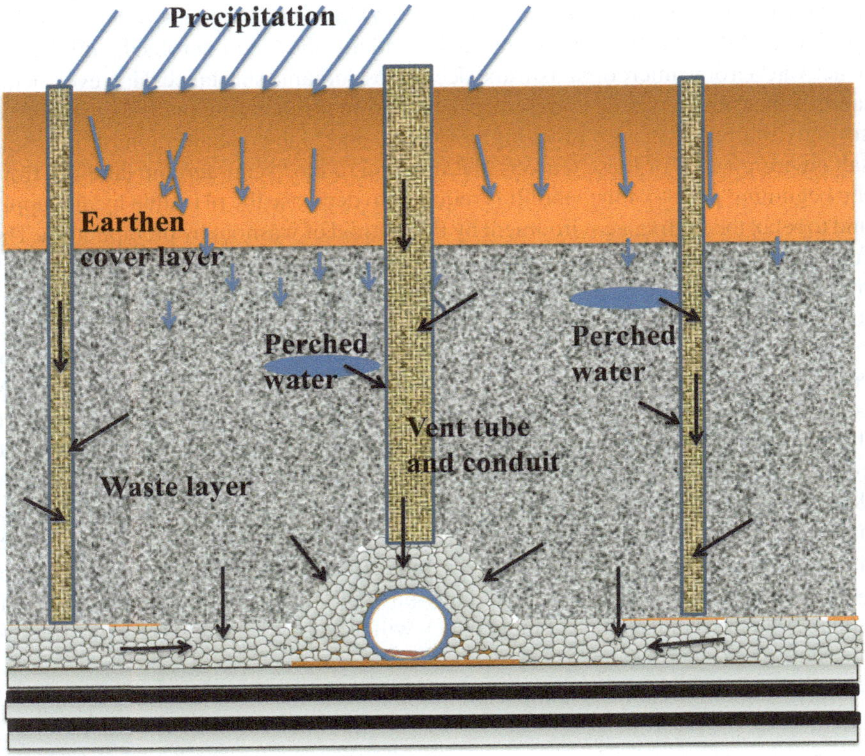

Fig. 3.4 Examples of Vent tube and conduit facility

measures to improve the byproduct salt recycling such like using disinfectants by eco sodium hypochlorite or neutralizers by acid-alkaline production and prevent the re-leach of byproduct salt.

Figures 3.5 and 3.6 show typical leachate treatment flow examples, including the measures to Ca and Cl.

3.2.4 Landfill Methods

In the landfill sites where mainly composing incineration residues, the concentration of the organic substance is reducing. However, the burden on leachate treatments such as Ca removal facilities, the dehydrating machines, and desalination treatment facilities are large. Based on the lime soda method, which is commonly used for the calcium removal facilities, the required amount of Na_2CO_3 is 2.65 kg to remove 1 kg of Ca^{2+}. Besides, 2.5 kg of calcium sludge is generated (Referring to the following equation).

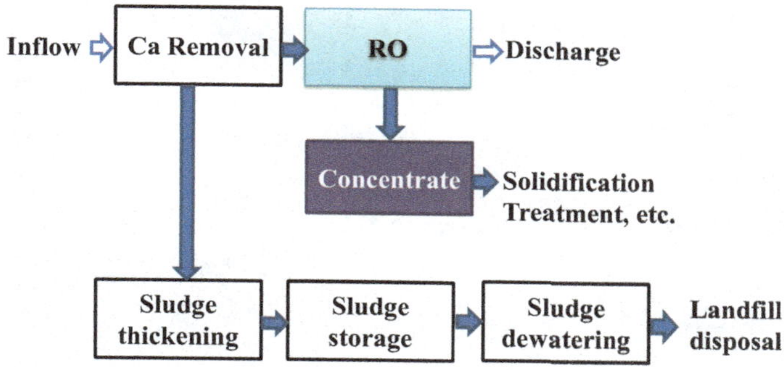

Fig. 3.5 Leachate treatment flow (RO method)

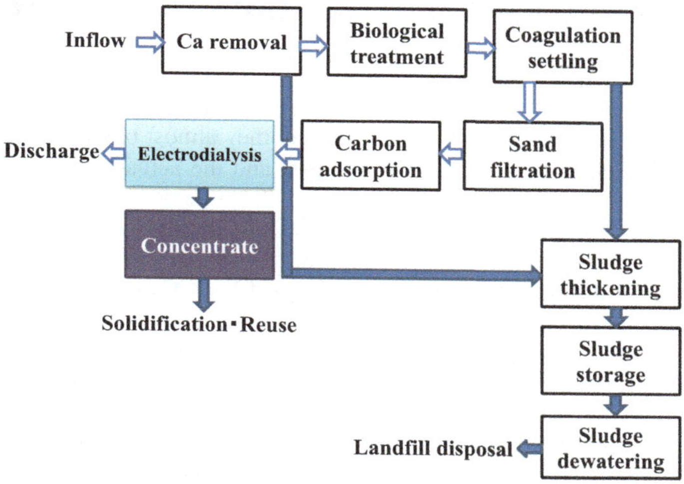

Fig. 3.6 Leachate treatment flow (ED method)

$$
\begin{array}{lllll}
Ca & + \; Na_2CO_3 & \rightarrow & CaCO_3 & + \; 2Na \\
40 & \quad 106 & & 100 & \quad 46 \\
1 & 2.65\ (106/40) & & 2.5\ (100/40) &
\end{array}
$$

Therefore, a large amount of lime soda and a large capacity dehydrating machine is required. In the case of reverse osmosis membrane method (RO) in the desalination treatment, since the concentration is 3–5%, the treatment efficiency decreases when the concentration of chloride ions in raw water exceeds 2%.

It is necessary to control the concentration of inorganic salt and make an efficient leachate treatment in the landfill sites where mainly composing incineration residues with landfill methods. The concentration of inorganic salt that the local governments usually set as design raw water quality is Ca^{2+} 2500–3500 mg/L and Cl^- 15,000 mg/L.

Fig. 3.7 Examples of adding Na_2CO_3 to landfill layers

Neither of them is based on scientific evidence; they almost based on the practical examples of leachate concentration in the past and the actual state of processing technology.

3.2.4.1 Ca^{2+} Inhibition by a Direct Addition of Na_2CO_3 in Landfill Sites

Since the amount of slaked lime blown into the exhaust gas varies depending on the incineration flue gas treatment performance, the Ca^{2+} concentration in leachate increases when the exhaust gas, especially HCl gas advances. As stated previously, when the Ca^{2+} concentration is set to a high value, numbers of treated chemicals and dehydrating machines are required. There is a method of producing $CaCO_3$ in the landfill layer by adding Na_2CO_3 directly to the landfill sites and percolating by precipitation. However, the $CaCO_3$ layer in the landfill layer becomes an impermeable layer, which becomes the factors causing ventilation inhibition and water flow. Therefore, when the landfill is performed on the area where Na_2CO_3 is added and $CaCO_3$ is produced, it is necessary to crush the $CaCO_3$ layer by heavy machine before the landfill. Figure 3.7 describes the examples of adding Na_2CO_3 to landfill layers.

3.2.4.2 The Dividing Landfill of Fly Ash

The fly ash usually contains about 20–25% of Ca (Waste Management and Recycling Department in MOE (Ministry of the Environment Government of Japan 2017), when the chelating treatment is carrying out in the waste incineration facility, the moisture content increases to 85% because of the mixture of chelate and fly ash. When the

Fig. 3.8 Solidified fly ash in the landfill layer

mixture of chelate and fly ash are transported in landfill sites, the plasticity is large due to the high moisture content, and there are problems in landfill operation, the bottom ash is mixed with earthen cover, etc. to ensure the ventilation and water flow. On the other hand, the Ca content and the moisture content is high in fly ash, it will be solidified by cement reaction, etc. in its present condition. Therefore, it is possible to prevent the leach of Cl^- and Ca^{2+} by landfilling the fly ash in each division and solidifying them. However, the leach of Cl^- continues because of long-term precipitation.

Figure 3.8 shows the example of solidified fly ash in the landfill layer.

3.2.4.3 Landfill by Stacking up

Inorganic salt such as Cl^- and Ca^{2+} are leached by advection and diffusion. Its' leaching behaviors are different depending on the content in the incineration residue and the thickness of the landfill layer. The thicker the landfill layer is, the higher the concentration becomes. Besides, the thinner the landfill layer is, the higher the leach speed it has. By using this property, the peak concentration decreasing by reducing the thickness of the landfill layer. Besides, accumulating incineration residues to the leached landfill layer will cause the lower layer becomes a buffer to decrease the peak concentration more. This kind of landfill method is adding the landfill method. Figure 3.9 shows an experiment in which incineration ash is filled in a simulated landfill layer with a diameter of 20 cm. When the thickness of the landfill layer is different and adding landfill is carrying out, the feature of Cl^- concentration in leachate can be confirmed. The thickness of the layer in RUN 1 is 2.0 m. In RUN 1, the artificial sprinkling is conducted and the Cl^- concentration of leachate is monitored.

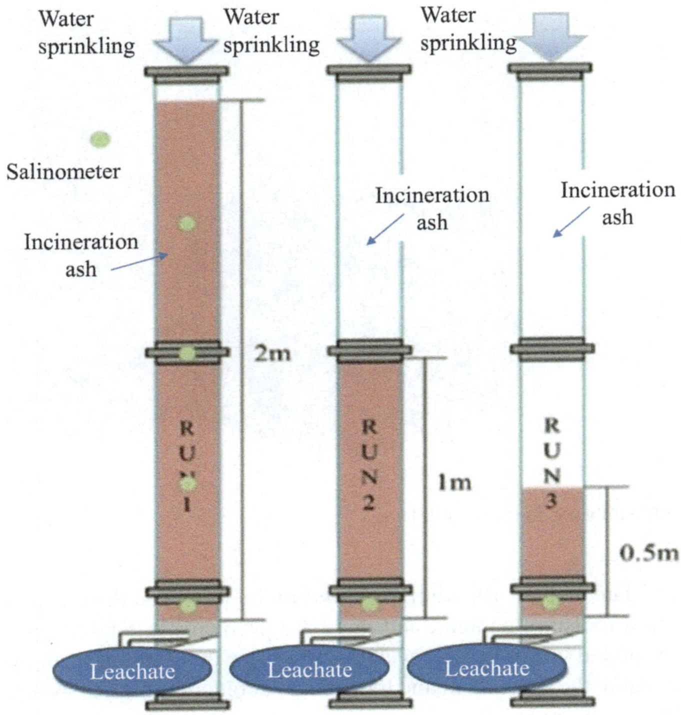

Fig. 3.9 Adding landfill by using incineration ash

The thickness of the layer in RUN 2 is 1.0 m. In RUN 2, the artificial sprinkling is also conducted and concentration in adding landfill when the Cl^- concentration below 1000 mg/L, 1.0 m of thickness will be added and the thickness of the landfill layer will become 2.0 m.

In RUN 3, the thickness is 0.5 m. The artificial sprinkling is conducted. When the Cl^- concentration in leachate is below 1000 mg/L, 0.5 m will be added until the thickness become 2.0 m.

The daily changes in Cl^- concentration of three RUNs are shown in Fig. 3.10.

By comparing the different thickness of the landfill layer, we can find that the greater the thickness is, the higher the peak value the Cl^- concentration has. Besides, comparing the values of peak concentration between three groups, it is lower in RUN 2 and RUN 3 than in RUN 1. It is because the lower layer serves as a buffer and leaching suppression is performed. By using a landfill method which considers the concentration suppression, it is possible to make an efficient and economic desalination treatment of leachate. Figure 3.11 shows the conceptual scheme of Cl^- concentration suppression which considers adding landfill. It is estimated that rainwater is used as dilution water in no dumping area. In the case of landfill sites mainly composed of organic substance; the rainwater discharge is actively carried out in

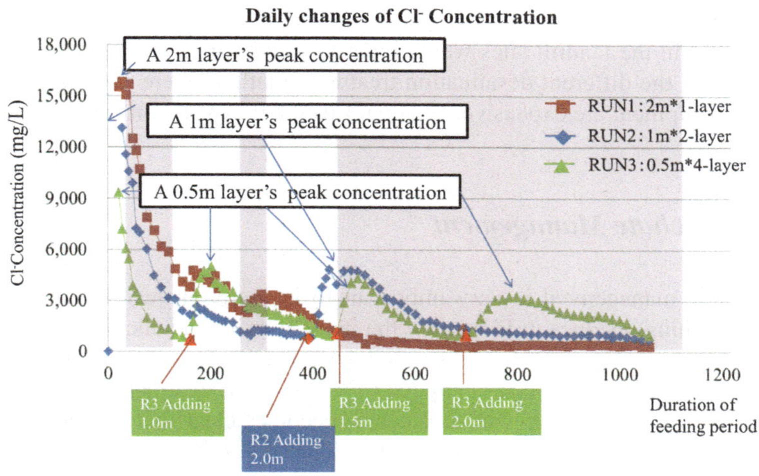

Fig. 3.10 The daily changes of Cl^-

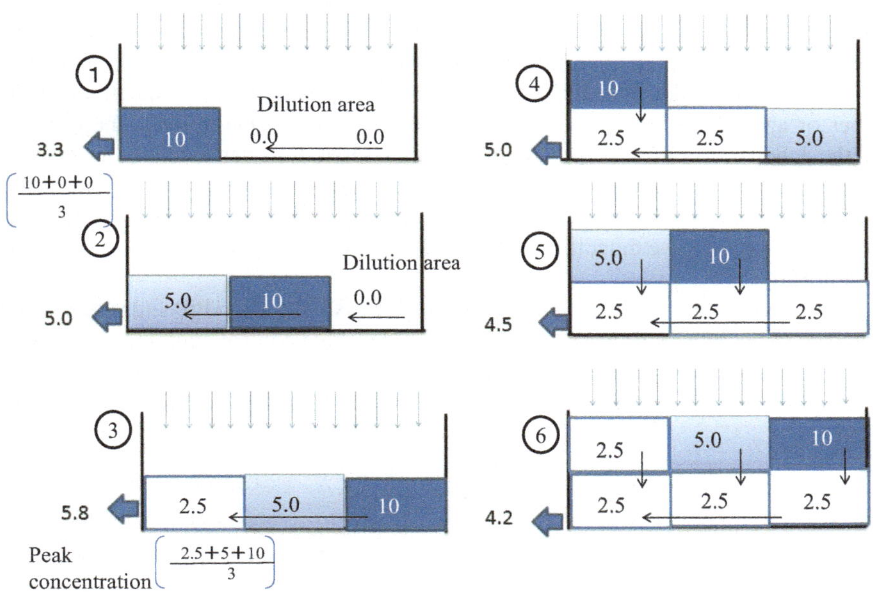

Fig. 3.11 The conceptual scheme in adding landfill

order to decrease the amount of leachate. However, when the desalination treatment is performed in the landfill sites which mainly composed of incineration residues, depending on the different desalination treatment method, there is a concentration limit and equipment are expensive. Rational treatment by dilution is suitable.

3.2.5 *Leachate Management*

The leachate management is very important to incineration residues and incombustible crushing residues. It is because the inorganic salt and alkaline components in incineration residues are easily dissolved when the leachate is liable to leached in the landfill layer. And the load is applied to the leachate treatment when the value of concentration becomes higher. Besides, it becomes clear that nitrification inhibition is caused by the influence of organic chelated residues (residual chelate) of stabilization of fly ash in recent days.

Residual chelate is easily dissolved in the leachate. Therefore, the internal shortage of leachate in landfill sites should be avoided as much as possible. In the case of incombustible crushing residues, although the landfill ratio is low, there isn't any heat treating performed. Therefore, when the impurities which against rule are mixed into the landfill, such like the motherboard for small appliances and fluorescent tubes or agricultural chemical such like herbicide for household used, the mixture of the landfill will still exist though crushing separation carried out. As results, mercury, lead, and rarely simazine is detected. Quality control of incombustible crushing materials is not carried out, and measures at recycling centers are difficult at present. Therefore, we must follow the discharge rules for residents and use leachate monitoring.

As shown in Fig. 3.12, the leachate treatment system for landfill sites consists of leachate collecting facilities, storage structure, leachate control equipment, leachate treatment equipment. Generally speaking, the capacity of the leachate control equipment and the leachate treatment facility is calculated by using the daily precipitation

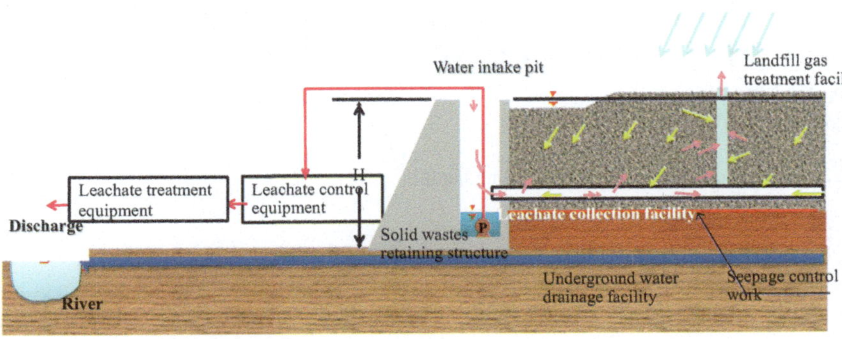

Fig. 3.12 Leachate treatment system

Fig. 3.13 Example of internal storage of leachate

time series of the largest precipitation year of the past 20 years, and then the capacity that water does not accumulate inside the landfill has been decided. However, under abnormal weather accompanied by recent climate change, the probability of occurrence of precipitation exceeding this value becomes higher. Therefore, to remain the leachate in the landfill (internal storage) so that the untreated leachate is not discharged, the height of the storage structure is often determined. When the period of internal storage is prolonged, inorganic salt and residual chelate in the incineration residue will be leached, the concentration will become higher and impede the leachate treatment. Furthermore, hydraulic pressure is applied to the seepage control work, which increases the risk of water leakage. Therefore, when internal storage state is reached, it is necessary to resolve this condition promptly. It is desirable to lower the internal water level within 1–2 weeks in summer or within 1 month in the winter when the solubility increases. Figure 3.13 shows a landfill case where it is in an internal storage state.

As a measure to avoid internal storage, it is effective to limit the landfilling operation area during the rainy season and typhoon season, capping measures to already landfilled areas. It is also a way to select a closed system landfill site in an area has heavy precipitation.

3.2.6 *Others*

In landfill sites of intermediate processing residues such as incineration residues and incombustible crushing residues, it is necessary to grasp the properties of the landfill waste mentioned above and to carry out facility planning and landfill management. The most important thing is to fully understand waste management such as collection transportation, intermediate processing policy, and to eliminate the communication gap between each department.

References

Explanation of Guidlines for the Structure of Waste Treatment Facilities (1978) Japan Waste Management Association, Japan.

Explanation of Guidelines for MSW Landfill Sites (1989) Japan Waste Management Association, Japan

Explanation of Guidelines for the Structure of Waste Treatment Facilities (1978) Japan Waste Management Association, Japan, pp 396–400

Noma Y, Kida A (1996) Influence of incineration ash on calcium capacity formation in landfill sites. High-salt problems and technology measures in landfill sites. A Report on Landfill Disposal of Waste. Research ethics committees in Japan Society of Waste Management Experts

Waste Management and Recycling Department in MOE (Ministry of the Environment Government of Japan) (2017) Annual report on emissions and processing status of municipal waste in Japan in 2015. https://www.env.go.jp/press/103839.html. Accessed 14 Aug 2018

Chapter 4
The Role and the Technology for the Future Landfill Site

Until now, we had discussed the peculiarities and issues of waste management and landfill site technology in Japan. In this chapter, we will describe the solutions to these issues and new essential development issues in the future.

4.1 Requirements for the Future Landfill Site

During the period of high economic growth, because of the exponential increase of waste, the climate and the topography, Japan's landfill site had developed independently in a different way from the West. The examples are as follows.

① There are no provisions for locating the landfill sites. (Japan is almost the only country which has no location avoidance provisions for landfill sites among all the developed countries.)

② By introducing intermediate processing such as incineration, volume reduction, detoxification, and stabilization are carried out. "Landfill disposing of residues" is conducted."

③ A landfill site on the sea surface is required.

④ The "semi-aerobic landfill," which makes the best use of natural organic matter purification function and its decomposition mechanism become the basic of landfill technology.

⑤ Because of the large precipitation, "washing out by precipitation" for inorganic substances becomes the basis of stabilization.

⑥ In maintenance management standards and structural standard, there is no landfill capping provisions after the landfill is completed.

(Capping is not prohibited, but it is not required.)

Among of these, "No location avoidance provisions," "The existence of waste landfill site at sea," "No capping provisions" are specially mentioned due to Japan's

S. Higuchi, *Municipal Solid Waste Landfill Technology in Japan*, Environmental Science and Engineering, https://doi.org/10.1007/978-981-16-2734-7_4

national conditions and waste management policies, they are also should be specially mentioned internationally.

Based on these facts, the requirements for future landfill sites are as follows.

4.1.1 Location Issues

When the landfill site becomes full, it is impossible to rebuild within the same premises as a waste treatment facility. Therefore, it is necessary to ensure a landfill site from a long-term standpoint. There was almost no land use regulation in conventional landfill sites. The main choices were mountainous areas with easily land acquisition and high landfill efficiency. Besides, it was selected and secured with symptomatic treatment when the landfill site became full. Since the landfill site is a necessary facility, it is desirable to incorporate it into urban planning and secure it for the plan for a hundred years. Therefore, it is one of the methods to position landfill sites as urban facilities or to define "Location avoidance areas" and "Location evacuation guide areas" ("Location feasible areas") which stipulated in Germany, etc. Considering the preservation of the living environment and natural environment, disaster prevention and regional characteristics, it is desirable to set the location avoidance areas separately. After setting the location avoidance areas, a candidate site is selected from the feasible areas, then determined through the evaluation and narrowing-down stages. This process needs to be opened to the inhabitants and ask for their opinions. It is important to make use of strategic environmental assessment actively, let inhabitants take part in the investigation from the planning stage.

4.1.2 Regeneration of the Existing Landfill Site and Its Early Stabilization

The waste quality in landfill sites has also changed in conjunction with these changes. For this reason, even after the landfill is completed, a high concentration of leachate and gas continues to be generated and the stabilization delays in some landfill sites. These landfill sites become a negative legacy of the areas. By repairing and regenerating finial landfill sites, it is possible to eliminate the negative legacy of areas and solve the problem of insufficient disposal capacity by prolonging the life of the landfill sites. Therefore, it is desirable to develop countermeasure technology and early stabilization technology capable of responding to such a situation. These technologies are as follows.

4.1.2.1 Contaminant Exploration Technology in the Landfill Layer

In the case of early stabilization and landfill regeneration, it is necessary to grasp the situation where the retardation of the decomposition has occurred in the landfill layer beforehand. Although a preliminary survey on boring is carried out, it is difficult to grasp the situation fully. There is a demand for exploration technology for locating perched water in either organic layers where stabilization is delayed or the landfill layer.

4.1.2.2 Promotion Technology for Stabilization of Organic Substance

The abolition standard established by the Ministerial ordinance in 1998 prescribes the leachate quality falls below the standard effluent values for more than 2 years, the gas generation, temperature, and subsidence. It is also an extremely strict standard towards the establishment of the elapsed years after the end of a landfill in each foreign country. The newly constructed landfill site will be assuming the decommissioning standard. For instance, with the introduction of pretreatment, it becomes easier to find the countermeasure towards only inert landfill waste. On the other hand, it is supposed that the landfill site constructed before the law regulation cannot be abolished semi-permanently in some cases. (For example, the landfill site where the effluent standard is much more strict than self-regulation and the landfill of active waste such as organic sludge is carried out) Therefore, maintenance cost after the landfill is completed may increase, and it may be impossible for the private landfill site to bear the cost. In this situation, it is thought to promote stabilization by an early stabilization technology for existing landfill sites. Many municipalities cite the slowing down of COD and T-N as a reason for the lengthening of the abolition period. We have considered a landfill layer as a type of bioreactor and promoted a semi-aerobic landfill that which can promote decomposition of organic substance. However, there is a demand for the early decomposition of the refractory organic substances, as typified by COD listed in the abolition inhibition factors. Therefore, the authors are developing a combining technology for promoting decomposition of an easily decomposable organic substance by the aerobic method and decomposition of the refractory organic substance by the chemical oxidation.

4.1.2.3 Countermeasures for Gas and Decomposition of an Easily Decomposable Organic Substance by the Aerobic Method

An air blowing method is mentioned as a measure for H_2S and CH_4 gas and decomposition of an easily decomposable organic substance by the aerobic method. In the air blowing method, there is a BIOPUSTER method in which air and oxygen are intermittently blown into a landfill layer under a pressure of 3 to 5 atmospheres and a SMELLWELL method in which air is continuously blown at a pressure close to normal pressure and an improved method thereof.

4.1.2.4 Decomposition of a Hardly Decomposable Organic Substance by the Chemical Oxidation

In general, a biologically easily decomposable organic substance in a landfill site is decomposed early, while it takes time for decomposing hardly decomposable organic substance. The chemical oxidation is used in these treatments, and it is a decomposition method of organic substance in waste. As oxidizing agents, hydrogen peroxide, potassium permanganate, etc. are used.

4.1.2.5 Regenerating Technology for the Landfill

In the case of the landfill sites which have been already landfilled or during the process of landfilling, since it takes a long time until abolishment, there is a concern over it will become a negative legacy, stabilization promotion and renewal are required. Different from other intermediate processing facilities such as incineration facilities, the landfill site cannot be rebuilt once it reaches the fully occupied amount, so it is necessary to construct a new one. From this kind of circumstance, the situation in some municipalities has already become worse. It is desirable to prolong the lifetime of existing landfill sites and to recycling use the sites which have been landfilled. In order to prolong the lifetime of landfill sites, conventionally, there is a history of the spread of intermediate processing facilities. However, more wastes have been landfilled without intermediate processing in the existing landfill sites. Through the changes in landfill sites described above, it is possible to solve the issues mentioned above by reprocessing and recycling these landfill wastes and ensuring the landfill disposal space. Besides, in the case of excavating the landfill layer, the wastes are certainly excavated, sometimes, it is used as a backfill material by detoxifying in situ. After the excavation work, it is sorted by trommel, and according to the need, coarse particles are crushed, carried out magnetic sorting, fine particles are carried out cleaning, used as a backfill material. In the landfill sites which mainly composed of incineration residue, when the dioxin content exceeds the reference value (3 ng-TEQ/g), after classification in situ, thermal decomposition becomes a method of making it detoxified.

4.1.3 A Newly Established Landfill Site

In order to let the landfill site to be accepted by the area, it is necessary to preserve the regional environment throughout the landfill and make sure it will not become a negative legacy such as early stabilization and being abolished after the landfill, making effective use of the site. For this purpose, it is certain to have a facility where the structure is safe, such as seepage control work, etc. It is also necessary to have a system that can make sure a low loading in the pre-treatment of waste and a backup in the post-treatment.

4.1.3.1 The Pre-treatment of the Landfill

The pretreatment of the landfill aims to secure a low loading of landfill waste and the aeration and water flow of the landfill layer and to activate microorganism and promote the washed-out effect. In the case of the former, though there is intermediate processing such as incineration because it is not consistent with landfill management and leachate management in a landfill site, it is necessary to reduce the load as needed. (For instance, the rise of pH value and salinity concentration will hinder activating microorganism.) In the case of the latter, waste such as sludge which has high plasticity hinders ventilation and water flow, causing perched water and making an anaerobic condition in the landfill layer. Therefore, it is necessary to reform the interior. In the pre-treatment which aims to reduce the load, there is a "washing" process. In this process, waste is forced washing before the landfill, through a stabilizing treatment, then landfilled. There are a machine-washing method and a water spray washing method. After the detergent drain is treated by membrane treatment, it is reused as washing water. (The volume of wash water is $1-10$ m^3/t). For machine washing, there is a spiral washing method, an ejector washing method, an air turbulence pipe method, etc. As washing waste is filled in the lysimeter and the long-term behavior is monitored, it is confirmed that stabilization is earlier by washing. For waste which organic substance with a high load such as organic sludge, the effect of pre-treatment by air treatment is confirmed. For pre-treatment which aims to secure ventilation and water flow, selection by trommel and removal of silt fraction by classification treatment are available. For waste with high plasticity, such as sludge, reforming by adding a moisture control agent or granulated solidification, are effective.

4.1.3.2 Landfill Operation

As factors that inhibit the stabilization, perched water in the landfill layer and water channel are in existence. Generally, there is a relation between the landfill height and leachate concentration. As the permeate water moves from the upper layer to the lower layer of the landfill layer, the leachate concentration also increases as the landfill height increases. On the other hand, there is also a kind of landfill site which is short passed by "water channel" and has no relation between each other. For this reason, the advancement of landfill operation technology is required. For instance, in order to avoid water channel and perched water, the increase of leachate concentration is inhibited by using a material with the void in daily soil covering or intermediate soil covering, the utilization of geosynthetic and biodegradable soil covering, or landfill by stacking up. It is also necessary to limit the average height and maximum height of the landfill layer from now on.

4.1.3.3 The Post-treatment of the Landfill

Even if the pre-treatment of the landfill is carried out, when the height of landfill becomes larger, it is thought to be hindered by ventilation and water flow and caused the delay of stabilization. Therefore, a post-treatment after the landfill should be carried out as necessary. The method of post-treatment is mentioned in the part of existing landfill sites. It is a compulsory air blowing method and chemical oxidation method. It can be handled by using pipes so that the post-treatment method can be taken before or during the landfill.

4.1.3.4 Leachate Management

Recently, due to a torrential guerrilla rain occurs in a short time, the leachate is unavoidable, and the untreated leachate overflows occasionally. Generally speaking, the water balance is calculated depending on the characteristics of regional precipitation, the precipitation which exceeds the probability of precipitation based on the statistical data is unpredictable. When leachate enters internal storage, water pressure is applied to the seepage control work, water leakage risk increases. Due to the anaerobic effect of the landfill layer, COD concentration in leachate increases and hinder the leachate management. Therefore, dealing with covered type waste disposal facilities is also an effective method in some areas.

4.1.3.5 Measures for By-Product Salt

It is said that salt discharged into the environment from the treatment process of municipal solid waste is about 220 thousand tons per year (Higuchi et al. 2012). Approximately 200 thousand tons of these are transferred to the landfill site. For this reason, the salt concentration in the leachate rises and the number of municipalities which incorporate desalting apparatus increases during the leachate treatment process. However, there is a problem of by-product salt accompanied by desalination treatment. In order to solve this problem, it is desirable to make efficient use of by-product salt. However, salt consumption enterprises, mainly soda industry, have a negative attitude towards the low price of imported salt and the use of by-product salt through the quality of by-product salt. At present, in addition to the soda industry utilization, road antifreeze, a fluxing agent for aluminum refining, sodium hypochlorite generation, etc. are conceivable. On the other hand, by-product salt is also recovered from incineration facilities. Recently, there was a case where sodium hypochlorite is generated by-product salt and recycled as a disinfectant at a sewage terminal disposal facility (Referring to Sect. 2.2 “Issue of high concentration inorganic salt”).

4.1.3.6 Cement Raw Material of Incineration Residues and Resource Storage Type Landfill System

When the salt content is removed from incineration residues, which accounts for the majority of the current landfill waste, cement raw material can be used as alternative clay material. For this reason, some municipalities are using cement recycling method of incineration residue. Besides, the old landfill sites are filled with plastic and metals. At present, these items are regulated to be taken out due to restrictions on the Waste Disposal Law. In this kind of situation, a resource storage type landfill system (repeatedly usable landfill site) with the aim of resource recovery should be suggested. As an example, in the case of incineration residue, the desalted washed ash is stored, and it is carried out to a cement plant at the time transport efficiency is high.

4.2 Concept of the Future Landfill Site

Figure 4.1 shows the conceptual example of the future landfill site. As mentioned above, the keywords which are the elements of landfill site function are "Early stabilization," "Early abolishment," "By-product salt," "Leachate treatment" and

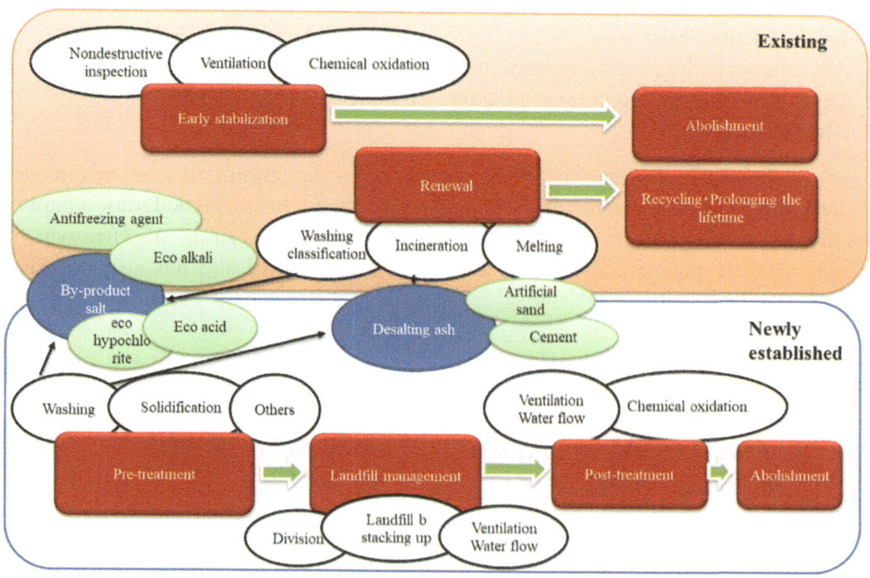

Fig. 4.1 The concept of the futurel and fill site and the required new technology

"Resource storage." "Pre-treatment" to make low landfill load, "Landfill management" to promote stabilization and "Post-treatment" become the constituent elements of the system to achieve the goals.

4.3 Location Selection of Landfill Site

The landfill site is a typical NIMBY (Not In My Back Yard) facility. Even now, its location is extremely difficult. There are various factors in those circumstances, and the representative factors are as follows.

- Site selection.

There is no scientific evidence for the reason why the landfill sites are built here.

- Residents' anxiety about environmental conservation.
- There is no reliability regarding security and safety for technology and operation management.
- Lack of information and communication
- Opinions which are unnecessary in a recycling-based society.

(A landfill site is unnecessary in order to realize a garbage zero society. Waste is put out because the facilities are built.)

- Fewer merits of the area.
- The image of the facility constructed in the past is bad.
- Distrust of the government and business people
- The others

In order to solve these issues, there are places for the opinion of residents in each process such as location selection based on scientific evidence, consideration papers, method books, preparatory documents in environmental assessment. However, in recent decades, the severity of the location of the landfill site has still not been improved.

4.3.1 Residents' Consciousness of the Location of the Troublesome Facility (Survey for College Students, 2001–2016)

As a part of the lecture series, the author conducts a questionnaire survey on the location of the waste disposal facility for junior and senior in college. The subjects are comprehensive clusters of subjects and can be taken regardless of humanities and science. The number of participants is about 80 people each year, and it continues from 2002 to the present. The contents of the questionnaire are as follows: ① "Do you agree or disagree if the construction facility for the landfill site near your house is put forward? What kind of action do you do?" ② "Where do you think the landfill site should be built?" The contents of the reply and opinion have not changed substantially for more than ten years since the questionnaire started. Figures 4.2, and 4.3 shows the results for 2012.

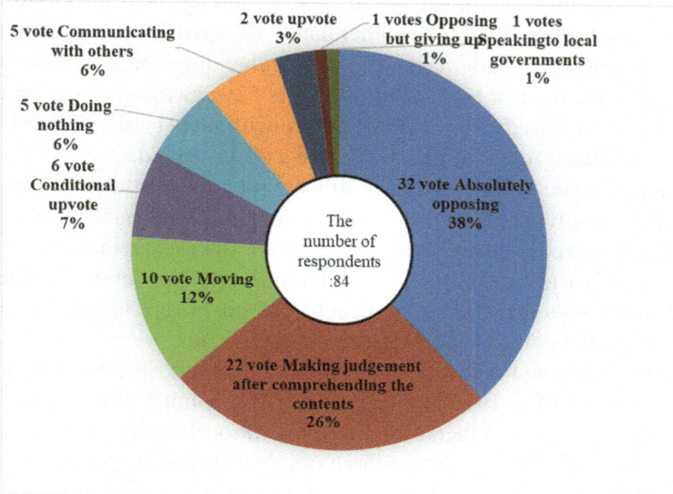

Fig. 4.2 Questionnaire for college students (2017)

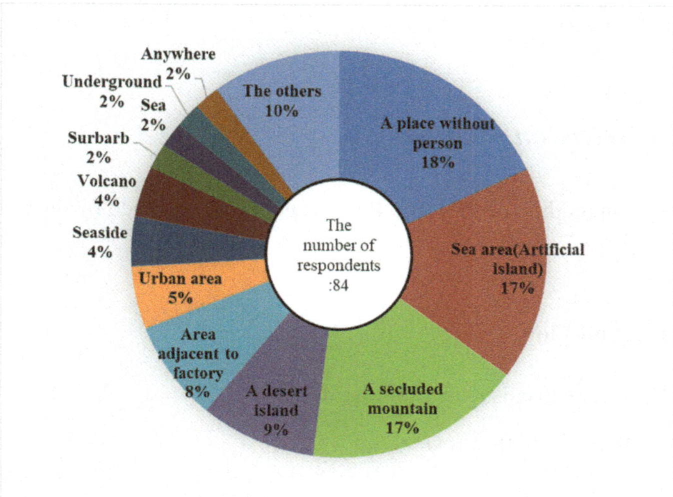

Fig. 4.3 Location of landfill site by questionnaire for college students (2017)

As a result, the total number of downvote is 39%, including 38% absolutely opposing and 1% opposing but give up. On the other hand, 3% upvote, 7% conditional upvote, and 22% vote making a judgment after comprehending the contents. Regarding the construction location of the landfill site, it is 52% away from the living-sphere (Multiple answers allowed). 18% choose a place without the human, 17% vote sea area, 17% prefer a secluded mountain, and 9% vote a desert island. As a result, the total number of downvote is 39%, including 38% absolutely opposing and 1% opposing but give up. On the other hand, 3% upvote, 7% conditional upvote, and 26% vote making a judgment after comprehending the contents. Regarding the construction location of the landfill site, it is 52% away from the living-sphere (Multiple answers allowed). 18% choose a place without persons, 17% vote sea area, 17% prefer a secluded mountain, and 9% vote a desert island.

In the questionnaire survey on the landfill site for college students, 39% of downvote and only 10% upvote (including conditional upvote). Because the waste incineration facility is covered with buildings, it is easy for townscape measures to be taken and the fact that waste can be seen at the landfill site also deteriorates the image, and many landfill sites constructed in recent years are covered type. Therefore, it is inferred that the elements of the landscape are also large.

On the other hand, as for waste incineration facilities, Prime Minister's Office in 1988 and Jiji Press Ltd in 1997 had carried out "An opinion poll concerning waste disposal." In the survey conducted in 1988, the ratio of downvote is 47.5%, while it was 40.7% in 1997. Though the research object is different, the ratio of opposing had decreased by about 7% in a decade. As the questionnaire for students is limited to the landfill site, though it is speculated, in recent waste incineration facilities, scenery consideration and the understanding of facilities is advanced. However, as the students had few opportunities to see the landfill site in general, the image has not been improved yet.

4.3.2 Site Selection Method

The site selection method of the landfill site is generally selected by the map overlay method.

4.3.2.1 Selection Flow

The flow of site selection is shown in Fig. 4.4.

① The first step (Primary filter)

In the first step, the selection range of the landfill site is set, and the areas which are subject to legal regulations on land use regarding the location of the landfill site are excluded. Then, we narrow down the areas which can be constructed. From

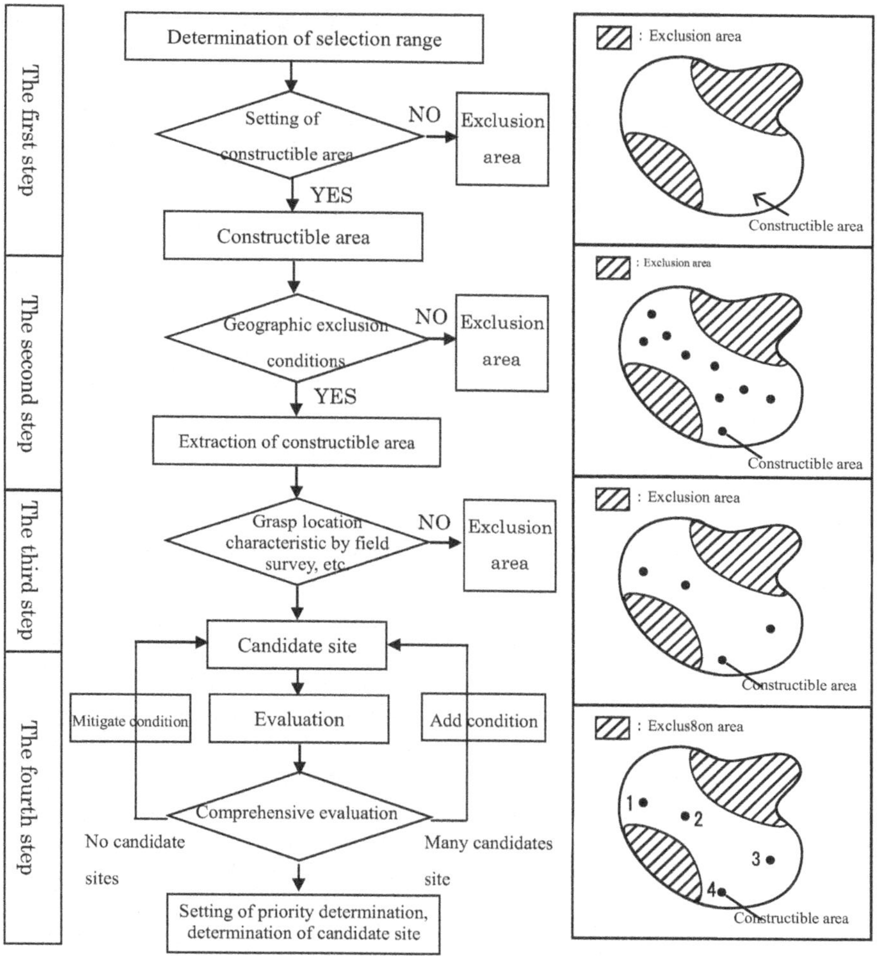

Fig. 4.4 The flow of site selection (Higuchi S 1999)

the situation of legal regulations on land use, it is challenging to release regulation, and excluding items which requiring permission from each ministry of a government in advance from the proper selection area. The regulations on land use depend on the contents of regulation, the right of regulation cancellation is entrusted to the competent minister of the country, the prefectural governor and the mayor of the Municipality. In the case of the natural environment conservation and disaster prevention, the regulation of high importance has the right of cancellation by the government and prefecture. Therefore, in the first step, the national land-use control areas are usually excluded from the selection area.

The working method is superposed generation by various regulations listed as exclusion conditions of the first step using GIS information.

Figure 4.5 shows the illustration of the work procedures.

In this figure, the quasi-national parks, the nature conservation area, the forest reserve is selected as a candidate area for exclusion in the first step. The area where these three exclusion areas are superposed on the map is set as the excluded selection area. The other area is set as the proper selection area.

② The second step (Secondary filter)

After narrowing down the selection areas in the first step, sites that can be concretely constructed from the selection areas are extracted in the second step. At this time, it becomes relatively smooth to extract the area if the geographical conditions for constructing the landfill site are clarified.

The following geographical conditions are mentioned as an example.

- A Mountainous area with a valley.
- No settlement in the vicinity.
- A place where the predetermined landfill capacity is possible to be secured.
- etc.

In the case of the working method, it is desirable to use a topographical map on a scale of 1/25,000 to 1/50,000.

The image is shown in Fig. 4.6. The construction area suitable for the landfill site is read from the contour lines in the appropriate site selection area. Since the small-scale landfill site cannot be read by contour lines of a topographical map on a scale of 1/25,000, this method is suitable for the landfill site exceeding medium-scale (above 100,000 m^2). There are many cases where work is done without considering the possession form of land at this point. The selected constructible area is plotted in a topographical map on a scale of approximately 1/200,000.

③ The third step (A field survey)

In the third step, a field survey is conducted on the constructible area extracted in the second step. Before the field survey, when a more precise topographic map is used, and landfill area, landfill capacity, drainage area is investigated, this confirmation will become easier. It is only a desk study until the second step. Since the topographic map is based on the aerial photograph, the precision of contour lines become coarse. Besides, since the land us regulation chart is generally shown on a scale of 1/200,000, it is necessary to confirm the area near the restricted area is outside on site. Based on these facts, items should be established in the field survey are geographical features, natures of the soil, the land category, river conditions, irrigation conditions, village locations, landscape, import access, traffic, the presence of urban facilities, other regulations, and comprehensive safety.

According to the field survey, the existence of new regulations and other development plans are confirmed, and it is revealed that the intake source is present at the area along the lower river, it will be excluded from the constructible area.

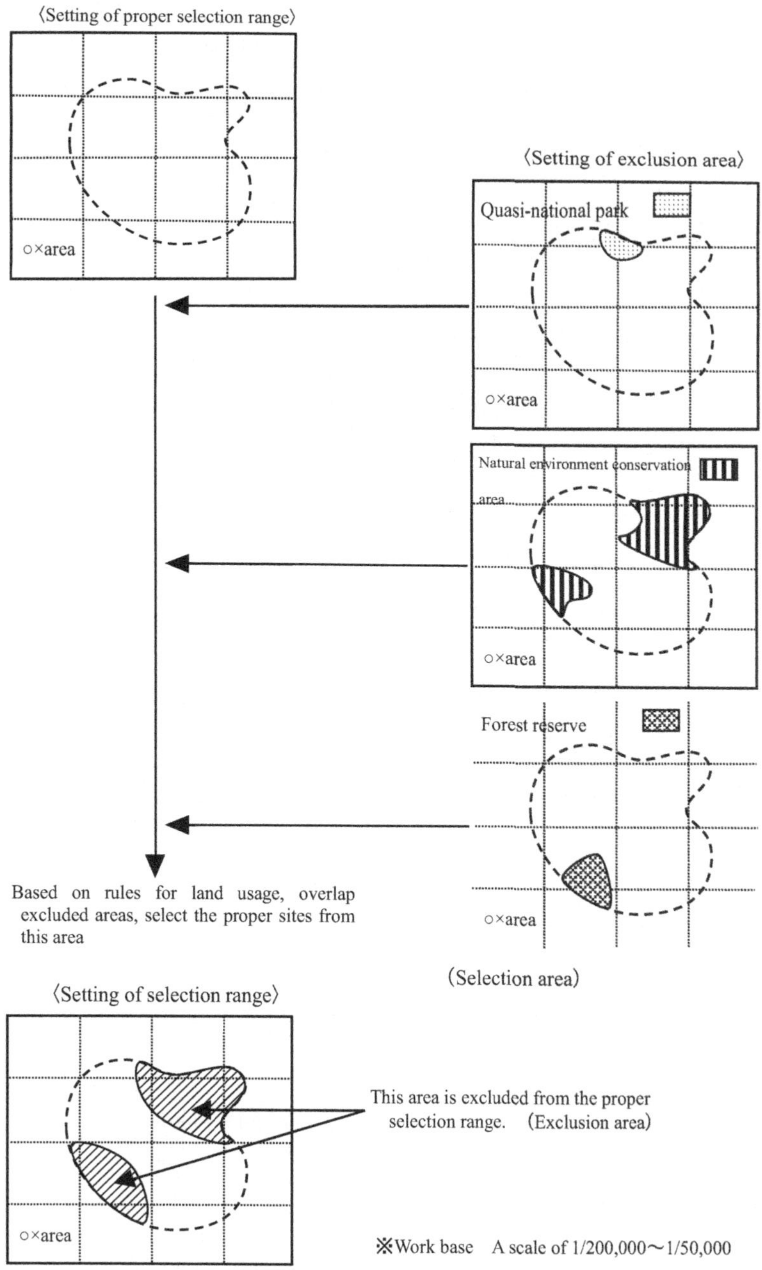

Fig. 4.5 The excluded area selection in the first step (Higuchi S 1999)

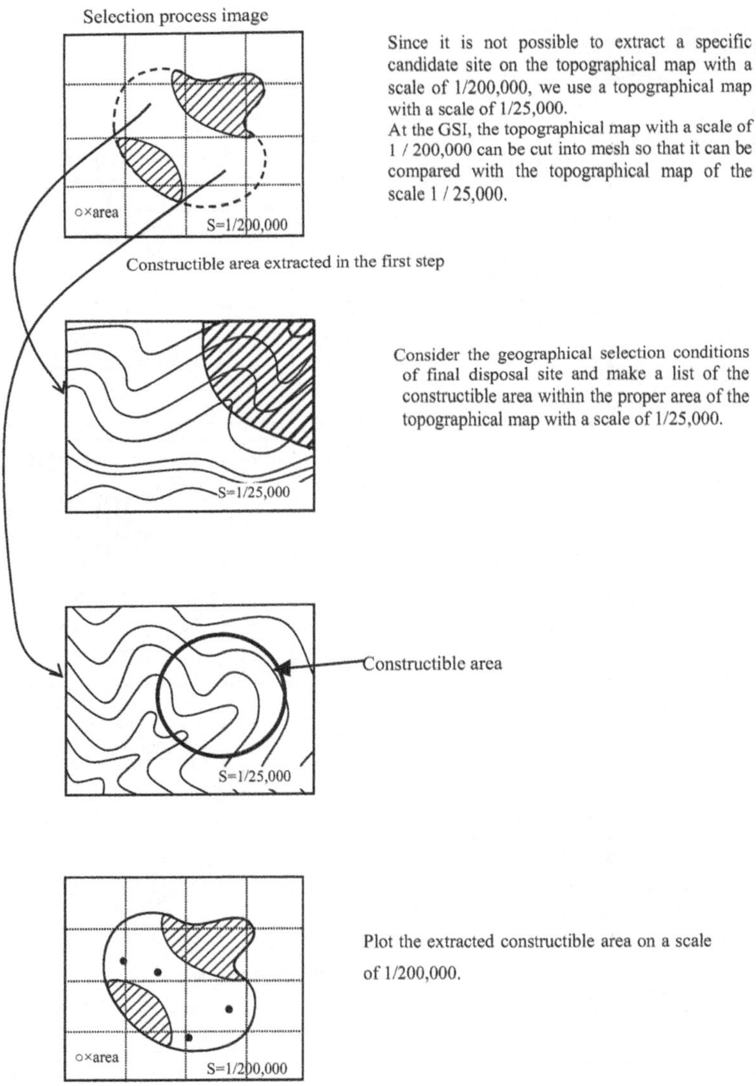

Fig. 4.6 The excluded area selection in the second step (Higuchi S 1999)

④ The fourth step (The evaluation of the candidate site)

Based on the field survey, in the fourth step, the evaluation of the candidate site which is narrowed down is conducted, the priority of the use of candidate site is set, and the final candidate site is determined.

The evaluation items and the evaluation reference are shown in Table 4.1. Evaluation items can be divided into site characteristics, environmental characteristics, and economic efficiency of the landfill site. The site characteristics of the landfill site include landfill efficiency, acquisition of soil cover material, an extension of approach roads, the utility value of the site, etc.

Landfill efficiency can be expressed as a value obtained by dividing the landfillable capacity by the landfill area. The capacity of landfill site-related facilities tends to be determined by the landfill area (such as leachate treatment facility, leachate collection/drainage facility, seepage control sheet area, etc.), and the landfill area also influences the costs for the construction of the facilities. Therefore, in the case of the plural number of landfill sites, when the landfill area is the same, the larger the landfill capacity is, the higher economic efficiency will be. It is possible to carry out the evaluation of the landfill site based on average landfill height obtained by dividing the landfill capacity by the landfill area. The cover soil is generally required for daily cover soil, an intermediate cover soil or final cover soil. It is an essential requirement whether the cover soil can be procured and used on the spot or must be purchased.

The shorter loading road is better, however, if the landfill site is set in the mountainous area, the extension becomes more prolonged, and the cost will increase due to the height difference.

In the case of the utility value of a site, after the reclamation, the wider the flat land obtained, the various uses of the site become and the higher the utility value of the land. It is effective to include the proportion of the flat area in the landfill as one of the evaluation items of the location characteristics.

In the case of environmental characteristics, a project can be smoothly carried out by a preliminary investigation into natural conditions, social conditions, and evaluation.

Location characteristics can represent economic efficiency because the relation of location characteristics is particularly close. Evaluation items include construction costs and maintenance costs.

The evaluation criteria are different depending on the regional conditions. Based on the evaluation, a comprehensive evaluation is performed, and the priority of use of the candidate site is determined. If there are no candidate site at this point, the conditions will be mitigated, and the candidate site will be reconsidered. On the other hand, if there are too many candidate sites, the requirements will be strict, and the number of candidate sites will be reduced.

Proper area selection of landfill site cannot be used unless an agreement is formed with landowners and surrounding residents even if it is selected based on the basis of scientific evidence. In many cases, since it comes in from Step 3, it becomes an individual discussion. And it is not possible to answer the residents' question "Why we build a landfill site here" clearly. Therefore, it is said that the proper area selection is essential according to the basis of scientific evidence.

Table 4.1 Evaluation items of candidate sites and examples of evaluation (Higuchi S 1999)

Evaluation item			Evaluation standard		
Large classification		Small classification	Rank 1 (High appropriateness)	Rank 2 (Normal appropriateness)	Rank 3 (Low appropriateness)
Location Characteristics	Location Characteristics (A)	Landfill efficiency	Landfill efficiency (Landfill capacity m^3/Landfill area m^2) is above 10	Landfill efficiency is between 5 to 10	Landfill efficiency is below 5
		Obtaining cover soil	Whole quantity available at the candidate site	Partly quantity available at the candidate site	Not available
		Extension of approach road	Extension from the public road to the MSW landfill site is below 2,000 m	Extension from the public road to the MSW landfill site is between 2,000 m to 3,000 m	Extension from the public road to the MSW landfill site is above 3,000 m
		Site use plan (Plain land/Landfill site)	Secure 75% or more plain land, and it is easy to use the site	Secure 50%–75% plain land, it is normal to use the site	Secure less than 50% plain land, it is difficult to use the site
Environmental Characteristics	Natural Condition (B)	Geology	No active fault within 1 km around the candidate site, heavy colluvial soil and old collapsed landform have not been found	No active fault within 1 km around the candidate site, heavy colluvial soil and old collapsed landform have been found	No active fault within 1 km around the candidate site, heavy colluvial soil and old collapsed landform have been found at the candidate site
		Animals and plants	The inhabitation of valuable animals and plants has not been confirmed within 2 km around the candidate site	The inhabitation of valuable animals and plants has been confirmed within 2 km around the candidate site	The inhabitation of valuable animals and plants has been confirmed at the candidate site
		Vegetation	Vegetation naturalness is 1–3	Vegetation naturalness is 4–6	Vegetation naturalness is 7–10
	Social Condition (C)	Land use	Afforestation area or cultivated land is below 1/3	Afforestation area or cultivated land is between 1/3 to 2/3	Afforestation area or cultivated land is above 2/3

(continued)

Table 4.1 (continued)

Evaluation item			Evaluation standard		
		Cultural property	Designated cultural properties are not confirmed within 1 km around the candidate site	Designated cultural properties are confirmed between 500 m to 1 km around the candidate site	Designated cultural properties are confirmed within 500 m around the candidate site
		Residence	There is no residence within 1 km around the candidate site.	There is a residence between 500 m to 1 km around the candidate site.	There is a residence within 500 m around the candidate site.
	Environmental Condition (C)	Irrigation condition	Effluent stream has not been used	Effluent stream has been used for agriculture	Effluent stream has been used for marine product industry
		Offensive odors · Noise · Vibration	No colony within 1 km around the candidate site	A colony between 500 m to 1 km around the candidate site	A colony within 500 m around the candidate site.
Economic Efficiency	Economic Condition (D)	Approximate construction cost (Excluding approach road)	Below ○○yen/m^2	Between ○○yen/m^2 to ○○yen/m^2	Below ○○yen/m^2
		Approach road construction cost	Below ○○yen	Between ○○yen to ○○yen	Above ○○yen
		Approximate maintenance cost	Below ○○yen/m^2	Between ○○yen/m^2 to ○○yen/m^2	Below ○○yen/m^2

4.4 Regeneration of Landfill Site

New technology has been developed year by year for the landfill site, and now it has reached the excessive facility level. For example, high reliable seepage control work technology, water leakage detection and restoration backup system corresponding to disasters and accidents, and leachate treatment technology approaching the level of distilled water past the tap water are beginning to be introduced. In order to stabilize the landfill site at an early stage or abolish it, various technologies such as pre-treatment of landfill have also been developed.

Therefore, in the case of the new landfill site, the groundwork is accepted by the areas as an environmental safety facility. These new technologies have been developed as symptomatic treatment and a post-treatment towards groundwater pollution caused by past water leakage in seepage control work or surface water contamination accident caused by leachate etc.

There are groundwater contamination risk and surface water contamination risk in some MSW landfill sites that have been landfilled. In the year of 1997, the Ministry of Health and Welfare announced five hundred places in the whole country as the improper landfill sites, and rationalizing work was carried out. Furthermore, waste has been landfilled directly in some of the MSW landfill sites that have been landfilled in the past. And gas is still generated, and high concentrated leachate is still discharged in some places. These have become a negative legacy, which is a problem in terms of regional environmental conservation. For the past improper landfill sites, though the rationalizing work is gradually carried out, it is still a negative legacy as it is not being abolished soon even if it is optimized. On the other hand, it is still difficult to secure a new landfill site. In some municipalities, the situation becomes even worse, and it is desirable to extend the life of existing landfill sites and to regenerate the sites that have been landfilled. Conventionally, there is a history that intermediate treatment facility is spread due to the difficulties in securing the landfill sites and extend life, however, landfill without intermediate treatment is carried out in many of the existing landfill sites, and these problems can be solved by reprocessing and reusing the waste to ensure the landfill space. In addition, when the amount of organic substance and dioxins in excavated waste and backfilled waste after sorting is high, it can be optimized by performing organic substance decomposition on-site during excavation or detoxication treatment.

Besides, when the breakage of seepage control work or the blockage of water collecting pipe or gas venting facility are found during the excavation of waste layer, it is possible to make the functional recovery of landfill site by restoration to prompt early stabilization and abolition. In this way, the regeneration of the landfill site can solve the lack of the landfill site. Besides, it is possible to reduce the pollution potential of existing landfill waste and minimize the risk of environmental contamination due to the reinforced structure.

4.4.1 Survey and Procedures for Regeneration

Figure 4.7 shows the example of the selected flow of renovating method and survey in the case of the MSW landfill site is regenerated.

4.4.1.1 Landfill History Survey

In the case of the regeneration of landfill sites, it is necessary to make a landfill history survey to determine whether to carry out a generation or not from the effect

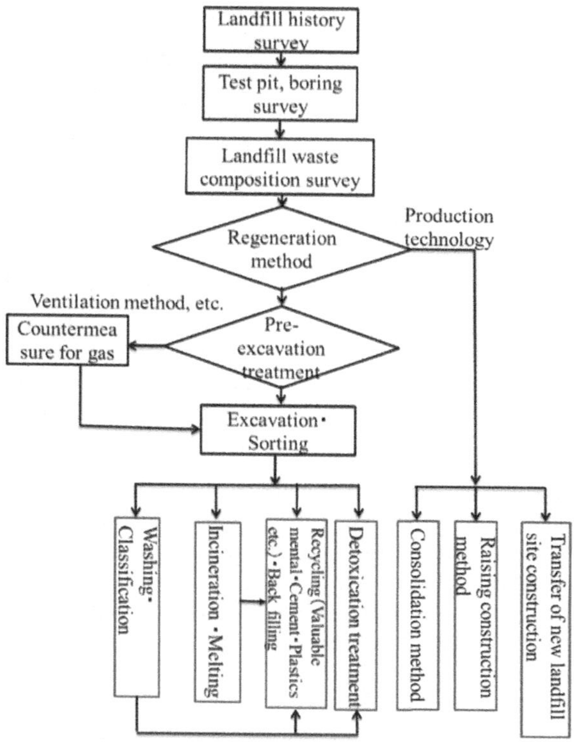

Fig. 4.7 The flow of generation

of regeneration. The landfill history survey is mainly focused on existing literature surveys such as landfill records. On the other hand, the exploration technique for waste quality (organic, inorganic, etc.) in the landfill layer or the contamination investigation on the underground of the landfill and surrounding soil is also required.

4.4.1.2 Test Pit, a Survey on Waste Quality (Higuchi Y et al. 2001)

When the accurate historical information on landfill does not remain, an investigation on landfill is conducted by test pit digging or boring. A test pit case is shown in Fig. 4.8.

In order to grasp the current conditions of the landfill waste layer, we investigate:

① Composition, density, moisture, the heat loss of landfill waste
② Presence of toxic substance
③ Stabilization of landfill waste
④ Status of landfill gas
⑤ Presence of perched water and its quality
⑥ Presence of leachate and its quality etc.

Fig. 4.8 Test pit case

Besides, in the case of planning to dug landfill waste, it is also necessary to investigate dust, offensive odors, poisonous gas, and combustible gas generated in digging up.

As a method of investigating underground conditions, the boring excavation method is common to be used. The waste layer is boring excavated, and collections are performed, such as a collection of landfill waste, a collection of samples for the survey on water quality which confirms the existence of perched water and leachate, a collection of landfill gas, etc. However, when the waster layer is boring excavated to collect the samples for composition and the three components analysis (combustible, moisture, ash), it is necessary to excavate waterless. It will become difficult to continue the excavation when the excavator hits concrete mass etc. in the waste layer, and it has no choice but to change the excavation position. In addition, in the case of the waste is collected as a core, when the waste is larger than core diameter, sometimes, it will be pushed down so that the accurate samples become impossible to be collected.

By using a test pit excavated the landfill waste layer with a backhoe, landfill waste can be collected as it is, and the density in landfill condition can be accurately measured. Besides, dust and gas which generated at the excavation of landfill waste also can be measured. In addition, as the landfill condition of waste can be directly observed, the kind of waste which is being landfilled can be directly grasped.

<Landfill density>

When the treatment residues of excavated waste are backfilled, the waste can be re-landfilled in the dup up space. Therefore, it is important to grasp the volume of this space clearly. However, when the waste is excavated, the volume will be generally increased by about 1.2 to 2 times. And the increase in volume depends on the different components of landfilled waste. Therefore, it is impossible to estimate the volume of excavated space from the volume of excavated waste, and it is necessary to measure the density in the landfill condition.

The density in the landfill condition of waste can measure the volume of the pit after digging up the waste and calculate the landfill density from the volume and the weight of the excavated waste. Methods for measuring the landfill density include boring excavation method and test pit method. In the boring excavation method, the volume is calculated from the length of boring excavation and the inside diameter of sampling tube, the weight of waste which is taken out as a core is measured, and the density is calculated. However, when the excavator hits tires which are difficult to cut off, metals or concrete mass, etc. during the boring excavation, all the obstacles will be pushed down, and it will affect the precision of density. In the test pit method, a replacement method using water and sand is used in public works. When water is used, a thin vinyl sheet is laid out in the excavation space, water is poured in, and the volume is obtained. However, if there are projections of waste in landfill sites or the excavation holes become deeper, the vinyl sheet may be torn by hydraulic pressure. As a case using sand is shown in Fig. 4.9. There is also a method for measuring the volume of the pit after the excavation of waste by using the laser volume measuring apparatus and calculating the landfill density from the weight of excavated waste. When the volume of the test pit is measured by the laser volume measuring apparatus, it can be measured with a precision of less than 5%.

As shown in Fig. 4.10, the laser volume measuring apparatus is an apparatus which attached a laser displacement meter at the lower end part of the movable rotating shaft. The volume of the pit is measured by the following procedure.

① Excavate a hole for in situ density measurement in the measurement site of the ground of waste.

② Install a measuring instrument in the just upper side of the hole and set the height of the laser displacement meter close to the upper side of the hole as much as possible.

③ Measure the distance R(θ) to the excavated face through laser displacement meter and record it.

④ Rotate the rotating shaft at angular pitch Δθ which is previously set and repeat the operations in ③.

⑤ When the laser displacement meter makes a circuit, the measurement of depth is terminated, and the cross-section area S_i based on the Formula (4.1) is calculated.

Fig. 4.9 Measuring the volume of the pit by a pre placement method

Fig. 4.10 The laser volume measuring apparatus

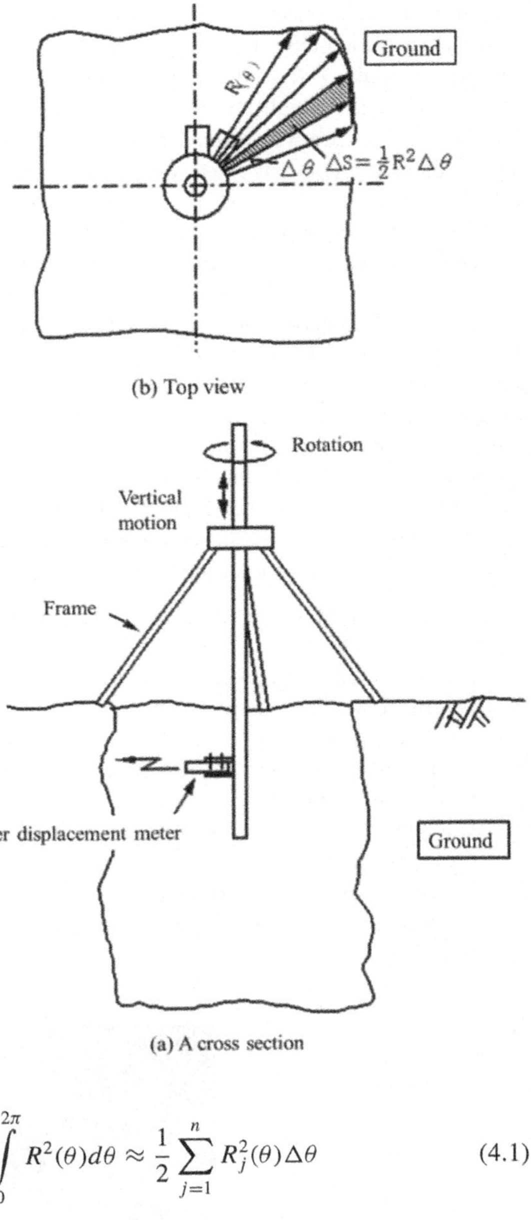

$$S_i = \frac{1}{2}\int_0^{2\pi} R^2(\theta)d\theta \approx \frac{1}{2}\sum_{j=1}^{n} R_j^2(\theta)\Delta\theta \tag{4.1}$$

⑥ The laser displacement meter is dropped by the depth Δh, which is previously set and repeat the operations ③ to ⑤.

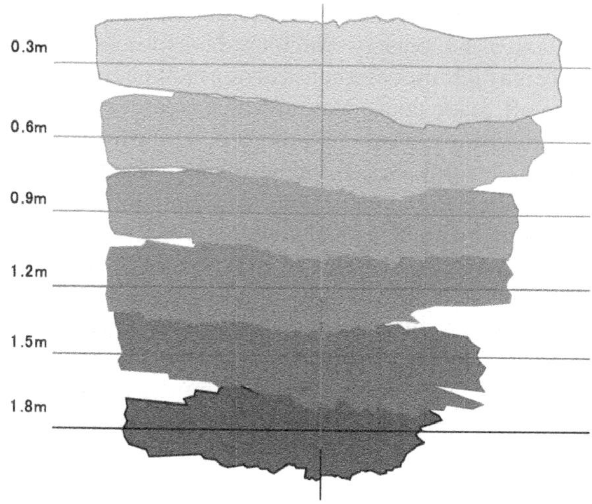

Fig. 4.11 Examples of laser volume measuring apparatus (Higuchi Y et al. 2001)

⑦ When the measured cross-section reaches the bottom, the measurement is terminated, and the volume V of the excavated hole is calculated by the Formula (4.2).

$$V = \frac{S_1 + S_2}{2}\Delta h + \frac{S_2 + S_3}{2}\Delta h + \cdots + \frac{S_{m-1} + S_m}{2}$$
$$\Delta h = \left(\frac{1}{2}S_1 + \sum_{i=2}^{m-1} S_i + \frac{1}{2}S_m\right)\Delta h \quad (4.2)$$

Regarding the pitch $\Delta\theta$ of the measuring angle of the laser volume measuring apparatus, as a result, verified by using the cylindrical container (the inside diameter is about 1.2 m) and the square container (about 1.2 m square), the measurement error is sharply decreased by fine adjustment of the measurement angle. Based on this result, the measurement is normally carried on with the measuring pitch angle of 6° (Higuchi 2001).

Figure 4.11 explains the cross-section examples of laser volume measuring apparatus.

4.4.1.3 An Investigation on the Regeneration Method

The judgment on whether it can be regeneration or not can be made according to the composition survey result of the landfill waste. The improper conditions for regeneration include the high ratio of organic substance in landfill waste, cases without economic value, being landfilled with fly ash and dust, etc. which causes the layer solidification and hard to excavation, and too much excavation cost. In these cases, life extension strategy which used creation technology is considered. For instance, there is a high-pressure construction method such as dynamic consolidation by weight drop off, static consolidation which creates space in the landfill layer by hydraulic, etc., a raising construction method, a transfer method which can transfer to new or existing landfill sites. Besides, as a regeneration method, excavated existing landfill sites, collect the valuables such as aluminum and iron by sorting technology, volume reduction by crushing and compressing plastics, methods of recycling vacant space for the utilize of landfill by reprocessing such as the incineration or melting of combustible waste, methods of after cleaning, classification and detoxification technology to make it harmless, the recycle it to cement, etc. Decides on the regeneration method is decided comprehensively based on the quality of landfill waste, such as life prolongation-effect, environment conservation, economy, etc. As an example, the evaluation criteria which focusing economy are listed as follows.

$$\mathrm{F/S} = (\mathrm{kD} - \mathrm{T} - \mathrm{V})\,\mathrm{a} \geq 0$$

In here,
D: Disposal cost (yen/m^3)
T: Transfer cost (yen/m^3)
V: Volumetric reducing treatment cost (yen/m^3)
a: Amount of target (m^3)
k: Constant

4.4.1.4 Pre-treatment of Excavation

Through the results of the boring survey, test pit survey and the investigation on waste quality, a blowout of methane gas and hydrogen sulfide gas, etc. are anticipated during the excavation. When the preservation of industrial safety and living environment is expected to be hindered, it is necessary to remove these troubles in advance. There are methods for removing troubles, such as ventilation method and oxidant injection method. These methods will be introduced in the part of "Early stabilization." As methods of investigating places where there is a possibility of gas generation in landfill sites, there are

① A survey on the temperature of the surface layer by an infrared camera
② A survey on remote sensing
③ A survey on surface gas
④ Temperature surveys at a depth of 1 m

In this section, we will introduce the Temperature surveys at a depth of 1 m

In the existing MSW landfill site, the raising of the temperature of waste landfill ground by the decomposition of the organic substance is focused, and through the temperature surveys at a depth of 1 m (5.0 m mesh measurement) and thermal conductivity analysis (G-HEAT/3D), it is possible to evaluate the three-dimensional distribution of the organic layer. Besides, by combination with high-density electric exploration used in the waste exploration of the past, it is possible to conduct research with higher precision.

a. Temperature surveys at a depth of 1 m (Kubo S 1988; Higuchi S 2005)

The temperature surveys at a depth of 1 m is a method used in the investigation on source of hot spring in the past, application to groundwater survey (Takeuchi 1983), it is used as an investigation method to grasp the planar presence situation of "water channel (fluidized layer of underground water)" in shallow ground (up to GL-15 m). The survey method is to measure and analyze the ground temperature of the one-meter underground with a 5.0–10.0 m mesh. The difference between the normal ground temperature at a depth of 1 m and the fluidized groundwater temperature is larger in summer and winter.

Through the temperature surveys at a depth of 1 m, the place where existing the flow of groundwater and the place where not existing the flow of groundwater can be divided by grasping the high and low temperatures.

In the case of the investigation on organic substance, the measurements (winter) and analysis are performed with 5.0 to 10 mesh. As a result, the distribution of high-temperature belt (above 13 °C) is detected because the places where the decomposition of the organic substance is processing becomes the heat source towards normal ground temperature at a depth of 1 m (about 10 °C).

Figure 4.12 explains the conceptual diagram of the temperature surveys.

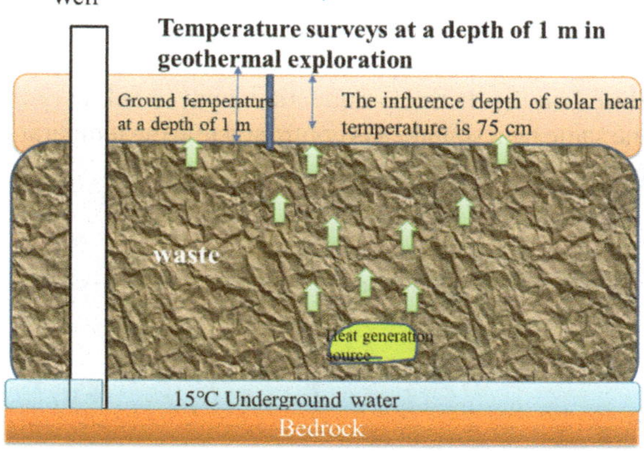

Fig. 4.12 The conceptual diagram of geothermal exploration

b. A three-dimensional distribution analysis by thermal conductivity analysis

Based on the results of the temperature surveys at a depth of 1 m, a three-dimensional distribution analysis of the organic layer is carried out by using thermal conduction analysis (G-HEAT/3D). An inverse analysis method by measurement results of physical properties such as thermal conductivity of ground has been proposed many times by research on tomography etc. On the other hand, as a method of identifying heat generation source, methods using integral equation method have been proposed (Kubo S 1988; Yamatani K et al. 1995), however, the current condition is that it is difficult to adapt the problem with complicated boundary condition and the condition in which thermophysical property is distributed, etc. Therefore, the authors attempt to estimate the position of the heat source by making a change of the position and the number of the heat source, conducting numbers of simulations and comparing the result with the measurement value. In this method, a finite element analysis is conducted by making a change of the position of heat source, and the temperature distribution is obtained when a specific heat generation source is assumed. Comparing the properly superimpose temperature distribution with the results of the temperature surveys at a depth of 1 m and the position of the heat source is estimated.

c. Thermal conductive analysis

In the analysis, a finite element model of a region including the target landfill site is created, a heat source is determined, and the temperature distribution of the region is obtained by steady-state analysis while sequentially moving. The thermal conduction equation is shown as follows.

$$\frac{\partial}{\partial x_i}\left(K_{ij}\frac{\partial T}{\partial x_j}\right) = \rho C\frac{\partial T}{\partial t}$$

K_{ij}, C and ρ represent the thermal conductivity, density, and specific heat of each ground. In addition, subscripts represent the orthogonal coordinate system and follow the sum-rule.

d. Identification of heat source

Based on the thermal conduction analysis while providing the same temperature to the same heat source, it is confirmed that the temperature is obtained in the measuring points as follows.

$$\begin{aligned}
t_1 &= \alpha_1 T_{11} + \alpha_2 T_{12} + \cdots + \alpha_j T_{1j} + \cdots + \alpha_n T_{1n} \\
t_2 &= \alpha_1 T_{21} + \alpha_2 T_{22} + \cdots + \alpha_j T_{2j} + \cdots + \alpha_n T_{2n} \\
&\vdots \\
t_i &= \alpha_1 T_{i1} + \alpha_2 T_{i2} + \cdots + \alpha_j T_{ij} + \cdots + \alpha_n T_{in} \\
&\vdots
\end{aligned}$$

$$t_m = \alpha_1 T_{m1} + \alpha_2 T_{m2} + \cdots + \alpha_j T_{mj} + \cdots + \alpha_n T_{mn}$$

t_i is the measured temperature at the i th measuring point, T_{ij} is the temperature at the i th measuring point generated by the j th heat source. α_j is a coefficient representing the degree of influence of the j th heat source on the measuring point.

In this analysis, the coefficient α is found by the least-squares method.

e. Analysis model

Figures 4.13 and 4.14 shows the finite element model used for the analysis. The analysis area is the ground including the landfill site, and the area size is determined to make sure the influence of the boundary condition on the temperature distribution is small enough. In addition, the temperature measuring point is made to match the node of the finite element model. Examples of thermal conductivity, such as waste are shown in Table 4.2. In the analysis, a heat source is assumed in the area of the waste, and the temperature is found at the measuring point. Furthermore, the heat source is moved in the area of the waste, the temperature of the measuring point is obtained in each case, and the magnitude of α is obtained according to the Eq. (4.2).

f. Examples of survey analysis (Higuchi S et al. 2005)

Figure 4.15 shows examples of geothermal exploration at the landfill site of industrial waste.

The soil temperature at a depth of 1 m is in a range of 10.0 to 28.8 °C, and it causes a magnitude of deviation. The soil temperature in the left side of the bank of

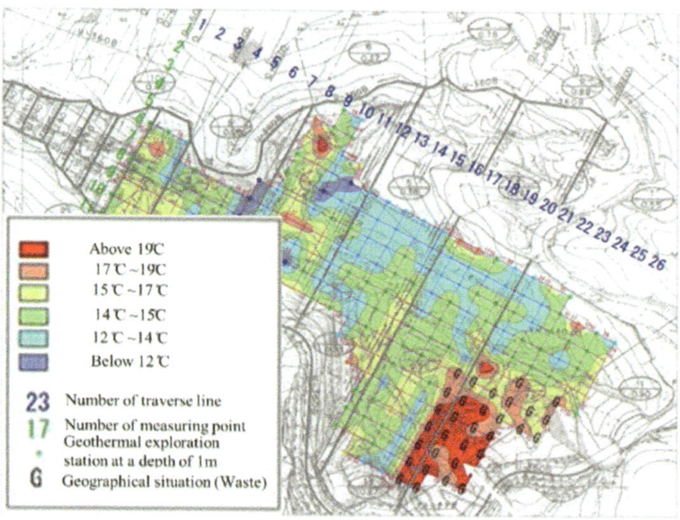

Fig. 4.13 Example of soil temperature measurement

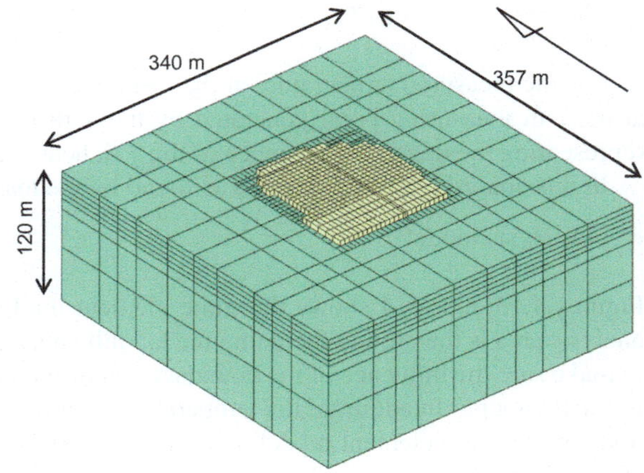

Fig. 4.14 The whole analysis model

Table 4.2 Examples of thermal conductivity in each material

Material	Thermal conductivity W/(mK)
Earthen cover	1.83
Waste	0.57
Bedrock	1.10

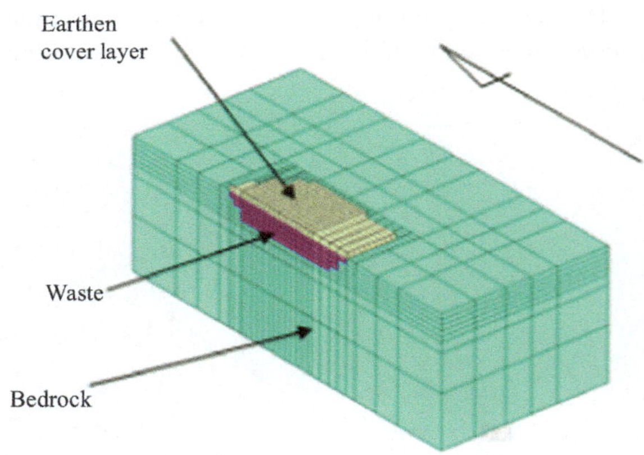

Fig. 4.15 A cross section of analysis model

the upper stretch is high because of the temporary placement of waste on the surface layer. The soil temperature at a depth of 1 m at the fixed point outside the landfill site is 12.8 °C on average. The groundwater is measured by using a gas pipe. The water temperature inside the hole is above 35 °C till GL-30.0 m, and it becomes lower to the depth. As a factor of forming the high-temperature section, there is a high probability that the organic substance contained in the waste is used as the heat source.

In the thermal conductive analysis, the distribution of heat source and the heat source temperature are estimated by creating a three-dimensional model and making an isothermal boundary identification analysis. The isothermal boundary identification analysis of heat source performs an inverse analysis based on the results of the temperature surveys at a depth of 1 m. A constant temperature is assumed in the underground heat generation source, a measured temperature at a depth of 1 m is used, and three-dimensional evaluation of organic substance is carried out by identifying isothermal. The necessary physical property values for the three-dimensional model creation are density, thermal conductivity, and specific heat. Besides, as the analysis condition to give the initial soil temperature, the soil temperature at a depth of 1 m and the geothermal gradient is necessary. In addition, considering the ground temperature while calculating the geothermal gradient, there is a possibility that it may be greatly affected by the external temperature. Therefore, in order to obtain a more accurate geothermal gradient, a geothermal gradient of intermediate (0.5–1.2 m) which relatively not affected by the external temperature is adopted. For analysis, the ground boundary surface is used as the heat transfer boundary between different phase, and the physical property value uses the external temperature and heat transfer coefficient (Table 4.3).

Figure 4.16 shows the distributional area in which heat generation source is above 48 °C. The depth of the heat source body is from GL-10 to 30 m. A high heat source body which above 48 °C distributed in the deep layer is found mainly in the (south side) old valley bottom of upstream. Old waste before standard revision is landfilled in the old valley bottom, and the rise in the temperature is thought to be influenced. In order to evaluate the calculation result, the boring of the heat source zone and the temperature logging are conducted (Refer to Fig. 4.17).

As a result, the temperature which is almost similar to the result of heat source analysis is detected, and organic sludge is sampled from the high-temperature boring core. Based on these facts, it is found that temperature surveys at a depth of 1 m are effective as a method of exploring the underground hot spot (undecomposed place of organic substance).

4.4.2 Sorting Technology (Higuchi S 2005)

Sorting technologies are generally classified into sieve sorting technology for size such as measurement, particle diameter, and magnetic sorting technology for metals, etc. The sieve technology is divided into primary sorting (coarse selection) and secondary sorting (trommel and screen for sorting by each particle size) by skeleton bucket. Besides, wind sorting is used for plastics and paper because of the difference in specific gravity. As excavated waste has a lot of moisture, it clogs the trommel and the screen and causes the malfunction. Therefore, some have a function of giving vibration or heating and drying. In the case of conducting a detoxication treatment or recovery of specific substances, the classification technology is used for fine fraction

Table 4.3 List of physical property values used for calculation

Thermal analysis physical property value		Adopted value	Notes
Density	Waste	1550	Unsaturated ground
(kg/m^3)	Foundation ground (Bedrock)	1900	Unsaturated ground
Specific heat	Waste	840	Unsaturated ground
(J/(kg K))	Foundation ground (Bedrock)	1400	Unsaturated ground
Thermal conductivity	Waste	0.57	Unsaturated ground
(W/(m K))	Foundation ground (Bedrock)	2.20	Unsaturated ground
Heat transfer coefficient (W/m^{2}K)		15	General value in outdoor (10–25 W/m^{2}K)
External temperature (°C)		12.8	The measurement of soil temperature in 5 days, the mean temperature from 9 o'clock to 17 o'clock
Geothermal gradient	Whole	2.25	Gradient by geothermal approximation curve between 0.5 and 1.2 m
	(Surface morphology) Bare land, Grass land	1.75	〃
	(Surface morphology) Waste	6.82	〃
Soil temperature at a depth of 1 m	Whole	15.2	
The mean temperature	(Surface morphology) Bare land, Grass land	14.4	
(°C)	(Surface morphology) Waste	20.6	

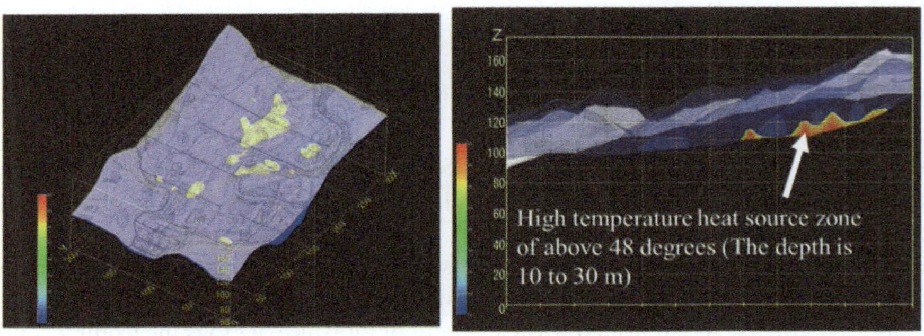

Fig. 4.16 Result of heat source analysis

Fig. 4.17 Temperature logging in heat source zone

sorting (below 100 μ). The classification is divided into dry type (sieve) and wet type (hydrocyclone, etc.).

We will introduce the empirical research which is carried out in synthetic research on regenerated reclamation (representation researcher: HIGUCHI Sotaro) by Environmental Science from the Ministry of the Environment.

① Target landfill site: The sorting of MSW landfill site in which the disposal of incineration residue, bulky waste, etc. is carried out

② Test machine: Excavator · Rough sorting machine · Transporting machine on rough terrain · Mobile sorting machine · Loader The sorting flow

③ Sorting flow
Figure 4.18 shows the sorting flow.

④ Excavation

A large number of wastes which is mainly waste plastics are moderately in wet conditions in the excavation sites. Besides, there are many clothes and sheet-like waste plastics in the case of over 200 mm by skeleton bucket. In addition, it is found that both the odor and the occurrence of gas is relatively little (Refer to Fig. 4.19).

⑤ Sorting

Figure 4.20 shows the condition of wastes over 200 m sorted by skeleton bucket. There are many clothes and sheets. The reburying wastes are transferred to the secondary sorting place by rough terrain vehicles. The excavated waste that transferred to the side of the secondary sorting machine (trommel) is thrown into sorting machine within the transferred day, and the sorting test is conducted in a wet condition to improve the scattering state to the utmost.

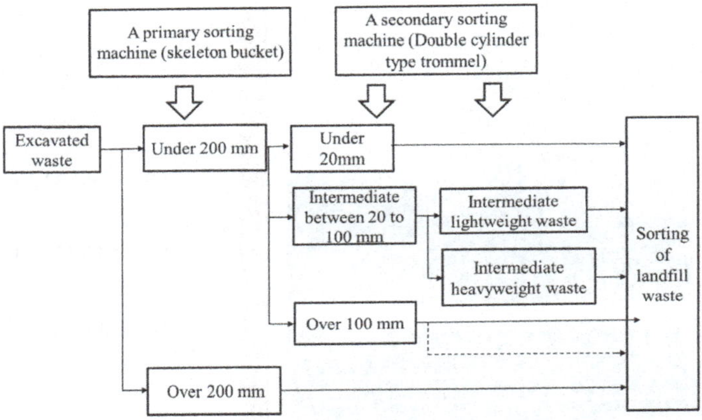

Fig. 4.18 The sorting flow

Fig. 4.19 Excavation condition

Fig. 4.20 Excavation condition (Waste over 200 mm)

Fig. 4.21 The outline of the secondary sorting machine

The throwing method is carried out by using a hydraulic shovel for excavation. Samples are thrown to make uniform the amount in a small quantity on the upper surface of vibration sieve of the secondary sorting machine, and the feed speed of feed belt conveyor of the secondary sorting machine is adjusted. The outline of the secondary sorting machine is shown in Fig. 4.21.

The secondary sorting is divided into 5 classifications as follows.

(a) Waste over 200 mm by vibration sieve (mainly for removing foreign substances in trommel)
(b) Waste by a magnetic separator
(c) Waste under 20 mm
(d) Intermediate waste between 20–100 mm
 d-1) Lightweight waste for wind sorting
 d-2) Heavyweight waste for wind sorting
(e) Waste over 100 mm

Because of the wet condition, it is possible to prevent the dust generation as much as possible and reduce the burden on the surrounding environment. In this test, though the wind sorting test on the waste over 100 mm is not carried out, it can be presumed that it is possible to maintain sorting precision of waste as the same equation as the intermediate waste by analyzing flight locus of waste inside the wind sorting machine for intermediate waste. Besides, it is possible to do the gravitational sorting of intermediate waste, over waste, lightweight waste, and heavyweight waste.

⑥ Characteristics of excavated waste

Table 4.4 explains the composition, the three-component test, and the caloric value test result of excavated waste. The moisture content in three-component tests is about 30–33% for "excavated waste" immediately after excavation, about 27–28% for "waste under 20 mm" in which organic substance is mixed, about 22–23% for "intermediate heavyweight waste" which mainly containing earth and sands, and about 27–38% for "intermediate lightweight waste" which mainly containing combustibles. In the case of ash content.

Table 4.4 Characteristics of excavated waste

Level name			First level			
Sample name			Excavated waste	Waste under 20 mm	Intermediate heavyweight waste	Intermediate lightweight waste
Sampling time			11:50	16:00	16:00	16:00
composition	Dry matter basis	Clothes, Paper (%)	1.0	0.3	0.8	4.5
		Synthetic resin, Gum, Leathers (%)	11.0	1.0	2.2	41.5
		Tree, Bamboo, Straws (%)	3.0	1.6	1.8	8.5
		Garbage (%)	0.0	0.0	0.0	0.5
		Incombustible waste (%)	23.6	53.0	89.3	31.0
		Others (%)	61.4	44.1	5.9	14.0
Three component	Moisture (%)		30.0	26.9	22.8	27.3
	Ash content (%)		–	68.0	73.9	36.1
	Combustible content (%)		–	5.1	3.3	36.6
Higher heating value (kcal/Kg)			–	–	–	2,240
Lower heating value (kcal/Kg)			–	–	–	1,920

Fig. 4.22 Intermediate waste between 20~100 mm

It is about 66–74% for incombustible system and about 29–36% for the combustible system. In the case of combustible content, it is 3–9% for incombustible system and 33–37% for the combustible system. In the case of the calorific value for the intermediate lightweight waste, the lower heating value is 1,740–1,920 kcal/Kg, and the higher heating value is 2,130–2,240 kcal/Kg. Figure 4.24 shows the result of the three components (Figs. 4.22 and 4.23).

Fig. 4.23 Waste under 20 mm

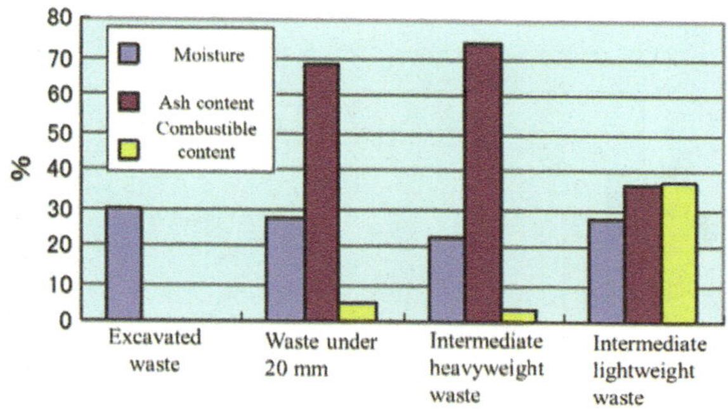

Fig. 4.24 Three components of excavated waste

⑦ Materials balance

The proportion of each sorting classifications in the whole buried object excluding soil is

(a) Waste over 200 mm skeleton 13–13.6%
(b) Waste under 20 mm 40.2–56.5%
(c) Intermediate lightweight waste 7.8–13.3%
(d) Intermediate heavyweight waste 16.4–22.3%
(e) Waste over 100 mm 4.9–10.5%
(f) Waste removed by selection separator 0.1–1.5%

About 50% of the waste is under 20 mm. The materials balance (the mean value) is shown in Fig. 4.25.

⑧ Composition in each classification

In the secondary sorting by trommel, we analyzed the composition of "intermediate lightweight waste," "intermediate heavyweight waste" and "waste over 100 mm." In this analysis, mixed samples around 50 kg are prepared from each sorting classification, and hand sorting is carried out. The results are shown in Fig. 4.26.

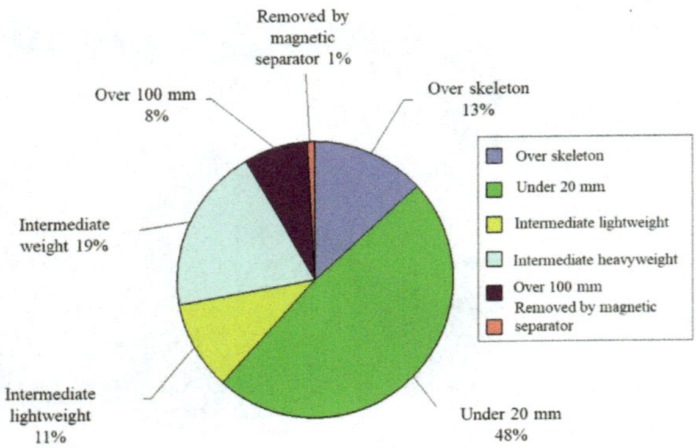

Fig. 4.25 Material balance (the average value)

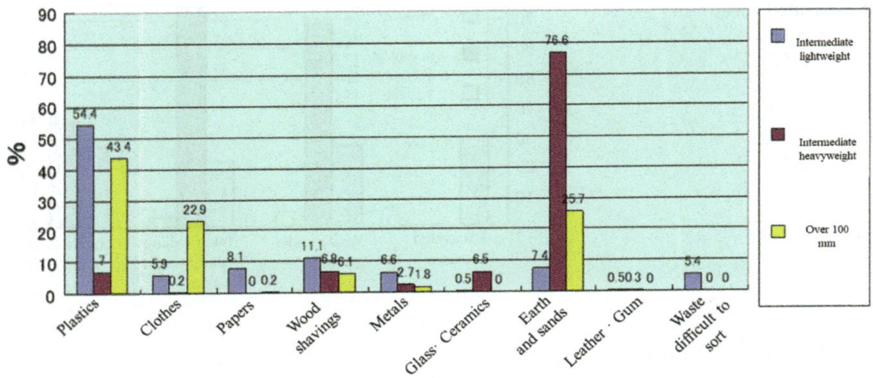

Fig. 4.26 Composition in each classification

(a) Intermediate lightweight waste

The number of wood shavings and waste plastics is overwhelmingly large, while the ratio of paper and clothes is low. Since it is a weight conversion, there is so much thin paper that lower weight ratio is counted. In addition, since many plastics are relatively thin, the weight ratio is less than that of waste over 100 mm.

(b) Intermediate heavyweight waste

There is an extremely large number of earth and sands and glass bottles. Besides, it is found that pet bottles with water are divided into weight side by wind sorting and wood shavings in a water absorption state are mixed in heavy things.

(c) Waste over 100 mm

There is a lot of waste plastics, wood shavings,

earth and sands. In the case of waste plastic, it includes a large number of thick plastics such as sheet-like uncrushed material, bottles, and domestic moldings. In the case of earth and sands, there are uncrushed materials such as bounding stone and others. It is said that through this kind of composition, it is easy to do the wind sorting of waste over 100 mm.

(d) Waste under 20 mm

There are many granular materials in the form of earth and sands. Some of them are the uncorrupted pruned branch. It seems that the sludge and incineration residues are mixed among them. Due to the wet condition, it is difficult to perform the sorting of combustible/incombustible waste efficiently on site. It is desirable to transfer it to full-scale intermediate treatment equipment separately and consider the drying process, etc.

(e) Waste removed by magnetic separator

Most of them are umbrellas, home electric appliances, and wires. The number of wastes which cause blockage of sorting machine when they are not removed is large.

(f) Waste over skeleton

It is a mixture of miscellaneous things, from oversized waste such as home electric appliances, bicycles, beds, blankets, carpets to demolition waste of the house burned down, compressed waste of beverage can.

The heavy machinery sorting is carried out by folk grabs on-site, and it is necessary to throw the wastes into the crushers which are suitable for their individual characteristics to reduce the volume.

⑨ Sorting precision

In the case of the intermediate materials used for wind sorting,

(a) The mixed amount of heavyweight material among lightweight material: 11–19.9%
(b) The mixed amount of lightweight material among heavyweight material: 9.1–14.2%

it is a practical value as a site in wet condition, and it can obtain a mostly good result.

⑩ The proportion of combustible/incombustible

Based on the results of this test,

(a) The combustibles in primary sorting waste are sorted by heavy machinery
(b) The wind sorting is conducted on the intermediate waste of the secondary sorting machine and waste over 100 mm, and machine sorting is carried out on lightweight waste and heavyweight waste. In this case, as an estimation from the composition of the site, the weight ratio is approximate as follows.

Incineration object 25%
Object to be landfill again 25%
The object can be utilized available in the soil covering 50%

Lightweight waste can be treated by incinerators or melting equipment, and heavyweight waste can be backfilled, some of the waste which is under 20 mm can be used as a soil covering material. Moreover, after sorting the lightweight waste, high-density rolling is available even if the heavyweight is re-landfilled. Furthermore, it is thought that it becomes helpful to the volume reduction due to the difference in rolling density. In fact, the ground density of the landfill is as low as 1.0 on-site, the vibration of the ground is strong due to heavy machinery traveling. Therefore, when the crushed landfill waste which is mainly composed of incombustible waste is sufficiently compacted, it is possible to shut down the ground density to around 1.5, and the effect of volume reduction is also great.

⑪ Conclusion

The results of the excavation and sorting test of the landfill site of B city in A prefecture are as follows.

(a) Waste with a moderate wet condition excavated in natural is thrown into the straight sorting machine without going through a drying process, and the excavation sorting testing is carried out on the same day. As a result, excavation sorting is possible in the condition where dust occurrence is extremely small.
(b) In spite of the landfill waste in a wet condition, three parts separated by size (under 20 mm: 40.2–56.5%, 20–100 mm: 24.2–25.6%, over 100 mm: 4.9–10.5%) is under a good condition, and separation of heavyweight waste/lightweight waste of intermediate materials of 20–100 mm by a wind sorting machine is also good (The mixed amount of lightweight material among heavyweight material is 9.1–14.2%; The mixed amount of heavyweight material among lightweight material is 11–19.9%). Besides, it seems that it is also possible to make a gravity separation on the waste over 100 mm.
(c) The preliminary sorting by the primary sorting machine of about 200 mm is effective for ensuring a stable treatment of foreign substances which is easily caused the blockage of the secondary sorting machine.
(d) Although the test has not carried out, when the sorting is effectively performed as mentioned above, since the waste suitable for the shear-type crusher such as waste plastics, clothes, wood shavings and the waste suitable for the impact type crusher such as concrete glass and natural stone can be separated, and it is considered to be effective for volume reduction.

4.4.3 Heat Treatment

As a reprocessing method of excavated waste, heat treatment such as melting, and incineration has the highest volume reduction effect. In this section, we will introduce the fluidized bed type gasification melting furnace, the indirect heating type kiln furnace, the stoker type furnace + the ash melting furnace and the cases which excavated waste with a high content of dioxins is detoxified (demonstration) by heat reduction treatment after classification.

4.4.3.1 Examples About Fluidized Bed Type Gasification Melting Furnace

① Outline

The landfill waste is various kinds of waste such as combustible waste, incombustible waste, oversized waste. These landfill waste may contain high concentrations of heavy metals such as mercury and lead. In addition, coarse metals and rubble which are inappropriate in the fluidized bed type gasification melting furnace may be landfilled. After excavating and selecting these wastes, a mixed combustion experiment is carried out with collected waste. The outline of the demonstration experiment is as follows.

(MSW landfill site for regeneration)

- Landfill area 43,500 m^2
- Landfill volume 144,600 m^3
- Landfill amount 104,000 m^3
- Landfill waste Mainly composed of municipal solid waste (combustible waste)
- Distance to incineration facility About 2 km

The specification of the excavation facility is as follows. The flow of excavation treatment is shown in Figs. 4.27 and 4.28 describes the facility.

- Name Pre-treatment facility for excavated waste

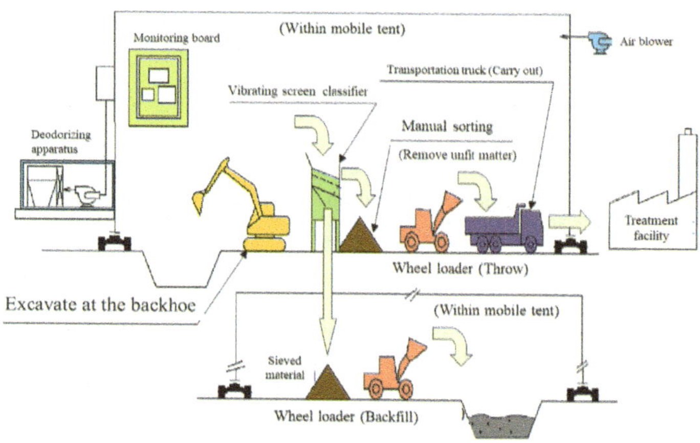

Fig. 4.27 The flow of excavation

Fig. 4.28 A mobile tent

- Method Heavy machinery with a mobile tent for excavation · Sieve selection method
- Processing capacity 100t/day (input amount of vibratory sieving machine)
- Processing time 8 h
- Heavy machinery for excavation

Excavation: a backhoe (bucket capacity: 0.4 m^3)

Loading: a wheel loader (bucket capacity: 1.4 m^3) · Vibratory sieving machine a water cooled diesel hydraulic screen (1 stage deck type)

- Transportation truck Two closed type 4 t dumps with canopies

② The composition of waste

Table 4.5 shows the composition of excavated waste, and Table 4.6 explains the composition of mixed combustion waste (20%) for collected combustible waste and excavated waste.

③ The result of mixed combustion

Figure 4.29 shows the material balance chart of collected combustible waste, and Fig. 4.30 describes the material balance chart of 20% excavated waste during mixed combustion treatment.

- Residues from fluidized gas bed type gasification melting furnace

It is confirmed that the concentration of dioxins shows a low value in the incombustible residues from gasification furnace and in the fluid sand, both are all less than 10 pg-TEQ/g. On the other hand, the lead content in incombustible waste and fluid

Table 4.5 The composition of excavated waste

Item		Excavated waste in surface layer		Excavated waste in deep laye	
		Oversize	Undersize	Oversize	Undersize
Three components	Moisture (%)	58.08	40.14	71.27	59.19
	Combustible content (%)	26.94	26.61	23.28	18.74
	Ash content (%)	14.98	33.25	5.45	22.07
The composition of combustible content	Carbon (%)	68.41	71.48	65.89	68.62
	Hydrogen (%)	8.91	8.91	8.89	8.80
	Nitrogen (%)	0.71	1.01	1.16	1.17
	Sulfur (%)	0.11	0.11	0.13	0.11
	Chlorine (%)	2.23	2.22	0.43	0.43
	Oxygen (%)	19.63	16.27	23.50	20.87
Lower heating value (kJ/kg) (Wet base)		6,540	7,160	4,780	3,680
Higher heating value (kJ/kg) (Dried base)		20,380	14,560	24,480	13,520
Unit volume weight (kg/m^3)		349	536	419	806

Table 4.6 The composition of mixed combustion waste (Mixed combustion ratio is 20%)

Item		Municipal combustible waste (Analysis value)	Mixed waste (Calculated value)
Three components	Moisture (%)	44.01	47.45
	Combustible content (%)	51.63	47.21
	Ash content (%)	4.36	5.34
The composition of combustible content	Carbon (%)	56.65	57.51
	Hydrogen (%)	8.19	8.25
	Nitrogen (%)	1.41	1.37
	Sulfur (%)	0.06	0.06
	Chlorine (%)	0.64	0.71
	Oxygen (%)	33.14	32.10
Lower heating value (kJ/kg) (Wet base)		10,380	9,593
Higher heating value (kJ/kg) (Dried base)		22,180	22,222
Unit volume weight (kg/m^3)		181	215

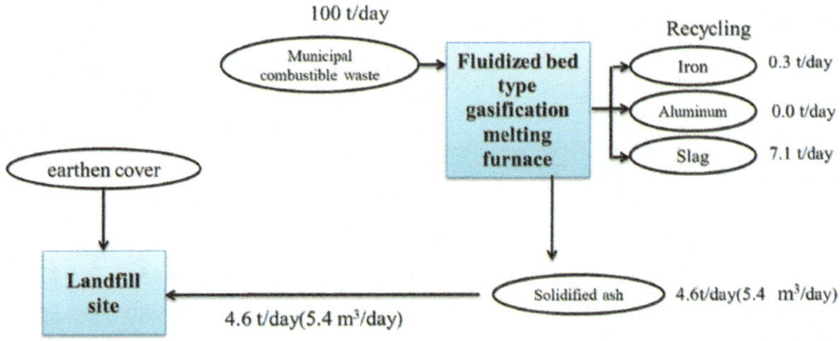

Fig. 4.29 The material balance chart of collected combustible waste

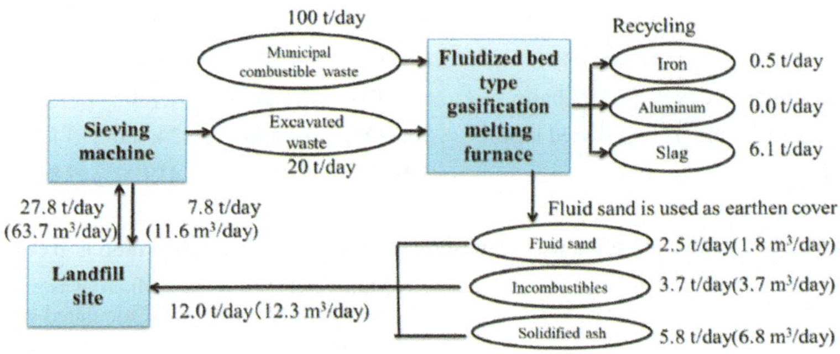

Fig. 4.30 The material balance chart of 20% excavated waste during mixed combustion

sand shows a higher value, and when it used for soil cover, it is necessary to use an intermediate soil cover according to the quality control for each lot and processed object.

Table 4.7 explains the composition of slag and Table 4.8 shows the result of elution test.

Though the value of lead content in the slag is sufficiently low, it exceeds the elution standard value (soil standard) in the case of the elution test. Although it has a quality that can be effectively used as a slag, considering the influences such as the contamination of granulation water used for slag cooling and in order to make effective use, it is necessary to perform the quality control for each lot and water quality management of granulation water and stabilize the quality.

④ Points of attention in mixed combustion treatment of excavated waste

- Even as the combustible waste, there are many unfit substances in gasification furnaces among all of the excavated waste. In the present situation, there are dry type vibration sieve sorting and visual-manual sorting, however, as the content of

Table 4.7 The composition of slag

Items		Incombustible waste	Fluid sand	Slag	Solidfied ash
SiO_2	(%)	–	–	32.0	9.5
CaO	(%)	–	–	21.0	39.9
ZnO	(%)	–	–	–	0.8
CuO	(%)	–	–	–	0.5
PbO	(%)	–	–	–	0.2
CI	(%)	–	–	0.1	13.3
Cadmium and its compounds	(mg/kg)	14	1	<1	–
Lead and its compounds	(mg/kg)	500	190	9	–
Chromium(VI) compounds	(mg/kg)	27	9	<2	<2
Mercury and its compounds	(mg/kg)	<0.5	<0.5	<0.5	–
Arsenic and its compounds	(mg/kg)	2	4	<1	–
Selenium and its compounds	(mg/kg)	<1	<1	<1	–
Metallic iron	(%)	–	–	0.17	–

Table 4.8 The elution test of slag

Item		Slag (Environment Report No. 46)	Solidified ash (Environment Report No. 13)
Cadmium and its compounds	(mg/l)	<0.005	<0.005
Lead and its compounds	(mg/l)	0.013	0.017
Chromium(VI) compounds	(mg/l)	<0.01	<0.01
Mercury and its compounds	(mg/l)	<0.0005	0.0010
Arsenic and its compounds	(mg/l)	<0.005	<0.005
Selenium and its compounds	(mg/l)	<0.005	<0.005

the dust bag cannot be sorted by the sieve, it may become a reason for the damage of crushers while transferring to gasification furnaces. In order to conduct a more effective sorting process, it is also necessary to consider the lightweight waste (waste plastic, etc.) and heavyweight waste by bag ripping and wet type sorting.

- During normal operation, the mixture ratio of excavated waste has enhanced from 10% to more than 16%. Even in the case of mixed combustion treatment, there is not any significant difference between measurement data of exhaust gas, and treatment can be conducted without any problems by some combustion aid due to the dryer with a lowering of the material of waste. Although there is a tendency to increase in fluid sands in operation control, a stable gasification operation can be performed without poor fluidization because of the fluid air pressure is monitored and sands are extracted.
- During mixed combustion treatment, the incombustible residue from the gasification furnaces is crushed and taken out from the bypass duct without being sent to the melting furnaces. During the melting treatment of incombustible residue, since the amount sent to the melting furnace is as same as generated slag, it is expected that there is a great increase in ash load which is treated in the melting furnaces, and it is also predicted an increase in water temperature of granulation water by the increase of the amount of slag. It is necessary to pay attention to these points while melting the whole quantity of residue.
- The waste pit used in this test is sufficiently wide, a part of the area inside the pit is provided to mixed combustion test, the incineration treatment could be conducted. However, in order to perform a more stable operation, it is necessary to ensure the receiving area for excavated waste in the waste pit, to secure the pit area for sufficiently mixing and stirring, dividing the crushed waste pit and stabilizing the material of waste.

⑤ Economic evaluations

- The target disposal amount and disposal period

Excavating 240 thousand tons of landfill waste which is landfilled at the existing MSW landfill site, receiving the plus sieve of combustible excavated waste (168 thousand tons) sorted by sieving machine by fluidized bed type gasification melting furnace and disposing 11.2 thousand tons of the excavated waste per year on the premise of treatment over 15 years.

- The disposal amount of waste in each case

The conditions of the case study for economic evaluations are set as follows.

Case1: Combustible excavated waste and combustible waste are the targets.

Case 2: Combustible waste is the only target.

Case 3: Combustible excavated waste is the only target.

It is used as an evaluation index of increment cost required for the disposal of excavated waste by obtaining the difference between the equipment costs and maintenance costs of Case 1 and Case 2.

The disposal amount of each waste in each case is shown in Table 4.9. And in the material of waste is shown Table 4.10.

Table 4.9 The disposal amount in each case

	Combustible excavated waste (t/year)	Municipal combustible waste (t/year)	Total disposal amount (t/year)
Case 1	11,200	56,000	67,200
Case 2	–	56,000	56,000
Case 3	11,200	–	11,200

Table 4.10 The material of waste in each case

Item		Case 1	Case 2	Case 3
Three components	Moisture (%)	50.0	49.0	55.0
	Combustible content (%)	37.5	42.0	15.0
	Ash content (%)	12.5	9.0	30.0
The composition of combustible content	Carbon (%)	57.51	56.56	67.24
	Hydrogen (%)	8.25	8.19	8.90
	Nitrogen (%)	1.37	1.41	0.92
	Sulfur (%)	0.06	0.06	0.12
	Chlorine (%)	0.71	0.64	1.40
	Oxygen (%)	32.10	33.14	21.42
Lower heating value (kJ/kg) (Wet base)		7,350	8,200	3,100

The annual operating days is 280 days, and each case has two furnaces. The facility scales in each case are set as follows.

[Case 1] 120t/day × 2 furnaces (the type of waste heat boiler · power generation by the steam turbine)

[Case 2] 100t/day × 2 furnaces (the type of waste heat boiler · power generation by the steam turbine)

[Case 3] 20t/day × 2 furnaces (a cooling type of water spray gas)

Table 4.11 shows the trial calculation result of disposal costs in each case. Referring to the data of construction cost in 2003, the unit price of construction is calculated.

Besides, 15 years simple amortization is calculated on the premise regardless of the interest rate.

The difference between Case 1 and Case 2 represents the increase in disposal cost while disposing of excavated waste (mainly composed of ash) for disposing of combustible waste. And this difference is compared with the single combustion cost of the excavated waste of Case 3 in Table 4.12.

Table 4.11 Cost in each case

Item	Unit	[Case 1]	[Case 2]	[Case 3]
The capacity of facility	t/day	120 × 2 furnaces	100 × 2 furnaces	20 × 2 furnaces
Construction cost	billion yen	108	94	30
Utilities cost	yen/t	2,485	2,018	10,357
Repair · Maintenance cost	yen/t	2,381	2,500	5,357
Labor cost	yen/t	2,396	2,750	10,000
Total maintenance cost	yen/t	7,262	7,268	25,714
15 years maintenance costs	billion yen	73.2	61.1	43.2
Construction cost + Maintenance cost	billion yen	181.2	155.1	73.2

Table 4.12 The evaluation of cost

Item	Unit	[Case 1]–[Case 2]	[Case 3]
Annual disposal amount	t/year	11,200	11,200
Construction cost	billion yen	14.0	30.0
15 years maintenance costs	billion yen	12.1	43.2
Construction cost + Maintenance cost	billion yen	26.1	73.2
Unit price of construction cost	yen/t	8,333	17,857
Unit price of maintenance cost	yen/t	7,202	25,714
Unit price of construction cost + Unit price of maintenance cost	yen/t	15,536	43,571

In the case of construction costs, the increased cost is 1.4 billion yen for mixed combustion treatment, which is nearly half of 3 billion yen for a single combustion facility for excavated waste. Besides, for the 15 years maintenance costs, the increased cost is 1.21 billion yen in the case of mixed combustion treatment, which is nearly 30% of 4.32 billion for single combustion treatment of excavated waste. When the total cost which is the sum of construction costs and 15 years maintenance costs are evaluated, it is 2.61 billion yen for the mixed combustion treatment and 7.32 billion yen for the single combustion treatment of excavated waste. The unit price per ton of disposal is divided by the disposal amount for 15 years. It is about 16 thousand yen for mixed combustion treatment, which is just over 1/3 of 44 thousand for the single combustion treatment of excavated waste. As mentioned above, it is found that the disposal costs are greatly reduced by the mixed disposing of combustible excavated waste and combustible waste.

4.4.3.2 Cases About Indirect Heating Type Kiln

① Outline

The outline of the facility where the demonstration is conducted is as follows.
(Treatment facility): E cleaning center

- Incineration capacity: 162 t/24 h (81t/24 h × 2)
- Incombustible waste: 23 t/5 h
- Disposing method: Kiln pyrolytic gasification melting furnace
- (Landfill site): E MSW landfill site
- Landfill area: 22,386 m^2
- Landfill capacity: 90,000 m^3

② The material of waste

The compositions of excavated waste and collected combustible waste are shown in Tables 4.13 and 4.14.

Table 4.13 The composition of excavated waste

	Each waste amount	
	Total amount (t)	Ratio(%)
Mainly ash	29,100	44.0
Mainly metals	18,800	28.4
After being crused	13,200	20.0
Mixed waste (4~5 years)		
Mixed waste(10 years ago)		
Kitchen waste 10 years ago	5,000	7.6
Total	66,100	100

Table 4.14 The composition of collected combustible waste

Type of waste	Each amount of waste	
	Total amount (t/day)	Ratio (%)
Collected combustible waste	80.60	49.9
Directly transported combustible waste	55.36	34.3
Sorting residue from recycle plaza	10.74	6.4
Sewage sludge	11.95	7.4
Excretion sludge	2.92	1.8
Total	161.57	100

Table 4.15 The mixed combustible experiment condition

	Item	Content
①	Mixing treatment of waste (mainly composed of crushed waste)	Municipal combustible waste+ Landfill waste which is mainly composed of crushed waste For municipal combustible waste, it is about 20% by weight ratio
②	Mixing treatment of waste (mainly composed of ash)	Municipal combustible waste+ Landfill waste which is mainly composed of ash For municipal combustible waste, it is about 20% by weight ratio

Among excavated waste, mixed combustion test is carried out on crushed waste and ash. The mixed combustion ratio for collected combustible waste is set at 20% (Table 4.15).

4.4.3.3 The Composition of Excavated Waste

Table 4.16 shows the composition of collected combustible waste and excavated waste.

Two kinds of excavated waste and combustible waste are sampled about 300 kg for each, and the condensation by quartering method becomes the analysis target. Compared with the combustible waste and excavated waste mainly composed of ash, the lead content in ash is more than three times in the excavated waste mainly composed of crushed waste. Besides, the dioxins amount in excavated waste is less than 0.1 ng-TEQ/g, and it is well below the standard of 3 ng/TEQ/g.

4.4.3.4 The Result of Mixed Combustion

Table 4.17 explains the composition of slag, and the result of the elution test is shown in Table 4.18.

The Pb content of molten fly ash during the disposing of excavated waste which mainly composed of crushed waste is more than during the disposing of excavated waste which mainly composed of ash, while the content of the solidified product of desalted residue and the content of slag are reversed. In addition, in the case of Pb content (Environment Report No.19), the value in 1 N hydrochloric acid extraction method is lower than that of in nitric acid extraction method. The ratio is 100:81 during the disposing of excavated waste, which mainly composed of crushed waste, and it is 100:41 in the case of an excavated waste which mainly composed of ash. There aren't any qualitative relationships depending on the kinds of target waste. All the results of the elution test are satisfied the landfill standard.

Table 4.16 The composition of waste in mixed combustible experiment

Item		Unit	Waste (mainly composed of crushed waste)			Waste (mainly composed of ash)		
			Waste (mainly composed of crushed waste)	Municipal combustible waste	Mixed waste (calculated value)	Waste (mainly composed of ash)	Municipal combustible waste	Mixed waste (calculated value)
Three components	Moisture	%-wet	31.0	31.7	31.6	40.9	36.4	37.3
	Combustible content	%-wet	25.1	62.1	55.3	1.0	47.1	38.3
	Ash content	%-wet	43.9	6.2	13.2	58.1	16.5	24.5
Apparent specific gravity		t/m^3	0.87	0.10	0.12	0.78	0.11	0.13
The floristic compostion	Fiber	%-dry	1.8	15.4	12.9	0.9	0.5	0.6
	Plastics	%-dry	18.2	17.1	17.3	2.1	23.2	19.4
	Gum · Leather	%-dry	0.2	<0.1	0.1	<0.1	2.2	1.8
	Wood · Bamboo · Straws	%-dry	8.7	1.6	2.9	0.1	2.2	1.8
	Papers	%-dry	1.6	63.7	52.1	0	34.1	27.9
	Kitchen	%-dry	<0.1	<0.1	<0.1	<0.1	19.0	15.6
	Metals (Iron)	%-dry	7.9	0.3	1.7	1.9	0.2	0.5
	Metals (Not iron)	%-dry	1.7	0.3	0.6	1.6	0.9	1.0
	Glass	%-dry	13.0	0.2	2.6	1.0	1.5	1.4
	Stone · Pottery	%-dry	11.0	0.2	2.2	0.1	2.9	2.4
	Impurities (below 5 mm)	%-dry	35.8	1.3	7.7	92.3	13.2	27.5
Element composition	Carbon C	%-dry	52.0	43.8	44.5	54.4	50.6	50.6

(continued)

Table 4.16 (continued)

Item		Unit	Waste (mainly composed of crushed waste)			Waste (mainly composed of ash)		
			Waste (mainly composed of crushed waste)	Municipal combustible waste	Mixed waste (calculated value)	Waste (mainly composed of ash)	Municipal combustible waste	Mixed waste (calculated value)
	Hydrogen H	%-dry	6.3	5.7	5.8	6.5	6.2	6.2
	Chlorine Cl	%-dry	1.4	0.2	0.3	3.0	0.6	0.6
Lower heating value		kJ/kg	4030	9100	8162	<200	7380	6005
Inorganic composition	Si	%-dry	24.6	22.0	23.6	18.6	24.0	21.5
	Ca	%-dry	5.0	7.3	5.9	7.9	8.2	8.1
	Al	%-dry	4.8	6.0	5.3	6.8	5.7	6.2
	Cu	%-dry	0.394	0.603	0.5	0.042	0.324	0.2
	Hg	mg/kg	0.088	0.16	0.1	0.86	0.078	0.4
	Cr^{+6}	mg/kg	2.5	6.9	4.2	<1	6.8	4.2
	Pb	mg/kg	1000	300	732	270	280	276
	Cl	%-dry	0.37	0.54	0.4	0.41	<0.05	0.2
DXNs (Official method)		ng-TEQ/g	0.054	※10.0021	0.012	0.087	※10.0021	0.018

Table 4.17 The composition of slag

Items		Unit	Mainly crushed waste				Incineration residue			
			Melting flying ash	Desalting residue solidified product	Slag	Pyrolysis residue	Melting ash	Desalting residue solidified product	Slag	Pyrolysis residue
Three component	Moisture	%-wet	–	–	–	<0.1	–	–	–	<0.1
	Ash content	%-dry	–	–	–	85	–	–	–	86.7
	Volatile component	%-dry	–	–	–	6.2	–	–	–	6.4
	Fixed carbon	%-dry	–	–	–	8.8	–	–	–	6.9
Apparent specific gravity		ton/m^3	–	–	–	0.84	–	–	–	0.71
Lower heating value		kJ/kg	–	–	–	3000	–	–	–	2900
Inorganic composition	Si	%-dry	17.7	8.9	22.3	–	13.1	9.9	25.3	–
	Ca	%-dry	6.2	8.7	6.1	–	6.1	8.1	4.9	–
	Al	%-dry	6.3	3.5	7.2	–	5.5	3.8	5.7	–
	Cu	%-dry	0.167	0.08	–	–	0.104	0.113	–	–
	Hg	mg/kg	31	10	<0.05	0.021	4.3	18	<0.05	0.061
	Cr^{+6}	mg/kg	21	0.86	<1	–	<1	1.8	<1	–
	Pb	mg/kg	6960	2340	270	850	2600	3700	340	1110
	Pb (Environment Report No.19)	mg/kg	–	–	220	–	–	–	140	–
	Cl	%-dry	8.2	11.2	0.1	0.45	5.2	11.3	0.1	0.34
DXNs (Official method)		ng-TEQ/g	–	0.92	0	4.4E-05	–	1.3	0	2.1E-07

Table 4.18 The elution test of slag

Item	Unit	Mainly crushed waste	Incineration residue	Reference value of landfill (Environment Report No. 13)
Cd	mg/l	<0.01	<0.01	<0.3
Pb	mg/l	0.018	0.028	<0.3
Cr^{+6}	mg/l	<0.05	0.06	<1.5
As	mg/l	<0.01	<0.01	<0.3
T–Hg	mg/l	<0.0005	<0.0005	<0.005
Se	mg/l	<0.01	<0.01	<0.3
Alkyl Hg	mg/l	No detection	No detection	No detection

4.4.3.5 The Volume Reduction Effect of the Landfill Site

Using the results mentioned above, we will consider the volume reduction effect while treating the excavated waste by a kiln gasification melting furnace. The mixed combustion ratio of excavated waste is 18.5% during the disposing of excavated waste, which mainly composed of crushed waste and 19.2% at the time of disposing of excavated waste which mainly composed of ash. In order to make a comparative investigation, we will unify it to 16.7%. (The emission rate of iron, aluminum, slag and the solidified product of desalted residue is applied as the result of this test is)

The investigation conditions are

The disposal volume of combustible waste: 100t/day

The disposal volume of excavated waste: 20t/day

The total disposal volume: 120t/day (the mixed combustion ratio is 16.7%)

The results of the investigation are shown in Figs. 4.31, 4.32 and 4.33. In order to collect the iron, aluminum, and slag as valuable substances, the solidified product of desalted residue becomes the only target to landfill. In the case of only disposing of combustible waste, the landfill amount is 5.8 m^3, the mixed combustion of excavated waste which mainly composed of crushed waste is 18.5 m^3/day while it is 14.3 m^3/day

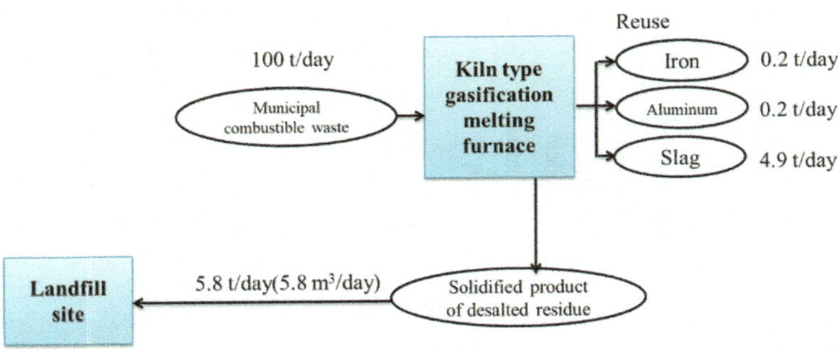

Fig. 4.31 The balance of collected combustible waste

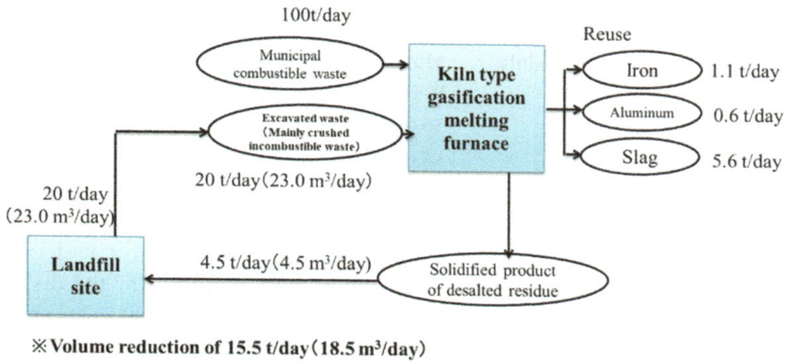

Fig. 4.32 The balance of excavated incombustible waste

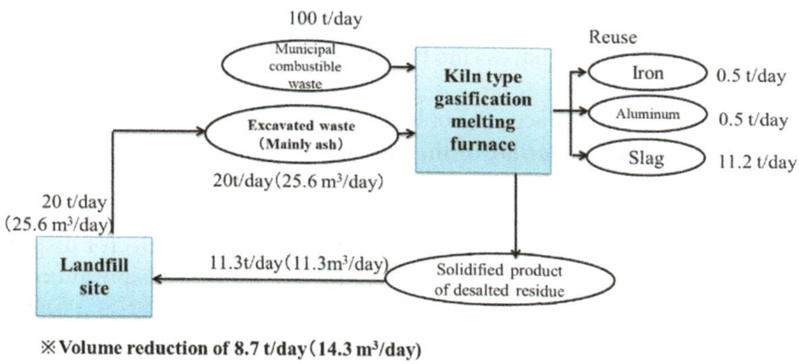

Fig. 4.33 The balance of excavated incineration residue

in the case of the excavated waste which mainly composed of ash. It is found that the volume reduction of MSW landfill site is achieved and it is effective to prolong the lifetime of the landfill site.

4.4.3.6 Economic Evaluations

Excavating 168 thousand tons of landfill waste which is landfilled at the MSW landfill site and assuming that disposing 1.12 thousand tons of the excavated waste per year on the premise of melting treatment over 15 years. The conditions of the case study for the amount of target waste for each case and their economic evaluations are set as follows.

Case 1: The disposal target is excavated waste (mainly composed of ash) and combustible waste.

Case 2: The disposal target is a combustible waste.

Case 3: The disposal target is excavated waste (mainly composed of ash).

Table 4.19 The disposal amount in each case

Item	Excavated waste (mainly ash) (t/year)	Municipal combustible waste (t/year)	The total throughput (t/year)
Case 1	11,200	56,000	67,200
Case 2	–	56,000	56,000
Case 3	11,200	–	11,200

It is used as an evaluation index of increment cost required for the disposal of excavated waste by obtaining the difference between the equipment costs and maintenance costs of Case 1 and Case 2. The disposal amount of waste in each case is shown in Table 4.19. The annual operating days is 280 days, and each case has two furnaces. The facility scales in each case are set as follows.

[Case 1] 120t/day × 2 furnaces (the type of waste heat boiler · power generation by the steam turbine)

[Case 2] 100t/day × 2 furnaces (the type of waste heat boiler · power generation by the steam turbine)

[Case 3] 20t/day × 2 furnaces (a cooling type of water spray gas)

Table 4.20 shows the materials of the target waste in each case. And Table 4.21 shows the trial calculation results of disposal costs by each method.

The difference between Case 1 and Case 2 represents the increase in disposal cost while disposing of excavated waste (mainly composed of ash) for disposing of combustible waste. And this difference is compared with the disposal cost of the excavated waste of Case 3 in Table 4.22.

In the case of construction costs, the increased cost is 1.8 billion yen for mixed combustion treatment, which is nearly half of 3.2 billion yen for the disposal facility for excavated waste. Besides, for the 15 years maintenance costs, the increased cost is 1.41 billion yen in the case of mixed combustion treatment, which is nearly 30% of 4.99 billion for disposing of excavated waste. When the total cost which is the sum of construction costs and 15 years maintenance costs are evaluated, it is 3.21 billion yen for the mixed combustion treatment and 8.19 billion yen for the disposing of excavated waste. The unit price per ton of disposal is divided by the disposal amount for 15 years. It is about 19 thousand yen for mixed combustion treatment, which is just over 1/3 of 49 thousand for the disposing of excavated waste.

As mentioned above, it is found that the disposal costs are greatly reduced by the mixed disposing of excavated waste (mainly composed of ash) and municipal

Table 4.20 The material of waste in each case

Item		Case 1	Case 2	Case 3
Three components	Moisture (%)	47.5	50.0	35.0
	Combustible content (%)	37.5	44.0	5.0
	Ash content (%)	15	6.0	60.0
Lower heating value (kJ/kg)		5,600	7,300	200

Table 4.21 The cost in each case

Items	Unit	Case 1	Case 2	Case 3
The capacity of facility	t/yen	120 × 2 furnaces	100 × 2 furnaces	20 × 2 furnaces
Construction cost	billion yen	120	102	32
Utilities cost	yen/t	2,803	2,221	12,000
Repairing · Maintenance cost	yen/t	2,679	2,679	5,804
Labor cost	yen/t	2,396	2,875	9,375
Total maintenance cost	yen/t	7,878	7,775	29,679
Maintenance cost for 15 years	billion yen	79.4	65.3	49.9
Construction cost + Maintenance cost	billion yen	199.4	167.3	81.9

Table 4.22 The cost evaluation in each case

Item	Unit	(Case 1)–(Case 2)	Case 3
Annual disposal amount	t/year	11,200	11,200
Construction cost	billion yen	18	32
Maintenance cost for 15 years	billion yen	14.1	49.9
Construction cost +Maintenance cost	billion yen	32.1	81.9
The unit price of construction cost	yen/t	10,714	19,047
The unit price of maintenance cost	yen/t	8,392	29,679
The unit price of construction cost +The unit price of maintenance cost	yen/t	19,106	48,726

combustible waste. Referring to the construction cost in 2003, the unit price of construction is calculated. Besides, 15 years simple amortization is calculated on the premise regardless of the interest rate.

The equipment costs for mixed combustion treatment is about 50% of the single treatment, and the maintenance costs for mixed combustion treatment is about 30% of the single treatment. And the total cost is just over 1/3 of the single treatment.

4.4.3.7 Stoker + Ash Melting

① It is possible to conduct a volume reduction of excavated waste by using stoker incinerator which has a lot of waste disposal experience as reproduction and surface melting type ash melting furnace, which is relatively easy to establish. At one MSW landfill site, we make a preliminary investigation, excavate landfill waste, sort and divide them, and conduct research on the component of waste. Then, the excavated waste after sorting and division are subjected to volume reduction treatment by incineration and melting by using adjacent existing continuous stoker incinerator and surface melting type ash melting furnace. We investigate its operating conditions and the property of waste matter.

② Survey outline

- MSW landfill site: F MSW landfill site (more than 20 years since the start of the landfill). Landfill waste: Incombustible waste, directly transport waste (it is rarely at the present time), incineration residues).
- Disposing facility: F clean center, continuous stoker incinerators 62.5 t/day furnace + Fuel type surface melting ash melting furnace 13 t/day

③ Excavation, sorting, and classification

The excavation point is centered on the boring point, and the shape is a truncated pyramid with a surface of 8.0 × 8.0 m, a bottom 2.0 × 2.0 m, and a depth of 5.0 m. The excavation was done at intervals of 0.5 m by using a 0.7 m^3 backhoe for the purpose of comparing the boring with the electrical survey. Solar drying is carried out after performing a coarse sorting of about 100 mm in the skeleton bucket and a magnetic sorting. After sorting by vibrating sieve and hand sorting, the waste is separated into over 20 mm and under 20 mm, then divided into soils and sands (including incineration residue), combustibles + plastics, bins, cans, stones, and rubble. Figure 4.34 shows an example of the composition of excavated waste. Table 4.23 shows the composition of the combustible waste and excavated waste.

④ Incineration test, Melting test

Figure 4.35 shows the processing flow of volume reducing the treatment of excavated waste. Among all of the sorted excavated waste, combustible waste + plastics are thrown into the existing adjacent continuous stoker incinerators, incineration test is conducted, and an investigation on the influence towards incinerators in the case of incineration treatment of excavated waste is carried out. Based on the analysis result of waste material of sorting combustible waste + plastics as stated below, the mixing ratio of combustible waste + plastics is 20wt% among all the thrown waste. Among all of the sorted excavated waste, the soils and sands which is under 20 mm are thrown into the existing adjacent continuous fuel type surface melting ash melting furnace, melting test is conducted, and an investigation on the influence towards melting furnace in the case of melting treatment of excavated waste is carried out. According to the result of melting lab test as stated below, the mixing ratio of soils and sands is 20 wt% among all the thrown waste. The existing of incineration facility and melting facility are used to understand the volume reduction of excavated waste. According to the analysis result of waste material of sorting combustibles + plastics, when the incineration treatment is 20 wt% of the total throwing amount, a stable volume

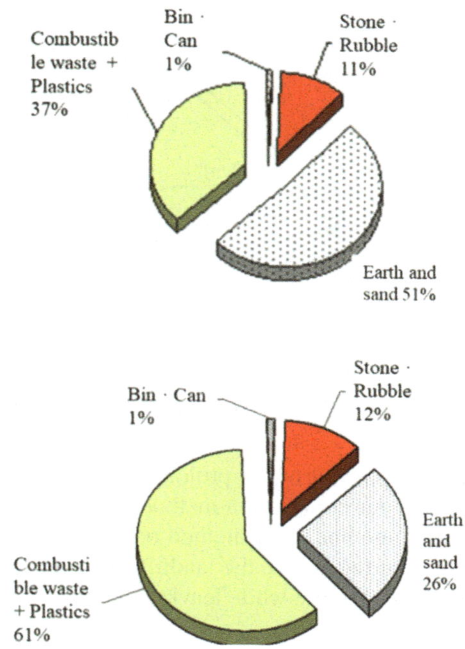

Fig. 4.34 The composition of excavated waste

Table 4.23 The composition of waste

Analysis item	Sample name / Unit	Sorting combustible waste + Plastics	General waste
Papers · Fiber	%-dry	1.3	78.1
Kitchen waste	%-dry	0.0	6.6
Plants	%-dry	0.3	6.0
Other combustible waste	%-dry	0.0	0.7
Plastics · Gum · Leathers	%-dry	44.2	7.3
Metal	%-dry	4.3	0.7
Glass · Crockery · Stone	%-dry	9.0	0.0
Earth and Sand · Other inncombustible waste	%-dry	40.9	0.7
Higher heating value	KJ/kg	12,000	8,080
Lower heating value	KJ/kg	10,300	6,110

reducing treatment can be performed in incineration test. And from the result of the melt lab test, when the melting treatment is performed by mixing 20 wt% of the total throwing amount, a stable volume reducing treatment can be performed. The melting slag discharged by the melting test satisfied the soil environmental standard, and the particle size is larger than that of the normal treatment.

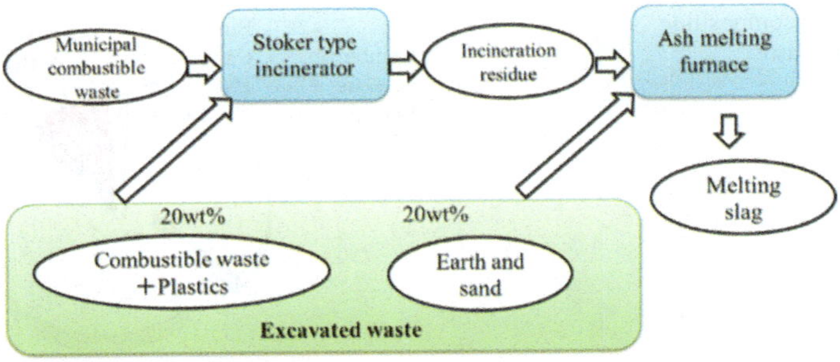

Fig. 4.35 Stoker+Ash

4.4.4 Life-Prolonging by Consolidation Method

For the purpose of life-prolonging, there are methods to make a higher in density such as transfer compaction method, dynamic consolidation method, and static consolidation method. This method regenerates the landfill site, while the transfer method does not regenerate the landfill site, it increases the density of the landfill layer and prolongs the life while leaving the function of the landfill site.

4.4.4.1 Transfer Compaction Method

In the landfill site where directly landfill is carried out since rolling compaction is not sufficiently, the void inside the landfill tank is large, and it is possible to secure the landfill capacity by compacting landfill ground. There is a method in which the old landfill site is excavated, and the landfill layer is rolling compacted with the thickness of about 50 cm. There is also a method in which combustible waste and oversized waste are removing by the sorting of excavated waste, the waste left behind is rolling compacted with the thickness of about 50 cm after reprocessing. In the case of a newly constructed landfill site, there is a case where the old landfill site is excavated and transferred to a new landfill site, and then a high-density landfill is conducted after sorting.

The volume reduction ratio by this method is estimated to be 30–50%. It is different depending on landfill waste.

4.4.4.2 Consolidation Method

Consolidation method includes dynamic consolidation method and static consolidation method. The dynamic consolidation method is a method of lifting weights such as iron by heavy machine and drop from altitude to compact landfill layer. There is some noise while weights are dropping. Therefore, it is necessary to consider the influence of noise and vibration at the MSW landfill site near the residential area.

On the other hand, in the case of static consolidation method, the underground is excavated by heavy screw auger, the auger is inverted, and excavated soil is pressed into the sidewall of hole wall which is formed by drilling with a special steel plate set in the auger. Depending on the material of the landfill waste, it is possible to landfill the waste in 10–30% drilling hole and prolong life.

The example of a demonstration experiment by static consolidation method (refuse press method) is introduced as follows (Higuchi 1999).

① The outline of the consolidation method

The refuse press method is developed as a method of volume reduction by statically compressing waste. The penetration body(φ850–1500 mm) is rotated and pressed into the waste and put the waste into the hole by backhoe from the top while compressing the waste into the wall hole, re-compaction work of waste thrown by penetration body is conducted, and volume reduction is devised. The features of this method are listed as follows (Ikeda et al. 2003).

A. Since rotation and press of the penetration body are driven by a hydraulic motor, there is only the engine sound of the heavy machinery such as backhoe and construction machine, and there is little vibration and noise generates.
B. Since the high torque hydraulic driving mechanism is used for rotation and press of the penetration body, it is possible to apply even the oversized waste is mixed. Besides, the penetration bodies (φ850–φ1500 mm as usual) are divided according to the type, composition and compaction condition of waste.
C. Improvement (compression) is performed reliably in order to let the penetration body to reach the planned depth. Figure 4.36 shows the construction condition and Fig. 4.37 shows the flow of construction.

② Method of demonstration experiment

The place where the demonstration experiment is carried out is an MSW landfill site established in the mountainous area, and the maximum thickness of the waste layer is 20 m. As a high-density landfill is performed by mixing crushed municipal solid waste which is left in the other place for two to three years and incineration residue, this landfill site is better tightened than other landfill sites. In the experiment, two types of penetrations of φ850 mm and φ1500 mm are used. And we will introduce the case using penetration of φ850 mm in Fig. 4.38. The construction is carried out, as shown in Fig. 4.39. Three points in alignment are set as one group. The procedure is constructing the side point in advance and backfilling the boreholes on two sides so that the penetration body will not be eccentric during the construction at the midpoint, and then constructing the midpoint. Then, after backfilling the borehole in the midpoint, the construction is conducted by the same method in the rear row. Figures 4.40, 4.41 and 4.42 describes the condition of excavation compaction.

③ Result of the demonstration experiment

A. The measured result of the volume reduction ratio

Fig. 4.36 The construction condition

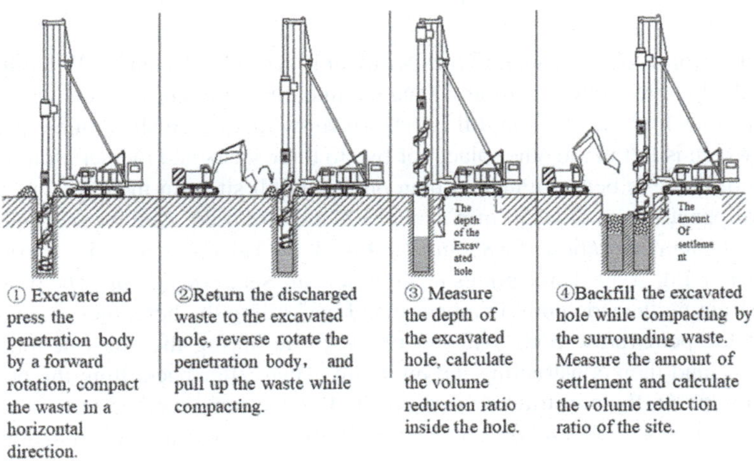

Fig. 4.37 The flow of construction (Ikeda et al. 2003)

Fig. 4.38 The penetration body(φ850 mm)

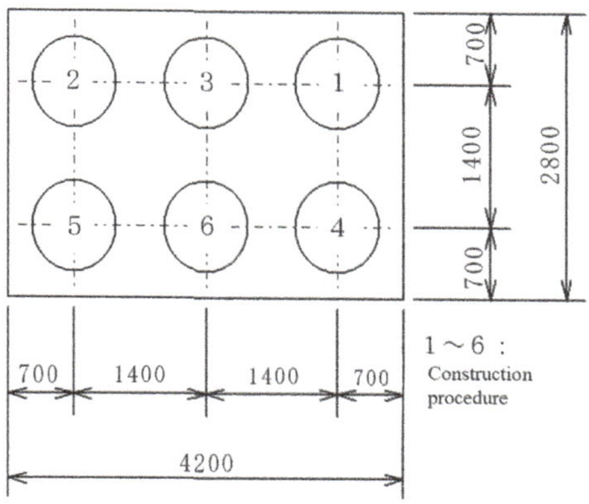

Fig. 4.39 Layout drawing of excavated points

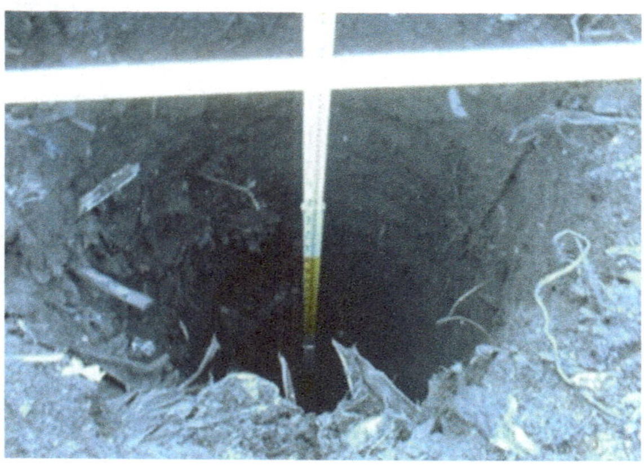

Fig. 4.40 The excavated hole (φ850 mm)

Fig. 4.41 The condition of compaction

Fig. 4.42 The obtained volume reduction space

After the compression at each excavation point, measuring the diameter and the depth of the shaped excavation hole, and calculating the volume reduction ratio inside the hole (the depth of excavation hole/the length of excavation is 10 m) and the capacity of excavation hole. Based on the results, the depth of the excavation hole is 4–4.4 m, and the volume reduction inside the hole is 43% on average. Besides, the diameter of the excavation hole is 1.15–1.2 as much as the diameter of the penetration body, and the capacity (the amount of volume reduction) of the excavation hole is 2.6 m^3 per hole on average.

The leveling is conducted before, and after construction, the quantity of a ground surface (hereinafter referred to as "the amount of settlement") is obtained from the average differences in elevation, and the volume reduction ratio of the site is calculated from this. According to the results, the amount of settlement is about 1.24 m, and the volume reduction ratio of the site is 12.4% despite the ground tightens well. When the diameter of the penetration body is the same, the amount of settlement depends on the excavation space. The wider the excavation space is, the smaller value it becomes. In addition, it becomes a higher value when the ground is soft.

B. The result of a boring survey

The standard penetration test by using borehole and the hole horizontally loading test (pressiometer method) is conducted before and after the construction, and the prediction of compressibility and the state of the strength changes are confirmed. Figure 4.43 shows the comparison of N value in the depth distribution map of the standard penetration test before and after the construction.

Based on the result of the standard penetration test, compared to before construction (○), the strength (N value) increases to the depth of excavation at the intermediate position between the four excavation points (●). Besides, it is a lower value as same as before construction at the hole excavating position (■). There is a difference in the volume reduction ratio due to the compaction inside the excavation hole, and it is necessary to pay attention to the compacting method (compaction time, input amount) inside the excavation hole. In addition, although there is an influence on the deeper bottom of the excavation hole, since there is no increase in strength at a depth of 1 m from the bottom of the hole, it is considered that there is less influence on seepage control work or the buried pipe if there is a clearance of about 1 m.

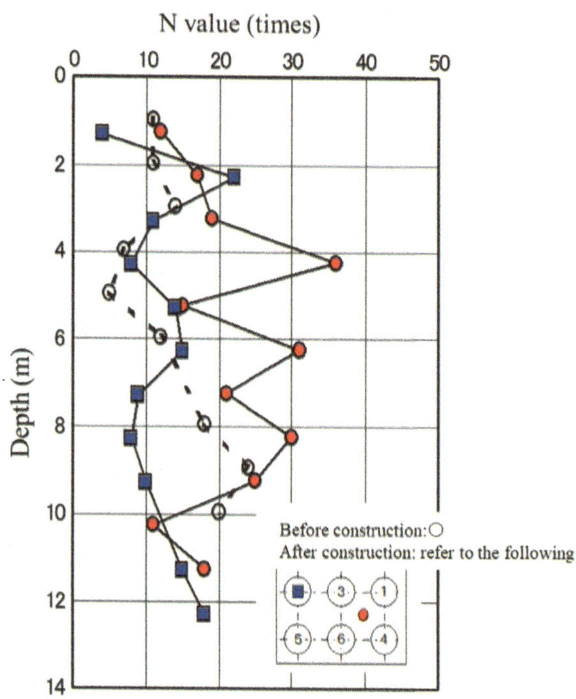

Fig. 4.43 N value before or after excavation

C. The measurement results of the surrounding environment

A distance attenuation measurement is carried out on the waste ground by using a pollution vibrometer and a noise meter. The measurement result is shown in Fig. 4.44.

Both the vibration level and the noise level are exponentially decaying towards the distance, and it is recognized that it has the same tendency as other ground improvement methods. However, the obtained values of the vibration and the noise level are quite lower than other ground improvement methods, and it indicates that the backhoe while backfilling has a larger value than other ground improvement methods. Therefore, it is confirmed that the RP method could be applied even in urban areas where nearby the residential area.

In addition, a measuring point is prepared for ground surface displacement measurement in the vicinity of a construction zone before construction. As a result of measuring the displacement of peripheral ground accompanied to execution, the horizontal and the vertical displacement is about 1 mm at the measurement point of 1.5 m from the excavation point. And when the distance is above 2 m, the displacement does not occur.

④ Influences on stabilization

When the consolidation method is used, the gap between waste layer becomes small, though it has an effect of life-prolonging, there is a concern that it prevents ventilation and water flow and becomes a hindrance to the stabilization of waste. For this reason, the authors filled crushed stones in a part of excavation holes and made a demonstration to verify the countermeasure effect. In the demonstration, 50 consolidation excavations are conducted in the 5 m × 7 m of the simulate landfill site, which

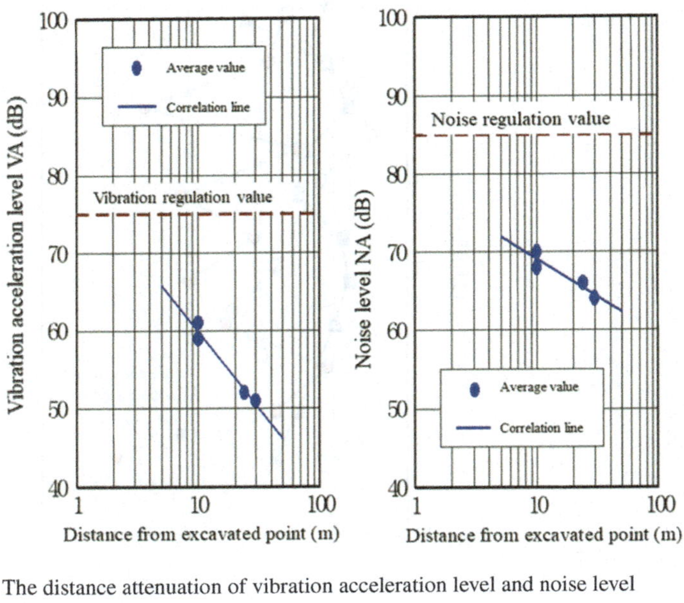

Fig. 4.44 The distance attenuation of vibration acceleration level and noise level

Fig. 4.45 The construction condition of sea surface landfill sites

is mainly composed of incineration residue, two of the excavation holes are filled with crushed stones after excavation. Besides, both generated gas (CH_4) around the crushed stone holes and BOD concentration in pore water are monitored. As a result, both the generated gas (CH_4) and the BOD concentration in pore water decreased remarkably. In the case of the setting of crushed stone holes, it is thought to be caused by the easily decomposable organic substance, preferentially leaching of salts and the decomposition by a microorganism in the waste layer.

It is possible to minimize the influence on stabilization by supplying air and water (precipitation) by the setting of crushed stone holes (Uryo et al. 2005, 2006).

⑤ Practical examples

As practical examples of static consolidation method, examples of sea surface landfill sites are cited. At the Minamihonmoku MSW landfill site in the city of Yokohama (the landfill area is 179,000 m^2, the landfill capacity is 4,270,000 m^3, 80% of landfill waste is municipal solid waste, incineration residue and incombustible waste, 20% of waste is cinders and industrial waste such as sludge), the construction area is 29,000 m^2 in 2014, and the volume reduction amount is 63,000 m^3 by RP method. After that, volume reduction work still has been carried out (Wada 2016) Fig. 4.45 shows the construction condition of sea surface landfill sites.

4.5 Technology for Early Stabilization

The landfill site shall continue the operation of the leachate treatment facility even the landfill is completed. And shall maintain it until the abolition. Based on the maintenance method for the landfill site, the period from the landfill is completed to the abolition is different. For instance, the landfill operation concerns the target waste, the landfill order, the height of landfill. Besides, the effluent standard can also become a factor. Since there is still a maintenance cost until the abolition, it is desired that it can be early abolished to protect the economy and preserve the living environment. Especially in the landfill site for industrial waste, it becomes no income by waste receiving when the landfill is completed, and the maintenance reserve fund, etc. cover the costs. In order to make an early stabilization of the landfill site, it is necessary to make sure a normal function of collecting pipe and landfill gas treatment facility. The landfill management technique which forms a landfill layer capable of ventilation and water flow is also required. In this section, we will introduce the pre-treatment technology, which is necessary for early stabilization, and the ventilation method and the chemical oxidation method which are carried during or after the landfill.

4.5.1 Pre-treatment of the Landfill

Though the Landfill site is constructed for landfill disposal of transferred waste, an efficient landfill and an early stabilization can be performed by landfill management and pre-treatment. The crushers and small incinerators were installed in old times. In current incineration facility and recycling facility, crushing sorting facility can be positioned as a pre-treatment facility for the landfill site. However, it takes time to stabilize the landfill layer because of the diversification of industrial waste and the sophistication of the incineration facility. Therefore, a pre-treatment of the landfill aiming to the early stabilization is necessary to be conducted. Table 4.24 explains the method of pre-treatment of the landfill. Based on the table, it is mainly intended to prevent the ventilation, water flow inhibition, and the formation of perched water in the landfill layer. Figure 4.46 shows a conceptual diagram of obstruction of ventilation and water flow by the establishment of a water channel and perched water in the landfill layer.

4.5.1.1 Moisture Control, Solidification

When the obstruction of ventilation and water flow occurs, the precipitation penetrated the landfill layer forms a "water channel," the function of waste washing out is damaged. Besides, in the landfill layer of waste that contains water like sludge and has a small particle size, an impermeable layer is formed, the perched water has occurred, ventilation and water flow are hindered, the decomposition of organic substance and washing out the pollutant are put off. Therefore, there is a method of adding moisture control agent to the sludge, reforming, and solidifying, promoting the efficiency of landfilling operation and stabilization. Figure 4.47 shows a reforming case where a neutral solidifying agent (mainly of paper sludge incineration residue)is adding to sewage sludge.

The pre-treatment can be carried out as part of landfill operation like the mixture of heterogeneous waste without a specialized facility.

Table 4.24 Examples of pre-treatment

Pre-treatment method	Aim	Target waste
Moisture control, solidification	Securing ventilation and water flow	Sludge (inorganic, organic)
	Prevention of perched water in the landfill layer	Flying ash, Soot (Granulated solid matter)
Ventilation oxidation	Reduction of load on leachate treatment plant by aerobic decomposition of high concentration organic matter	Organic sludge, Animal/plant residues
Mixture of different kinds of waste	• Securing ventilation and water flow by controlling particle size of the waste layer, and blocking prevention towards flying and dust • Suppression of hydrogen sulfide generation by using waste properties	Example: ash + Main ash Soot + Slag Waste gypsum board + Dissolution Dust + Slag
Washing	Desalting by washing, pollutants removal for an early stabilization	Flying ash, Main ash, Soot, Burnt residue, incombustible waste, Others

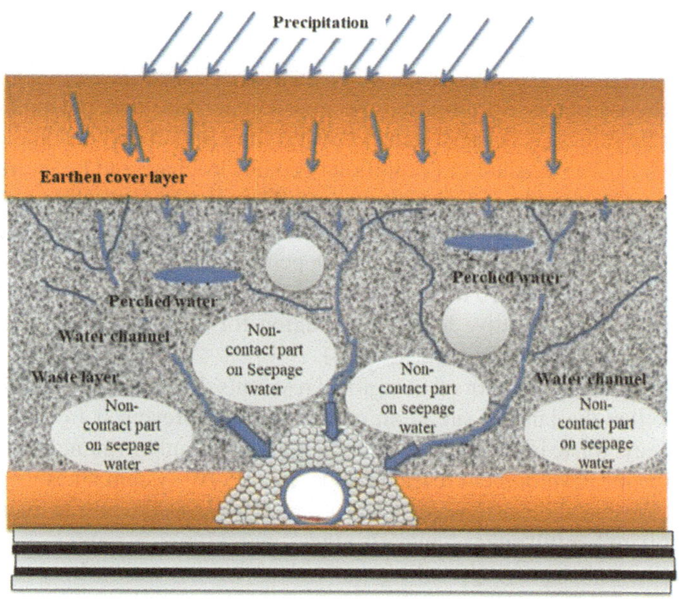

Fig. 4.46 Conceptual diagram of obstruction of ventilation and water flow

Fig. 4.47 A reforming case where a neutral solidifying agent is adding to sewage sludge

4.5.1.2 Washing Out

① Concept of pre-treatment washing out

The landfilled waste is decomposed by microorganisms, and the pollutants are washed out by rain. It is being stabilized over a long period and become extremely close to the soil. The washed-out pollution becomes leachate and being treated by a leachate treatment facility and then discharged. Though it is generally used for about 15 years, as the leachate will be generated even after the landfill is over, the leachate treatment facility will continue to operate. Depending on the various kinds of conditions, the duration is from 10 to 30 years or even longer. Therefore, the Landfill site will be existing as a negative legacy in a region for 30 to 50 years, including its construction period (Referring to Fig. 4.48).

It becomes one of the factors that regional residents dislike the landfill site. Therefore, to stabilize the landfill site at an early stage and prevent it from becoming a negative legacy, waste is washed out before landfill. The authors name it as the WOW system and make development research on it. The WOW system takes the initials of "Wash Out Waste," the waste is washed out, recycles as a cement raw material, and aims to stabilize landfill sites at an early stage.

There are a water spraying washing method, an immersion washing method, and a mechanical washing method. The washing apparatus is set in a waste incineration facilities or landfill sites. Recently, the old landfill sites are restored, and prolonged life by re-excavating, reprocessing, detoxifying, the landfill sites are regenerated. Through this regeneration project, the WOW system can also be used (Referring to Figs. 4.49 and 4.50).

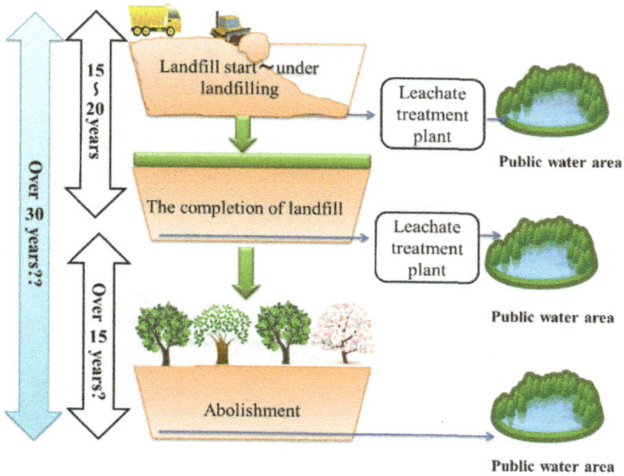

Fig. 4.48 Life of a landfill site

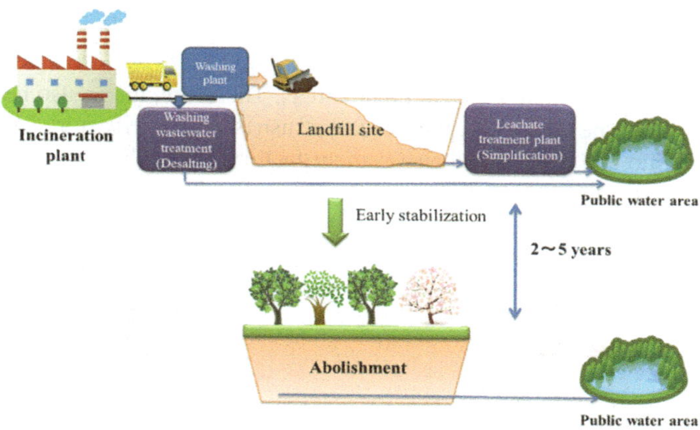

Fig. 4.49 WOWs ystem in Landfill site (open type)

Furthermore, incineration residues can be reused as cement raw material. In that case, it is necessary to remove the salt contained in incineration residues, as the salt is removed by washing, the washed-out incineration residues can be recycled without landfill disposal. The washing wastewater is treated in the leachate treatment facility or the special treatment facility, and the treated water is possible to be reused as washing water. For cement recycling of incineration residues, basically, it is a reverse compensation. With the reduction of chlorine content in incineration residues, it is possible to reduce the cost of receiving by the cement facility. There is a problem

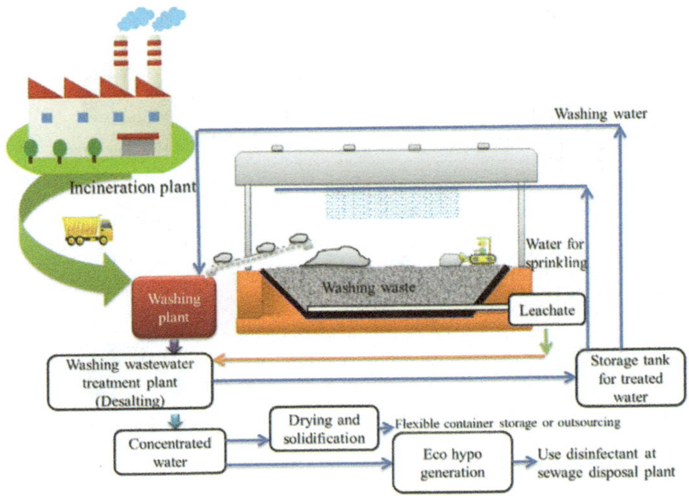

Fig. 4.50 WOW system in Landfill site (covered type)

with transport cost in small cities in inland areas. Therefore, it can also be used as a resource storage type landfill site where washed ash is temporarily stored at a covered type landfill site and transported to a cement factory with enhanced transportation efficiency (Referring to Fig. 4.51).

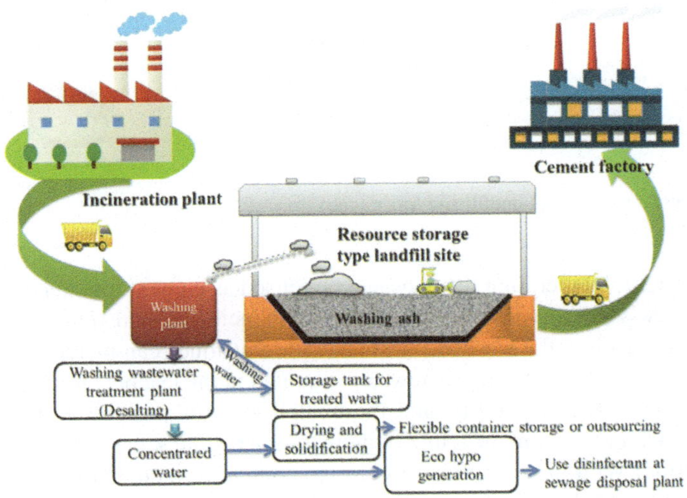

Fig. 4.51 Resource storage type landfill site using WOW system

② Washing out method and its effect

There are mechanical washing method, water spraying washing method, immersion washing method. Washing classification performed, which is a combination of washing, is also useful for the separation of heavy metals. Table 4.25 shows the types of washing methods and their properties.

The washing out method varies depending on the properties of the target waste. Generally speaking, the immersion washing is useful towards the waste which has a larger specific surface area. The amount of water used for washing also varies depending on the properties of the target waste and the washing effect, in the case of using as a pre-treatment of the landfill, the liquid: solid ratio is about two to eight as usual. The larger the liquid: solid ratio is, the higher the washing effect it will become, meanwhile, the higher the treatment cost of washing wastewater it will be. Therefore, it is necessary to select it considering the reducing effect the leachate treatment cost obtained by the early stabilization.

Table 4.25 Types of washing method and its property

Washing method	Example of apparatus	Outline	Target waste
Machine-washing	Stirring apparatus	Stirring washing by stirring apparatus at the liquid-solid ratio of 3 or more	Whole waste
	Spiral	Sorting washing used in mine at the liquid-solid ratio of 5	Incineration residue
	Ejector	Washing by ejector at the liquid-solid ratio of 5	Fly ash
	Air turbulence compressor	Inject the high pressure air from the bottom, grain refine the waste by collision plate in the upper	Whole waste
	Jig	Washing while rotating the jig	Incineration residue
	Wet vibrating filter, Sprinkler nozzle	Sprinkler washing is conducted while the waste is moved on the vibration sieve	Whole waste
Sprinkler washing	Nozzle, Spray gun	Sprinkling water with nozzle or spray gun from the upper of the waste and washing	Whole waste
Immersion and stirring washing	Stirring apparatus	After being immersed in the water of a certain period of time, stirring washing is conducted	Incombustible crushed waste, Fly ash

③ The washing effect (incineration residue) (Misumi 2001, 2003, 2004; Tanioka 2002; Minakata et al. 2003).

The bottom ash of MSW incinerator (stoker furnace) which is being washed (spiral washing and ejector washing) and those without washing are packed in a lysimeter with a capacity of 1 m^3, the quality of the leachate produced by the precipitation is monitored. Figures 4.52 and 4.53 indicate the washing machines. Table 4.26 shows the composition of ash before and after washing. Table 4.27 mentions the packing condition of lysimeter.

The experiment was conducted over 885 days and the analysis of leachate generated by precipitation was performed almost every week for five months from the beginning of the experiment. Then, it was conducted once a month, and it became once every two months from the time of two years later. Based on the Amedas weather data in the vicinity, the precipitation in 885 days was 4,500 mm, and it was 155 mm per month on average. The leachate volume was 60 to 70% of precipitation. Figure 4.54 shows leachate quality. The leachate of non-washing ash is unstable in

Fig. 4.52 Ejector washing machine

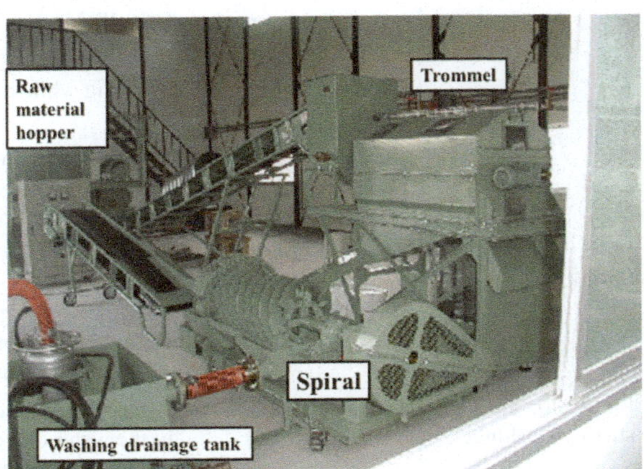

Fig. 4.53 Spiral washing machine

Table 4.26 The composition of ash before and after washing

Items	Unit	Non washing ash	Washing ash	
			Ejector washing machine	Spiral washing machine
Moisture	%	26.1	22.9	38.8
Ig-loss	%	6.8	3.3	2.9
Pb	mg/kg	960	580	123
Cu	mg/kg	38,600	1,910	6,690
Zn	mg/kg	13,800	3,410	3,060
Cr	mg/kg	124	76	142
Fe	mg/kg	55,400	57,600	7,060
Mn	mg/kg	4,380	728	740
Na	mg/kg	18,900	5,150	6,590
K	mg/kg	12,100	6,200	7,840
Ca	mg/kg	8,380	5,740	9,210
Mg	mg/kg	8,340	7,780	10,500
Al	mg/kg	44,200	47,200	44,400
Cl	mg/kg	16,500	2,140	644
DXNs	ng-TEQ/kg	1.1	0.4	0.0074

Table 4.27 The packing condition of lysimeter

Type of ash	Non washing ash	Ejector washing ash	Spiral washing ash
Packing capacity (m^3)	0.759	0.694	0.773
Packing weight (kg)	1160	989	876
Packing density (t/m^3)	1.53	1.43	1.34

every item. It needs to take a long time to stabilize at a low level compared with the leachate of washing ash. It stabilized at a relatively early stage in the case of BOD concentration while it needed to take time to become stabilize from low concentration in the case of COD concentration.

In terms of T-N concentration, it was unstable and became 20–30 mg/l after 750 days in every leachate. Moreover, T-N in leachate was most existing in the form of nitrate-nitrogen (NO3-N).

Regarding SS concentration, it increased when the precipitation became higher, or it was generally below 3 mg/l.

For the inorganic salts, the leachate of washing ash showed a low concentration from the beginning of the experiment, and it stabilized at an early stage, while the leachate of non-washing ash needs a long time to become stabilize from low

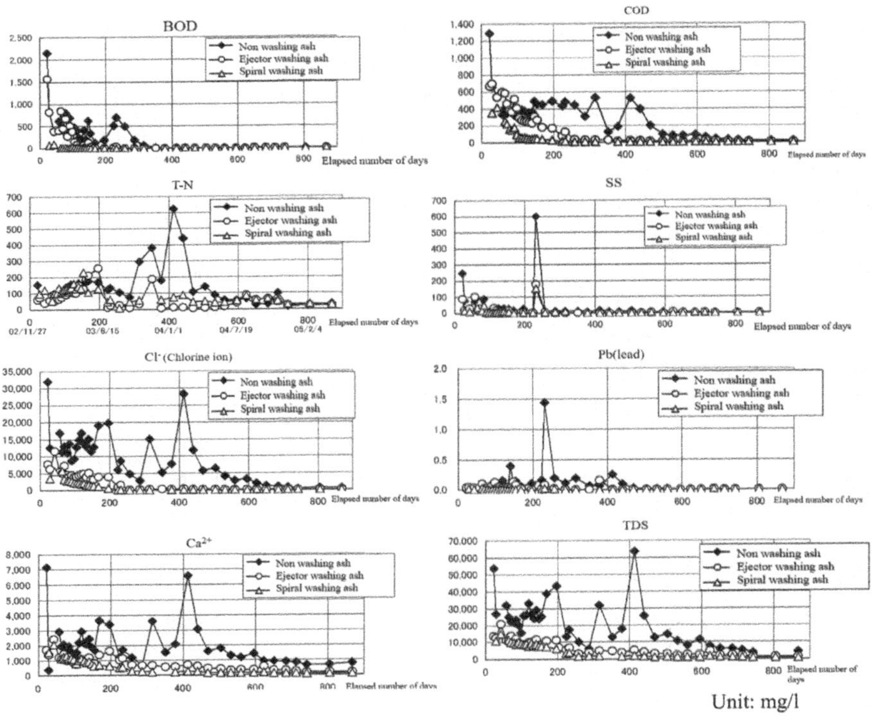

Fig. 4.54 Leachate quality of washing andnon-washing ash

concentration. In regards to the concentration of Ca^{2+}, it is 100–200 mg/l in washing ash and 700–800 mg/l in non-washing ash after 600 days.

In non-washing ash, lead was detected occasionally until 500 days after the beginning of the experiment, after then, it could not be detected.

The leachate of washing ash stabilized at a low concentration at an early stage than the leachate of non-washing ash. According to the experiment result, the washing was evaluated as pre-treatment. As evaluation methods, based on the leachate quality data, pollution load per unit weight of washing and non-washing waste were calculated, and leachate quality was assumed. In the case of washing waste, leachate quality was assumed by using the data of ejector washing ash. Table 4.28 shows the investigation conditions.

In regards of BOD, COD, and T-N among assumed water quality, the pollution load quantity per unit weight of waste during the landfill period was calculated from the data on precipitation of 885 days of the experiment and the monthly mean water quality of leachate. The landfill period of examination conditions was 15 years, and the experiment period was 885 days (nearly 2.42 years). Since the pollutant load quantity at the time of 800 days after the landfilling is extremely small for BOD and COD, the value of calculation result was assumed to be rounded. For T-N, the value

Table 4.28 The leachate quality investigation conditions of washing ash

Landfill period	15 years
Height of landfill	9.7 m
Earthen cover (Volume ratio)	20%
Unit volume/mass of waste	1.0 t/m^3
Annual precipitation	1,800 mm
Leachate ratio	0.7

has not reduced yet, therefore, 1.3 times as much as the calculation result was assumed as the expected value. Table 4.29 shows calculated load quantity during the experiment period and the assumed leachate quality fitting the investigation conditions in Table 4.28.

Besides, the leachate quality of Ca^{2+} and TDS was calculated from the approximation of annual changes, which is based on the leachate quality data. Through the approximation, the average leachate quality for 15 years was obtained, the load amount was calculated as same as BOD. The results show the concentration of Ca^{2+} is 2,696 mg/l in non-washing ash and 952.2 mg/l in washing ash. On the other hand, TDS was 20,890 mg/l in non-washing ash and 6,990 mg/l in washing ash. Table 4.30 shows the planned leachate quality when the washing is performed as a pre-treatment or not performed.

Leachate is stabilized at an early stage by washing the waste as a pre-treatment. It means the period from the end of the landfill to the abolition can be shortened. In addition, it can be assumed that the leachate quality during the landfilling is low. Therefore, the load of the leachate treatment facility is reduced, and the operation cost spent for the facility is suppressed, and as a pre-treatment of the landfill, the washing process is thought to be effective.

Table 4.29 The calculation result of the load quantity and the assumed leachate quality

	BOD		COD		T-N	
	Washing	Non-washing	Washing	Non-washing	Washing	Non-washing
Load amount during the experiment (g)	518.1	137.4	573.5	204.1	249	132.1
Weight of filler (t)	1.16	0.989	1.16	0.989	1.16	0.989
Load amount (kg/t)	0.4466	0.1389	0.4944	0.2063	0.2147	0.1336
Load amount (kg/t)	0.5	0.15	0.6	0.25	0.279	0.174
Estimated water quality during the landfill (mg/l)	206.2	61.9	247.4	103.1	115.1	71.6

Table 4.30 Leachate quality by washing

	Washing	Non-washing
BOD (mg/l)	65	210
COD (mg/l)	105	250
T–N (mg/l)	75	120
Ca^{2+} (mg/l)	1,000	2,700
TDS (mg/l)	7,000	21,000

Figure 4.55 shows the lysimeter conditions in about 60 days after the beginning of the experiment.

From the top, they are the non-washing ash lysimeter, the washing ash lysimeters (ejector washing, spiral washing). Through this photo, we can find although the facilities grow thick in the washing ash at the beginning, it is estimated that it influenced by the difference in salt-containing in the incineration residue.

④ Washing effect (incombustible shredded residue) (Lei X et al. 2014; Lei X 2014)

With the spread of the recycling centers, the landfill ratio of incombustible shredded residues is lower than that of incineration residues. As mentioned in Chap. 2, the incombustible shredded waste is sorted by household and then brought to recycling centers. There is only a crushed sorting treatment conducted. Therefore, fluorescent tubes, small household electric appliance, domestic pesticides may become possible to mixed therein, and lead or mercury may be detected by the elution test. BOD and COD may also be recognized at a high concentration due to the organic substance adhering to the plastics. For these reasons, an early stabilization should be carried out by the pre-treatment such as washing at the landfill sites which are exclusively used for incombustible wastes and incombustible shredded waste. Whether the washing process is performed or not is depending on the sorting method and the type of crushing sorting machine, therefore, an elution test is conducted to make a decision. However, the quality of incombustible waste is different according to the regional characteristics (for instance, rural area, urban area, etc.) at each collection day, and it is difficult to grasp the average quality of waste. For this reason, it is desirable to use much more data of the elution test as possible. Three municipalities (city A, city B, city C) conducted an elution test on incombustible shredded waste, the washing experiment and the lysimeter experiment of washing waste were also carried out. The compositions of incombustible shredded waste and the results of the elution test in three cities are explained in Tables 4.31 and 4.32. Figure 4.56 shows the incombustible shredded waste in city A.

A. Washing experiment

As a pretreatment of landfill, washing experiment was conducted using incombustible shredded residues. In city A, water spraying washing is carried out assuming a wet vibrating filter. The liquid: solid ratio of washing was set as 0.5,1.0, and 16 kg of each

Fig. 4.55 Lysimeter

Table 4.31 The composition of incombustibles hredded waste (Lei X 2014; Lei X et al. 2014)

	City A	City B	City C
Waste plastics	61	13.2	53.5
Styrene foam	0.4	–	0.7
Glass Pottery	18	42.6	24.6
Metals	5	30.5	18.9
Woodchips	6	1.5	1.6
Papers	1	–	0.2
Fiber	2	0.9	–
Others (earth and sand)	6.6	11.3	0.5
Unit: Weight%			

Table 4.32 The elution test of incombustible shredded waste

	Unit	City A	City B	City C
PH	–	9.3	8.4	8.5
BOD	(mg/I)	210	240	39
COD	(mg/I)	230	230	72
T–N	(mg/I)	18	14	5.5
Cl	(mg/I)	60	46	8.5
Pb	(mg/I)	<0.01	0.22	<0.01
Hg	(mg/I)	0.002	<0.0005	<0.005

Fig. 4.56 Incombustible shredded residues in City A

is washed. The immersion washing is performed in city B and city C. The liquid: solid ratio is set as 0.5,1.0,2.0, and 1.0 kg of each was washed out.

(City A): In order to do a complete washing, it was paved with the same thickness. The thickness was approximately 1–3 cm, and a high-pressure washing machine (a

Table 4.33 The washing waste water quality of incombustible shred waste in City A

Item	Unit	Liquid: solid ratio: 0.5	Liquid: solid ratio: 1.0
PH	–	7.1	7.1
BOD	mg/l	465	341
COD	mg/l	1,046	1,040
T–N	mg/l	61	88
CI	mg/l	518	767
Pb	mg/l	0.45	0.21
T–Hg	mg/l	0.14	0.068
F	mg/l	0.7	0.5
B	mg/l	2.8	1.8

Kärcher product) was used. When the liquid: solid ratio is 1.0, it is able to provide water to the front. On the other hand, when the liquid: solid ratio is 0.5, water spraying is not possible on the whole surface, and very slight unwashed portions still remained. Besides, soils and sands were deposited on the bottom of the washing tank. Table 4.33 shows the water quality analysis result of washing wastewater. From the table, we can find that the value of Pb and Hg exceed the effluent standard. Through the visual observation on the composition, it is thought to be caused by the mixture of the board of the electronic toy and the shredded residues from fluorescent tubes.

The results of the elution test of the sample before and after the washing process are shown in Table 4.34.

From this table, we can get a conclusion that the larger the washing liquid: solid ratio is, the higher removal ratio of each water quality item becomes. Besides, the

Table 4.34 The results of elution test before and after washing process

Item	Unit	Elution test before washing	Elution test after washing (Liquid: solid ratio: 0.5)	Elution test after washing (Liquid: solid ratio: 1.0)
PH	–	7.7	7.7	7.7
BOD	mg/l	170	160	96
COD	mg/l	180	180	110
T–N	mg/l	22	21	17
CI	mg/l	49	46	28
Cd	mg/l	<0.01	<0.01	<0.01
Pb	mg/l	0.02	0.02	0.02
Cr^{6+}	mg/l	<0.05	<0.05	<0.05
As	mg/l	<0.01	<0.01	<0.01
CN	mg/l	<0.1	<0.1	<0.1
T–Hg	mg/l	0.0026	0.0025	0.0009
F	mg/l	0.2	0.2	0.2
B	mg/l	0.4	0.3	0.2

removal ratio of BOD, COD, T-N by washing is low. In the case of water spraying washing, it is difficult to wash out the organic components adhered to the waste due to the short contact time with washing water, and washing water cannot contact with the backside of the waste with a large specific surface area such as plastics. Based on the data in Table 4.34, in Fig. 4.57, the effect of spraying washing (liquid: solid ratio is 0.5 and 1.0) of incombustible shredded waste in City A is summarized from water quality load quantity. It is divided into each different item of water quality.

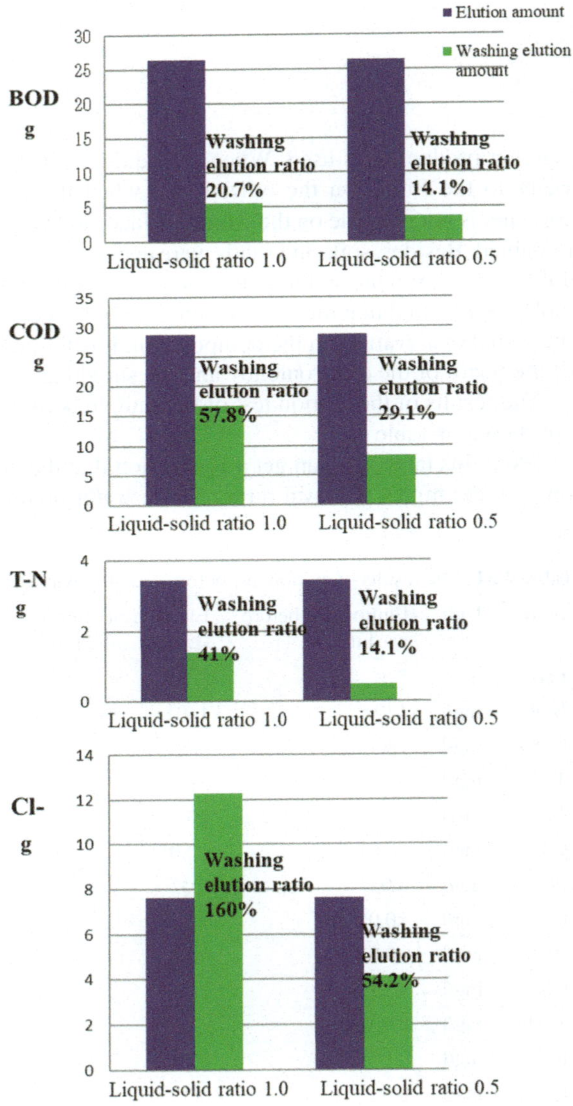

Fig. 4.57 Removal ratio of water spraying washing

The water quality load quantity was obtained by multiplying the amount of washing water in the elution test result. Based on the data in this figure, we could find the higher the liquid: solid ratio is, the higher washing elution ratio becomes, while the lower the removal ratio of BOD is.

(City B, City C): Water spraying washing experiment was conducted in City B and City C, moreover, immersion washing experiment was also carried out in these two cities. The immersion washing is a method of adding the water of a predetermined liquid: solid ratio and immersing for a certain time without stirring, it is the same method as leaving the washing materials to soak in a house and then washing. By using this method, it is possible to remove the extraneous substance adhered to soils and sand by water spraying washing, however, since adhesion of organic substance containing oil cannot be removed in a short time, there is a method of taking a certain time to immerse this kind of organic substance in water. From the washing experiment result of City A, the water spraying washing in this time was conducted at the liquid: solid ratio of 1.0 (the sample is 2.0 kg). The immersion washing was conducted at the liquid: solid ratio of 2.0 (the sample is 1.0 kg) under the condition of checking the waste is completely immersed. The immersion time is set as one hour, two hours and three hours. And after reaching the immersion time, the waste is stirred for 5 min. This outline is shown in Fig. 4.58. Besides, the immersion can be carried out even at the liquid: solid ratio of 1.0 in City B. In City C, it is impossible to completely immerse at this ratio due to a large number of plastics. When the liquid: solid ratio becomes 2.0, complete immersion is possible. Therefore, the liquid: solid ratio of immersion washing needs to change the immersion condition (liquid: solid ratio) according to the composition of incombustible shredded waste.

Table 4.35 points the washing wastewater quality of incombustible shredded waste by immersion washing in City B and City C. According to the data in this table, we can find the BOD concentration at the liquid: solid ratio of 1.0 by water spraying washing in City B is 380 mg/l. And it is 656 mg/l at the liquid: solid ratio of 2.0 by immersion washing for one hour. The higher the liquid: solid ratio is, the higher the BOD concentration in immersion washing becomes. It means immersion washing is more effective in washing. A similar tendency is seen in City C. As a factor, incombustible shredded waste has various kinds of shapes and qualities, and it is a larger shape than incineration residues. Therefore, it is thought that a complete washing cannot be performed because there is a dead angle zone where water cannot contact with the backside during the process of water spraying washing.

Comparing the water spraying washing and immersion washing in City B, the effect of immersion washing is two to three times more than that of water spraying washing in the case of BOD and COD. Besides, since there is no significant variation with the immersion time, it is considered that sufficient washing effect can be achieved with a 1-hour washing.

B. A simulation of reclamation

A landfill experiment in which the landfill site was simulated by using the incombustible shredded waste in City A was carried out. A transparent plastic lysimeter

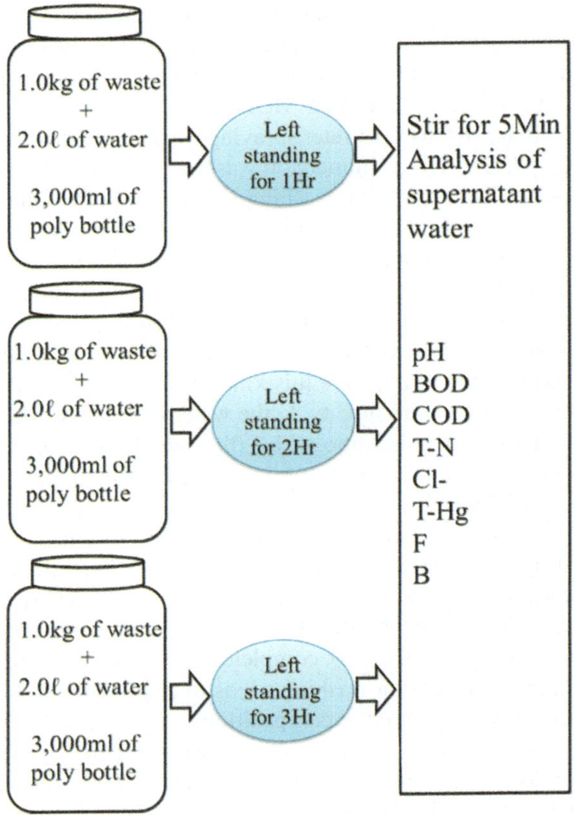

Fig. 4.58 The outline of immersion washing experiment

with a diameter of 200 mm and a height of 1,800 mm is used in this experiment. The experiment conditions were three cases, non-washing waste, washing waste at the liquid: solid ratio of 0.5, and washing waste at the liquid: solid ratio of 1.0. The packing amount and packing density are shown in Fig. 4.59. The soils and sands deposited in washing wastewater during the washing are simulated in each lysimeter. An artificial water sprinkling was conducted once a week for the lysimeter, and the analysis of leachate was carried out. The volume of water sprinkling was 940 ml, and it equals to a week rains in the average annual precipitation of 1,563 mm from 1993 to 2012 in City A. The simulation of reclamation is conducted for 300 days. Figure 4.60 shows the changes of the COD during the experiment. The removal ratio of COD by washing and by artificial water sprinkling is explained in Fig. 4.61.

Figure 4.60 confirms the reduction of water quality becomes faster by washing. And the result of the experiment shows the removal efficiency was higher at the liquid: solid ratio of 1.0. According to Fig. 4.61, it becomes clear that the removal effect by washing is higher than the removal effect by artificial sprinkling.

Table 4.35 The washing waste water quality of incombustible shredded waste by immersion

Item	Unit	City B				City C			
		Water spraying	Immersion			Water spraying	Immersion		
		Liquid: solid ratio: 1	Liquid: solid ratio: 2 (1H)	Liquid: solid ratio: 2 (2H)	Liquid: solid ratio: 2 (3H)	Liquid: solid ratio: 1	Liquid: solid ratio: 2 (1H)	Liquid: solid ratio: 2 (2H)	Liquid: solid ratio: 2 (3H)
PH	–	6.9	6.9	7.0	6.8	7.6	7.9	8.5	7.7
BOD	mg/l	380	656	419	434	285	267	359	207
COD	mg/l	720	1,027	824	846	559	557	782	426
T–N	mg/l	26	17	41	32	32	47	49	48
CI	mg/l	328	343	1,070	503	146	108	132	188
Pb	mg/l	0.43	0.61	0.49	0.43	0.40	0.46	0.39	0.44
T–Hg	mg/l	0.003	0.007	0.007	0.004	0.001	0.002	0.002	0.001
F	mg/l	0.36	0.41	0.39	0.35	0.25	0.20	0.24	0.19
B	mg/l	1.2	0.84	0.89	0.99	0.65	0.45	0.49	0.37

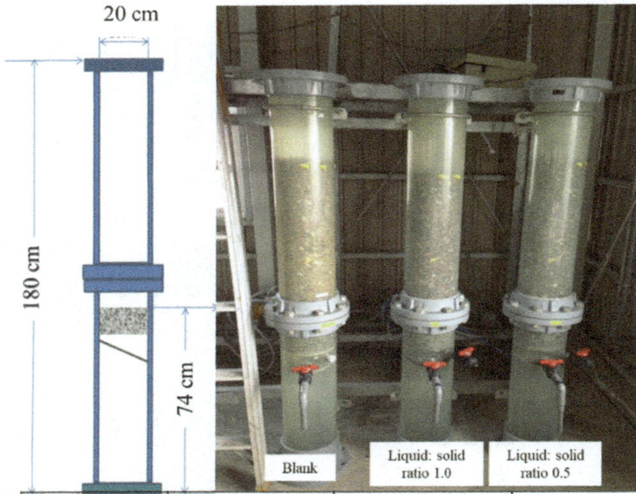

	Non-washing (blank)	Washing liquid: solid ratio 1.0	Washing liquid: solid ratio 0.5
Packing amount	16,000g	18,920g	18,500g
Moisture content	0	29.4%	29.1%
Height of packing materials	73.14cm	67.7cm	74.1cm
Capacity	23,000cm^3	21,000 cm^3	23,000cm^3
Packing density	0.696g/cm^3	0.761g/cm^3	0.696g/cm^3

Fig. 4.59 A simulation landfill experiment assuming City A

The concentrations of Pb and Hg, which are exceeding the effluent standard in washing wastewater are also exceeding the effluent standard in leachate even after washing. On Day 49, the concentration of Pb became lower than the standard, and on Day 35, the concentration of Hg was below the standard. In the case of non-washing, on Day 91, the concentrations of both were below the effluent standard.

Table 4.36 shows the result of the liquid: solid ratio calculated from the total amount of washing water and artificial sprinkling water and the amount of waste for stabilization. The value of effluent standard was set as BOD 20 mg/l, COD 30 mg/l, T-N 10 mg/l, Pb 0.01 mg/l, T-Hg 0.005 mg/l.

As shown in this Table 4.36, when we only pay attention to the total liquid: solid ratio, we can find the target water quality can be achieved more easily at a small value of liquid: solid ration in the case of non-washing than washing. As mentioned above, the washing water cannot contact the backside of waste which has a large specific surface area such as waste plastics in the case of water spraying washing,

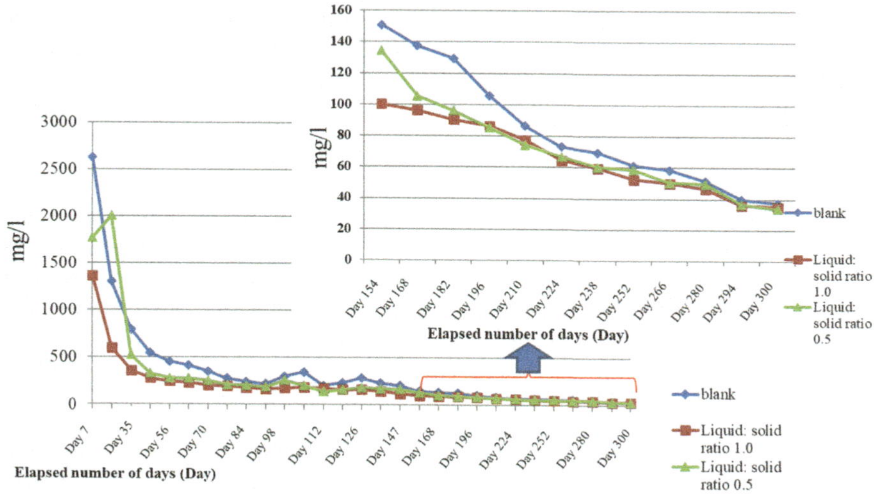

Fig. 4.60 The changes of COD in a simulated landfill layer of City A

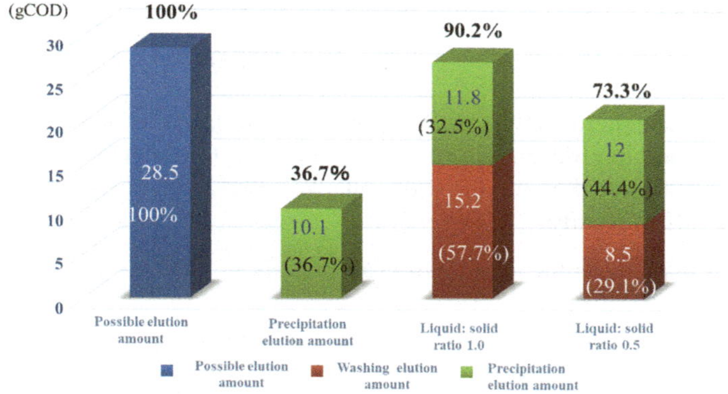

Fig. 4.61 The removal ratio of COD in a simulated landfill layer of City A

Table 4.36 The required liquid: solid ratio to wards the stabilization of incombustible shredded waste

	Effluent standard, etc.	Blank (non-washing)	Liquid: solid ratio: 1.0	Liquid: solid ratio: 0.5
		Total liquid: solid ratio	Total liquid: solid ratio	Total liquid: solid ratio
BOD	20 mg/l	0.63	1.53	1.07
COD	30 mg/l	2.52	3.13	2.68
T–N	10 mg/l	1.65	2.2	1.9
Pb	<0.01 mg/l	0.76	1.35	0.85
T–Hg	0.005 mg/l	0.77	1.29	0.79

and it shows the same tendency in the case of artificial sprinkling. It also can be seen from the results of elution tests of washing waste and non-washing waste in Table 4.34 that it is closer to true stability when the "water channel" in landfill layer is formed by washing.

C. The washing method of incombustible shredded waste

The washing method was to conduct an immersion washing with the aim of improving the washing effect from the results of water spraying washing, assuming a wet vibrating filter. The liquid: solid ratio of water for washing was set as 0.5, 1.0 (water spraying washing) and 2.0 (immersion washing). As a result, the following washing characteristics and issues have been found.

- Vary widely in waste quality.

Incombustible shredded waste varies widely in waste quality, and it shows different properties in three cities. Different from incineration residues, since it varies widely in waste quality, during the collecting or the sorting, in order to grasp the waste quality, at least to grasp the seasonal variation, it is necessary to collect multiple kinds of samplings while conducting this kind of experiment.

- Waste quality differs according to the shred sorting method.

Depending on the sherd method (impact, shearing, etc.), shapes and dimensions are different. Furthermore, depending on sorting, according to whether the fine-grain soil and sands are separated or not, the waste quality differs, and the quality of washing wastewater differs greatly.

- A high possibility of the mixture of heavy metals

Even the mixture quantity of small home electric appliances and fluorescent lamp fragments is very small, there is a possibility of mercury elution. Even the sorted collection is carried out, a minimal mixture is inevitable.

- The organic substance load of adhered substances is large.

Based on the results of the elution test, the adhesion quantity of organic substances (BOD, COD) is larger than expected, and the necessity of carrying out a peak-cut by washing is recognized.

- The leachate load becomes higher when the washing liquid: solid ratio is 0.5.

As a result of the water spraying washing experiment which using the waste of City A, it is difficult to perform a complete washing at liquid: solid ratio is 0.5, and the non-washing part is generated. The elution test was performed once again on the washing waste subjected to water spraying washing at the liquid: solid ratio of 0.5, a high concentration of the organic substance is provided. Therefore, the liquid: solid ratio is desirable to above 1.0 in the case of water spraying washing.

- Water spraying washing and immersion washing

Comparing with the washing effect of water spraying washing, immersion washing has a higher washing effect. A water spraying washing at a liquid: solid ratio of 1.0 and an immersion washing at a liquid: solid ratio of 2.0 was performed in City B and City C. Although the target waste in the experiments is different, the washing water amount is twice in the case of the liquid: solid ratio is 2.0. If it has the same pollution load, the concentration of washing water becomes lower when the liquid: solid ratio is 2.0. However, in this experiment, the concentration became higher when the liquid: solid ratio is 2.0. That is, the washing effect becomes higher. For incineration residues, etc., the smaller the liquid: solid ratio is, the higher the concentration of washing water becomes. On the contrary, the larger the liquid: solid ratio is, the lower the concentration of washing water becomes. The inorganic substance adhering to waste is possible to separate by water relatively quickly. However, it is inferred a certain contact time is necessary to peel off and wash when the organic substance strongly adhering to waste.

According to these results, it is considered that the immersion washing method is a suitable method for washing incombustible shredded waste.

- The time of immersion washing is 1 h.

The time of immersion washing was set as 1 h, 2 h, 3 h, and the quality of washing wastewater was analyzed. As a result, it can be considered that 1 h is enough for washing. The washing effect slightly increases when the washing time is set above 2 h. However, the volume of washing water tank becomes larger when the washing time is increased. In the case a site without a margin area, it is thought that the maximum washing effect can be obtained with the minimum requirement of 1 h. Regarding immersion washing, in order to immerse the whole waste, the liquid: solid ratio needs to above 1.0.

- By immersion washing, it may be below the effluent standard at the time of landfill disposal.

The washing effect of immersion washing is high. When the washing waste is landfilled, The leachate is generated by precipitation and artificial water sprinkling. There is a possibility that it may be lower than the effluent standard of leachate, and it may higher than the standard. When the leachate treatment is performed, it has the possibility to reach the effluent standard with a simple and minimum capacity of the effluent facility.

D. Practical examples

We will introduce the installation examples of pre-treatment of incombustible shredded waste before reclamation. The target waste at covered type landfill site is the incombustible shredded waste.

The capacity of the pre-treatment washing facility is 31.3 t/day

The capacity of the leachate treatment facility is 38 m^3/day

After taking an advanced treatment, the washing water and the leachate from an artificial sprinkling can be recycled as washing water and sprinkling water. The washing method is immersion washing + aeration stirring, the washing liquid: solid ratio is 1.0.

Figure 4.62 explains the washing flow of incombustible shredded waste, and Fig. 4.63 shows the photo of the washing facility.

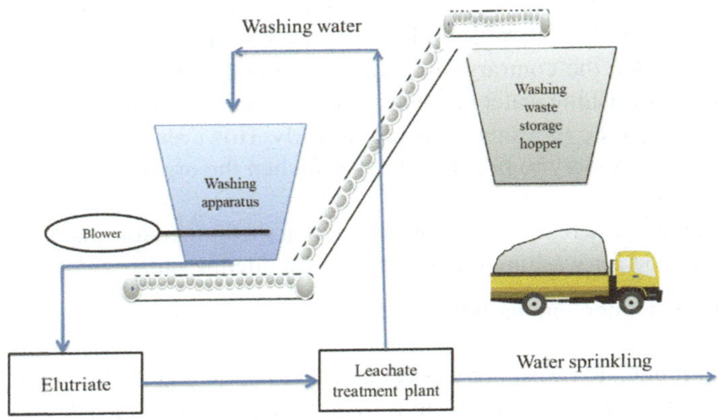

Fig. 4.62 Washing flow of incombustible shredded waste

Fig. 4.63 Washing facility for incombustible shredded waste

4.5.2 *Post-treatment of the Landfill*

Maintenance management such as the operation of the leachate treatment facility is performed at the landfill site from the end of the landfill to its abolishment period. However, when a place where the obstruction of ventilation and water flow exists in the landfill layer, it may be impossible to achieve the early stabilization even by long-term maintenance and management. Besides, even the pre-treatment is conducted, the obstruction of ventilation and water flow will be caused by the increase of the height of the landfill, and it leads to the postponement of the stabilization. Therefore, a post-treatment is necessary to be performed as needed. Many technologies for post-treatment are similar to the techniques for contaminated soil treatment. In this part, the ventilation method, the chemical oxidation method, and the combined method of both two methods will be introduced. All of them are applied to the landfill site.

In this case, effective countermeasures can be taken by pipe arrangement to make sure the measures of post-treatment can be taken before landfill or under landfilling.

4.5.2.1 A Survey for Post-treatment

To stabilize the landfill layer after the end of the landfill, we need to find the reason for delaying the stabilization, the target place in the landfill layer, and the method used for conducting a treatment. A survey is necessary for this condition. The factors which are thought causing the delaying of stabilization are listed as follows. By the obstruction of ventilation towards burial sites of the refractory organic substance and the organic sludge, an anaerobic atmosphere is made, and hydrogen sulfide gas and methane gas are generated. Water channel causes the inhibition of water flow and the delay of washing out. The decomposition of substance, which becomes the reason for delaying the stabilization can be promoted by identifying the location of burying and water channel. The survey methods include the heat source survey which investigates the heat source generated by the decomposition process of organic substance and the high-density electric exploration, which identifies the place where washing out of inorganic salts is delayed.

In regard to the heat source survey, as introduced in **4.4.1.4** Survey and procedures for regeneration in **4.4**. The Regeneration of the landfill site, there is a geothermal exploration at a depth of 1 m and an infrared camera method.

4.5.2.2 Ventilation Methods

Both the BIOPUSTER method developed in Austria and the SMELLWELL method are the ventilation methods. By air supplying, they are used as a countermeasure to hydrogen sulfide and methane gas and a decomposing method towards easily decomposable organic substance. The BIOPUSTER method is a method of blowing air and oxygen into the landfill layer intermittently under a pressure of 3 to 7 atmospheric pressure, and its corresponding depth is 10 m. The SMELLWELL method is a method of blowing air continuously under a pressure of 1 to 2 atmospheric pressure, and its corresponding depth is about 3 m. Figures 4.64 and 4.65 are the photos of BIOPUSTER method and SMELLWELL method.

The results of the demonstration experiment using BIOPUSTER method is shown as follows.

① The principle of BIOPUSTER method (Higuchi S 2005)

As described in Fig. 4.66, the main principle is that air is compressed intermittently and pressed in the waste with about a pressure of 3 to 7 atmosphere (adjustment) as a pressure flow at a near-sonic speed. Thereby, air and oxygen are supplied to the coarse stomatal part and the densely tight part of the waste. Besides, pure oxygen and nutrients reaching 40% can be injected with compressed air to the waste to promote

Fig. 4.64 BIOPUSTER method

Fig. 4.65 SMELL WELL Method method and SMELL WELL method

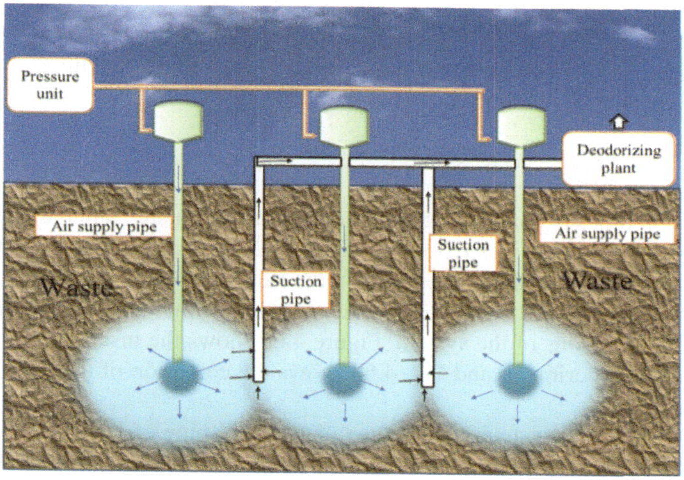

Fig. 4.66 The principle of BIOPUSTER method

more aerobic decomposition. During the aeration, the injected air is discharged from the landfill layer by suction pipe and sent to the filter system. Since the discharge amount is about 30% more than the injection amount, the condition inside the waste tank shows a negative pressure condition than that of the ground surface. Therefore, there is not any gas discharge from the ground surface during the purification. Generally speaking, after 4 to 16 weeks 'aeration (according to the condition of the distinctive landfill site), it is possible to stabilize malodorous gases and methane gas.

② Outline of the demonstration experiment site

The demonstration experiment is conducted at the landfill site where buried the untreated waste (basically domestic waste except papers, cardboard, bins, hardware) in the 1980 s.

The average moisture content of waste quality is about 30%. The average disposal density (wet density) of the landfill site is nearly 1.25 t/m^3. The maximum depth of the landfill site is 23 m, and the thickness of soil cover is 1 m to 4 m.

- The landfill area is 14,007 m^2 (on the north-eastern part of the site which is approximately 900 m^2 is used for demonstration test)
- The land capacity is 190,338 m^3 (the target is 2,000 m^3 which is until the depth of 5 m)
- The construction period is from 1979 to 1980
- The landfill period is from 1982 to 1991
- The landfill structure: The rubber sheet construction is up to the bottom of the slope. There isn't any sheet construction in the bottom due to the bedrock.

Five gas vent pipes are set inside the site.

- Characteristics: Many organic wastes such as kitchen waste and excretion treatment sludge are landfilled.

③ Outline of the demonstration experiment

Nine aeration pipes (P) are arranged at the demonstration site, and the intervals are 8.0 m. Twelve suction pipes (Sv) are placed at the outer periphery at intervals of 6.5 m in order to cut off the influence from the surroundings. Four suction pipes (S) for gas measurement are set near the center to monitor the operating status and the purification conditions of the system. Figure 4.67 shows the installation map of the demonstration experiment, and Fig. 4.68 shows the condition of the demonstration experiment.

The pressure pipe system uses electro-plated steel pipe of DN2" PN10, and the air is supplied to each BIOPUSTER by the deliver apparatus disposed at the center. For convenience, a high-pressure hose (1/2"PN10) is used between the delivering apparatus and the pipes. The suction pipe system is sealing as usual, and it is used when measuring the chemical composition of the suction gas. After the suction gas is treating with an activated carbon filter or biofilter inside the facility, it is released into the atmosphere.

The compressor compresses and supplies air at a high pressure of 4 atmospheres. The air is dried and purified before pumping into the storage tank.

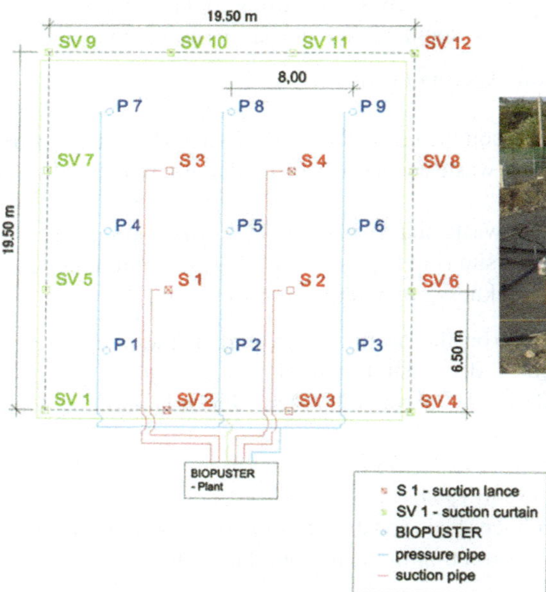

Fig. 4.67 The installation map of demonstration experiment of BIOPUSTER (Higuchi S 2005)

Fig. 4.68 The installation photo of BIOPUSTER (The container in this side is a pressure unit and a management apparatus)

Figure 4.68 The installation photo of BIOPUSTER (The container in this side is a pressure unit and a management apparatus)

Then, it is introduced to blast pipe through a pressure system network, stored in the main body (27ℓ) of BIOPUSTER, and discharged into waste through an insertion pipe at a pressure of 2 to 3 atmospheres. The aeration ratio and volume can be adjusted separately according to the suitable pressure and injection frequency. The suction volume through the suction pipe is 20 to 30% more than the volume of air injected into the landfill site. Therefore, the pressure prevents gas leakage from the surface layer of the landfill site.

The volume of air for aeration is 27ℓ for a main body of BIOPUSTER. Compressing air at a pressure of 2 to 3 atmospheres, it becomes about 200ℓ in a minute at the intervals of 15 to 20 s, and it becomes 12 m^3 in 1 h.

Considering the maintenance time, the operation coefficient is 0.85 by working 8 h a day, 80 m^3 of air can be injected per unit. It is possible to inject 720 m^3 of air in a day by 9 units. In the case of working for 24 h, the injection volume is 2,200 m^3. The operation time for air blowing is 8 h in the first 12 days. Then, it becomes 24 h in the next 12 days. As a result, 35,000 m^3 of air is inserted to the 2000 m^3 (20 m × 20 m × 5 m) of target waste amount. (17.5 times of air than the waste amount is inserted in 24 h)

④ The result of the demonstration experiment

By forcedly shifting to an aerobic environment, odor and methane gas can be reduced by air blowing about 17.5 times than the waste amount for 20 days.

Besides, since the state where the concentration of carbonic acid gas is 1.5% and the oxygen concentration is above 18% achieves the working environment measurement standard from Industrial Safety and Health Act, the measurement values at the final confirmation during aeration is similar to the atmosphere, the aerobic environment can be created. Through the data in Table 4.37, we can find the unit of methane gas is changed from vol.% to ppm, and the odor concentration is reduced to one-fourth.

Table 4.37 The demonstration of BIOPUSTER

Item	Unit	Before enforcement	After enforcement	Notes
Methane gas	V/V%	30.0	0.11	Malodor Prevention Law
Odor concentration	–	980	230	Malodor Prevention Law
Humidity	°C	17.6	16.0	–
Ammonia gas	V/V%	28.5	0.0	Simply measuring instrument
Carbon dioxide	V/V%	5.9	1.2	Simply measuring instrument
Oxygen	V/V%	6.8	17.8	Simply measuring instrument

4.5.2.3 Oxidizing Agent Injection and Combined Method of Ventilation and Oxidizing Agent (Yoshizaki K et al 2006)

The leachate continues to flow outside over a long period of time, COD and nitrogen (Higuchi S 2003) among the leachate are thought as the main factors for inhibiting abolishment of the landfill site, the refractory organic substance in landfill layer is considered to be the generation source. The authors believe that it is possible to reduce COD and nitrogen in leachate with the promotion of the decomposition of this refractory organic substance. They use a large-sized tank to conduct a demonstration experiment which concerns the Water equivalent to 1–2 times (1,700–3,400 mm/year) the national average precipitation is sprayed to five tanks. During the time, forced ventilation ($4.7L_{-Air}/min/m^3_{-Waste}$) is continued conducting in the Tank No.3 and No.4, and the chemical oxidation is carried out two times in No.2 and No.3. In the first time, hydrogen peroxide and phosphoric acid are injected into the simulated waste. The total addition ratio of hydrogen peroxide is $2.3\,kg_{-H_2O_2}/m^3_{-Waste}$, and the total addition ratio of phosphoric acid is $0.7\,kg_{-Phosphoric\ acid}/m^3_{-Waste}$. In the stabilization promotion (chemical oxidation method) by the decomposition of refractory organic substance using forced ventilation and oxidizing agents. We will introduce the results of this experiment.

(A) A large-sized demonstration facility

The photograph of the external view of the facility used for the demonstration experiment is shown in Fig. 4.69. The schematic drawing of a large-sized tank inside the facility is shown in Fig. 4.70. The size is 1.75 mW × 7 mL × 1.5 mH (18.4 m^3/tank), and four similar tanks are used in this experiment. The addition of oxidizing agent and continuous water sprinkling is carried out towards the simulated waste, and a mixture of incineration residue 90% w/w, compost 5% w/w, crushed stone 5% w/w is used in the simulated waste.

Fig. 4.69 A simulated landfill tank

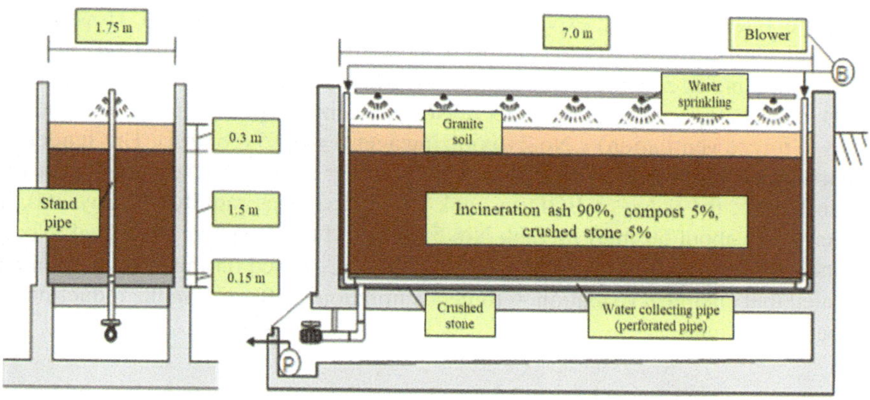

Fig. 4.70 Schematic drawing of a landfill simulation tank

(B) The outline of the experiment

The experimental tank is composed of five tanks, Tank No.1 is intentionally used as a hot spot, and it aims to bury organic sludge in the layer. The purpose of each tank is as follows.

Tank No. 1: hot spot survey and purification
Tank No. 2: oxidizing agent
Tank No. 3: oxidizing agent + forced ventilation
Tank No. 4: forced ventilation (aerobic tank)
Tank No. 5: control (semi-aerobic landfill)

Water equivalent to 1–2 times (1,700–3,400 mm/year) the national average precipitation is sprayed to five tanks. During the time, forced ventilation

($4.7L_{-Air}/min/m^3_{-Waste}$) is continued conducting in the Tank No.3 and No.4, and the chemical oxidation is carried out two times in No.2 and No.3. In the first time, hydrogen peroxide and phosphoric acid are injected into the simulated waste. The total addition ratio of hydrogen peroxide is $2.3\,kg_{-H_2O_2}/\,m^3_{-Waste}$ 2, and the total addition ratio of phosphoric acid is $0.7\,kg_{-Phosphoric\ acid}/\,m^3_{-Waste}$. In the second time, the addition ratio is the same.

(C) The result of the experiment

Figure 4.71 shows the daily changes of pH in leachate. At the beginning of the water sprinkling, pH in leachate is more than 9, though it has an upward tendency, it gradually decreases with the increase of the water volume, and after 200 days, it remains almost below 8. Although the simulated waste contains a large amount of incineration residue and it is an unsuitable condition for the growth of the microorganism, it is inferred that the state inside the tank is kept in a state suitable for biodegradation because the pH is below 8. Figures 4.72 and 4.73 shows the daily changes of COD and BOD in leachate.

At the beginning of the experiment, the COD is 1,600–1,800 mg/l. However, at Day 460, COD in Tank NO.5 (control) reduced to about 90 ml/g and COD in Tank NO.3(oxidizing agent + forced ventilation) reduced to about 30 mg/l. The value of COD in leachate becomes in the order of No. 3 (oxidizing agent + forced ventilation) < No 4 (forced ventilation) < No.2 (oxidizing agent) < No.5 (control). This tendency is particularly remarkable in the case of BOD. After adding the oxidizing agent for the first time, the value of BOD in Tank No. 3 decreases rapidly. At Day 460, it decreased to about 50 mg/l in Tank No. 5 tank, and to about 5 mg/l in Tank No.3. A similar result can be obtained in the case of TOC. Based on this result, it can be revealed that chemical oxidation + forced ventilation is effective in the reduction of refractory organics.

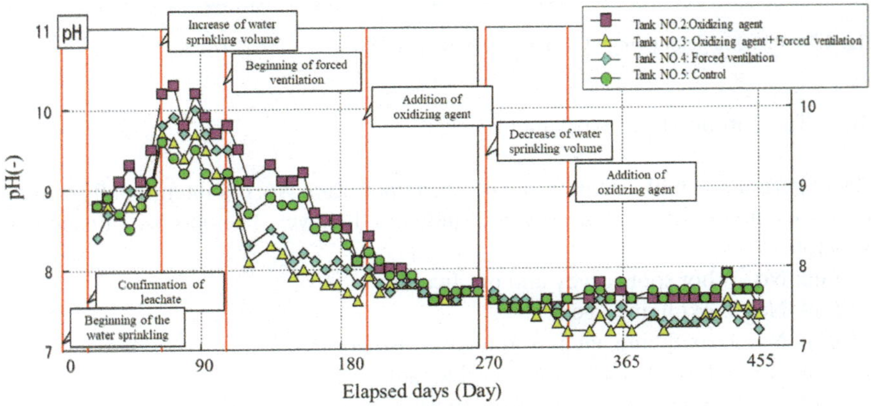

Fig. 4.71 The daily changes of pH

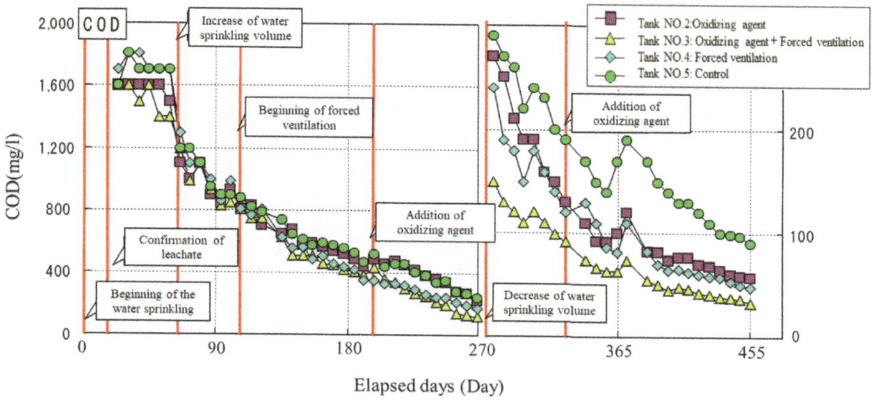

Fig. 4.72 The daily changes of COD

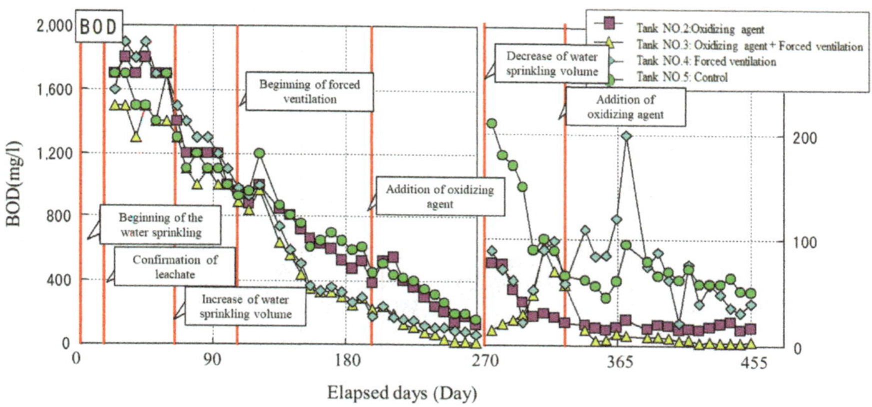

Fig. 4.73 The daily changes of BOD

Figures 4.74, 4.75 and 4.76 explains the daily changes of T-N, NH_4-N, NOx-N in leachate.

In regard to T-N in leachate, the value reduces in order as same as COD, and it becomes 60 mg/l in Tank No.5 and 30 mg/l in Tank No.3 at Day 460.

The forms of nitrogen are different in each tank (in different conditions). In Tank No.5, the principal form of nitrogen is NH_4-N. In Tank No.3, NOx-N is produced particularly, and NH_4-N has almost not been observed.

In Tank No.4, NO_3-N is observed, it is suggested that the nitration is conducted in the tank where forced ventilation has been performed.

Although NOx has been observed in Tank No.3 and Tank No.4, as NOx has almost not been observed in Tank No.5(control tank) about 90 days after the beginning of

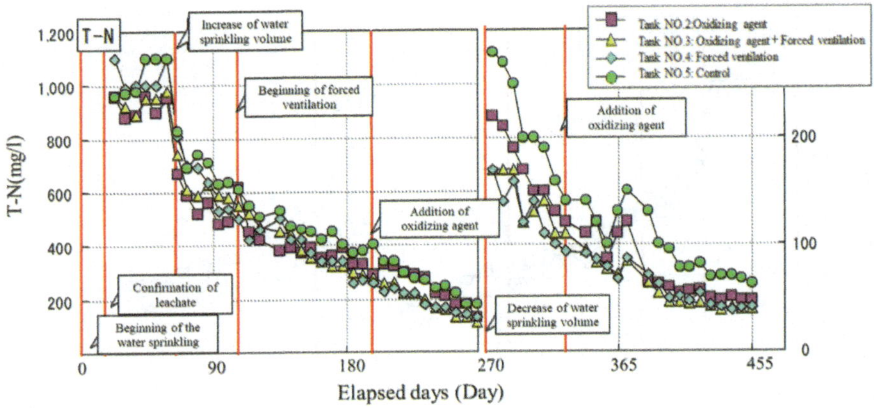

Fig. 4.74 The daily changes of T-N

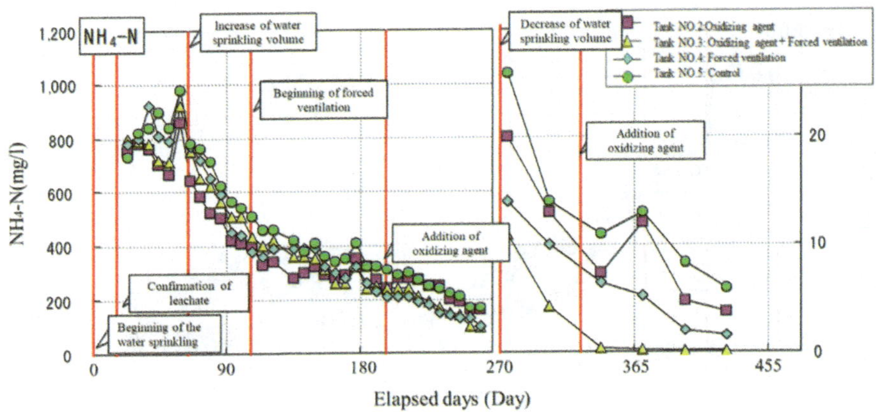

Fig. 4.75 The daily changes of NH4−N

the forced ventilation. It is estimated that forced ventilation affects the growth of nitrification bacteria in the waste, and the nitrification is promoted.

Furthermore, comparing with the nitrogen between No.3 and No.4, the concentration of NOx-N in No.3 is higher than that of in No.4. It can be inferred that the oxidizing agent has some influence on the growth of nitrification bacteria.

Figure 4.77 shows the balance of nitrogen in leachate. As described above, the nitrogen in each tank shows a different form, and it is about 30–60 mg/l as T-N. As the result of No.5 control tank, after the end of reclamation, T-N changes to NH_4-N. Due to the nitrification does not carry out, the post-treatment makes it possible to reduce the load of the nitrification process of the leachate treatment facility by the oxidizing agent and forced ventilation. Besides, the leachate which nitrification was

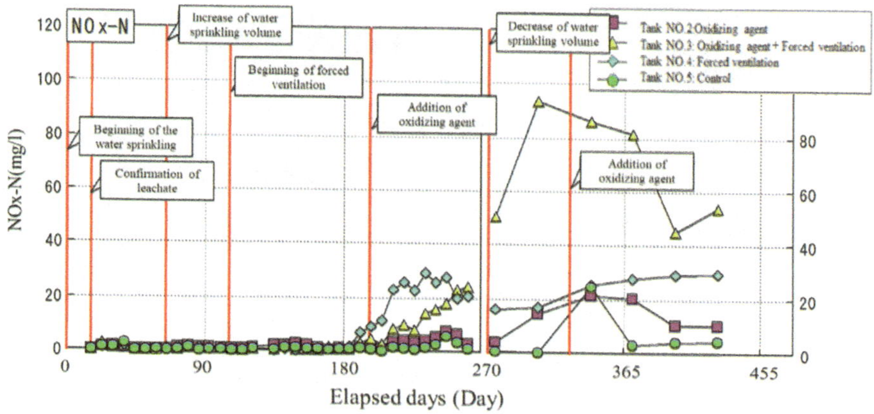

Fig. 4.76 The daily changes of NOx−N

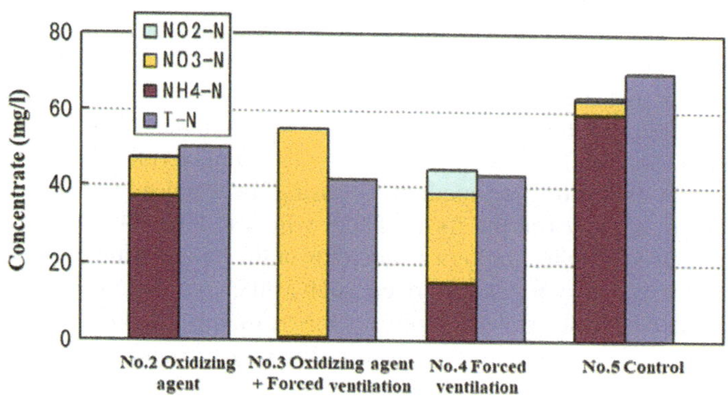

Fig. 4.77 Form of nitrogen in each tank

carried out can be expected to have a denitrification effect by returning to the landfill layer.

(D) Practical example

We will introduce the case of COD reduction using ozone water in the oxidizing agent. The geothermal exploration confirms the high-temperature heat source zone (hot spot) underground at a depth of 1 m and the boring survey. Along with it, the injection hole of ozone water is set, and ozone water is injected to make the COD reduction. Figure 4.78 shows the ozone generation apparatus, ozone dissolving tank, and the injection state of the ozone water.

Fig. 4.78 Ozone generation apparatus, ozone dissolving tank and the injection state of the ozone

4.5.2.4 Injection Method of Mist Oxidizing Agent

In the case of the chemical oxidation method using an oxidizing agent, with the injection of the oxidizing agent into the underground, the oxidizing agent is introduced into the "water channel" in the waste layer as same as the process which precipitation changes to leachate. And a large amount of oxidizing agent is consumed due to a short path is made without contacting the hot spot. Besides, the liquid oxidizing agent has a low efficiency of gas-liquid contacting with the methane gas and hydrogen sulfide gas underground, and the treatment efficiency of gas is also low. To solve these issues, the authors developed an injection method of mist oxidizing agent. The mist oxidizing agent uses a dry fog method. By using a special nozzle (Fig. 4.79), the oxidizing agent atomizes in the form of fog, which is 10 to 30 μ. The dry fog is below 10μ. The cooling technologies which lower the temperature by using the heat of vaporization when dry fog is vaporized, such as the cooling tower is used. By the injection method of mist oxidizing agent, as the oxidizing agent is made in the form of mist from the nozzle and then sprayed to the underground, hydrogen peroxide water, and ozone water is used as oxidizing agents. Besides, by injecting two liquids into the nozzle, it is possible to show a stronger oxidation effective due to the effects of accelerated oxidation. In the fogs, dry fog sews the gaps in the waste layer, and the oxidizing agent enters in every detail. Since it is in the form of mist, it has a better efficiency of gas-liquid contacting and the oxidation decomposition of methane gas and hydrogen sulfide is also possible. The fog of 10 to 30μ is possible to oxidization

Fig. 4.79 Dry fog

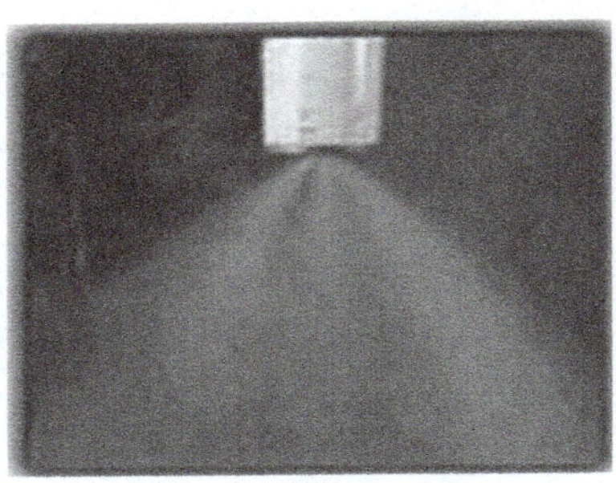

decomposed or easily decomposed organic substance adhering to waste. Figure 4.80 shows the conceptual diagram. Figure 4.81 shows the conceptual diagram of injection. Comparing with the ventilation method, which the depth limit on air supply is 10 m, the characteristic of the injection method of mist oxidizing agent is no depth limitation because the fog drops by gravity. However, when a large amount of methane gas which is lighter than air is generated, it is necessary to adjust the location of the nozzle as it is pushed back to the surface of the ground by uplift pressure. Besides, the diffusion in the horizontal direction is about 5.0 radius from the injection point.

① Practical example (A countermeasure against hydrogen sulfide in illegal waste dumping areas)

Hydrogen sulfide gas shows a high concentration of up to 950 ppm in the illegal waste dumping site of construction waste. As it is necessary to reduce the hydrogen sulfide concentration in the current recovery project, hydrogen peroxide (0.1–0.3%) is injected into the borehole by the injection method of mist oxidizing agent, and it decreases to 0 ppm after 16 days of injection. A geothermal survey of the illegal dumping site is conducted, hydrogen sulfide is removed by the injection method of mist oxidizing agent in the high-temperature section, and a recovery project is carried

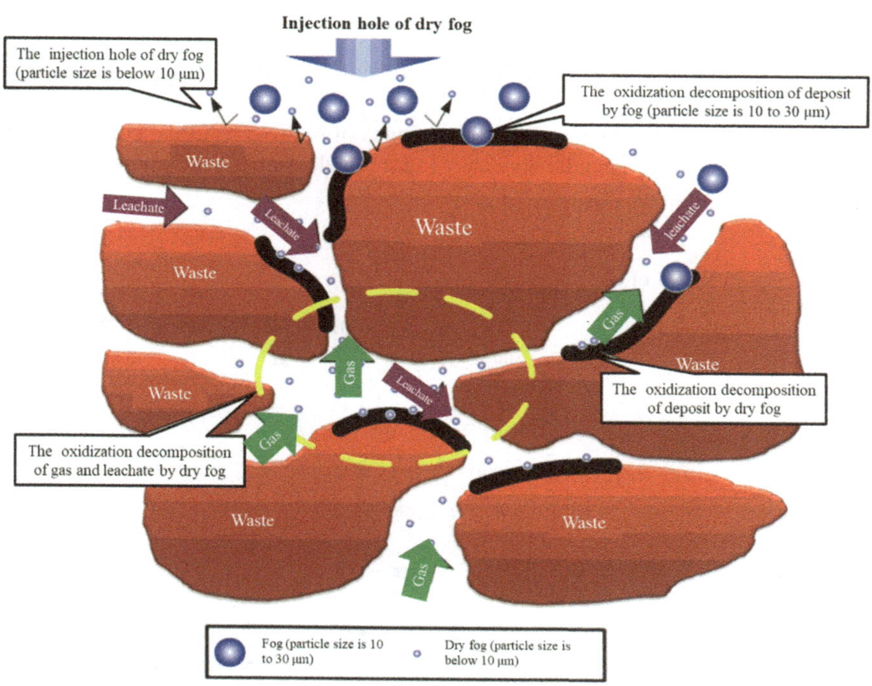

Fig. 4.80 Conceptual diagram of the injection of mist oxidizing agent

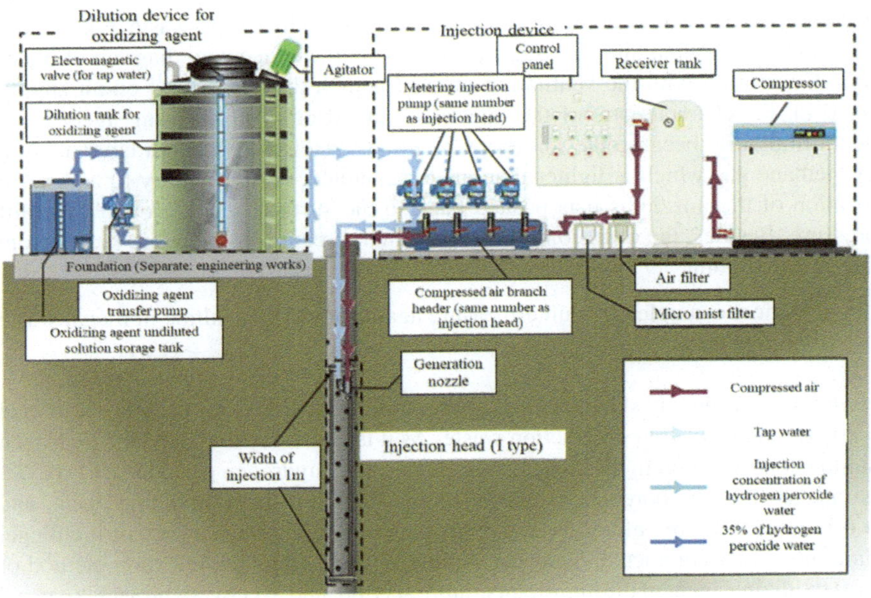

Fig. 4.81 Conceptual diagram of the injectionplant of mist oxidizing agent

out. Biological toxicity test on leachate is performed while the safety assessment is conducting.

Figures 4.82, and 4.83 show the daily changes of hydrogen sulfide concentration and oxygen concentration in the borehole. Figure 4.84 describes the injection facility of mist oxidizing agent.

Fig. 4.82 Daily changes of hydrogen sulfide concentration (Tameda K et al. 2010)

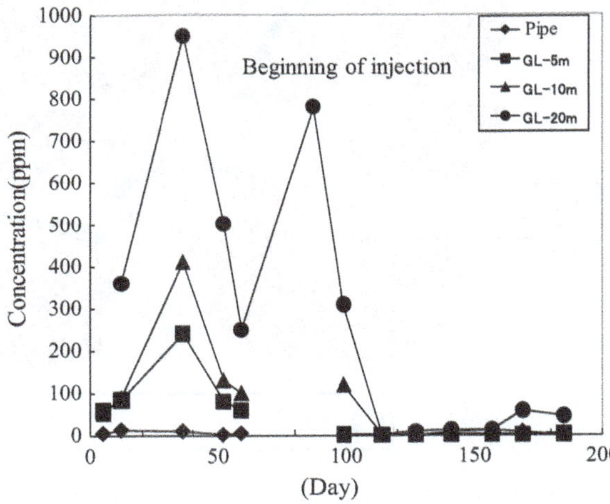

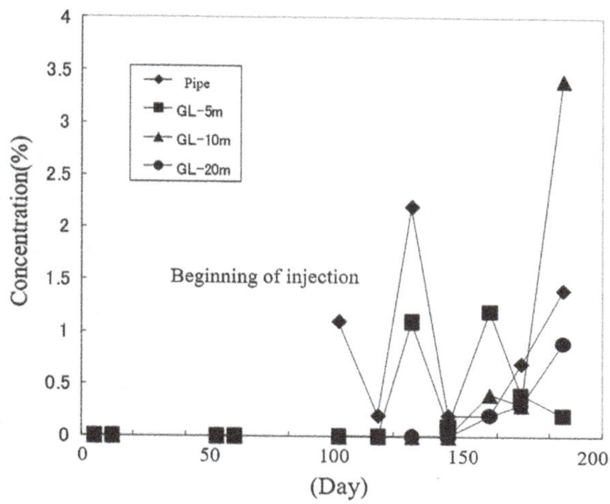

Fig. 4.83 Daily changes of oxygen (Tameda K et al. 2010)

Fig. 4.84 Injection plant of mist oxidizing agent

<Countermeasures against early stabilization>

Nearly 25 years have passed since the MSW landfill site of City B is over landfilled. Since it was constructed before the establishment of the structural criterion, there is no seepage control work, and many organic wastes are landfilled. Therefore, methane gas and hydrogen sulfide gas are generated even at the present time. By using the injection method of mist oxidizing agent, countermeasures against gas were carried out, and former site redevelopment was performed.

As the oxidizing agent,the hydrogen peroxide water (the concentration is 3%) is injected and the injection is stopped when the concentration of gas decreases. After a little while, the concentration of gas increases, the injection is repeated, and the concentration decreases gradually. Besides, as the concentration of methane gas increases with the rises of the internal water level, to make a stabilization, at the same

Fig. 4.85 Injection condition of mist oxidizing agent

time as lowering the position of the intake pump, an improvement of the leachate treatment capacity when the water level rises is performed, and the water level is controlled. The injection condition is shown in Fig. 4.85.

4.6 Technology for Future Landfill

As described above, the disposal technologies have been developed for a long time according to symptomatic treatment and post-treatment. As a result, Japan's landfill technology can be divided into those that were originally developed from the characteristics of waste management and those that were enhanced internationally regardless of waste management. The leachate treatment, early stabilization of landfill site, and technologies for abolishment can be listed for the former. The latter is a case of seepage control work. The method of selecting the type of seepage control sheet differs depending on the topography and geological conditions. Table 4.38 summarizes the facilities configuring the landfill site and their design factors. The topography and the geology conditions are undoubtedly the determinants of the capacity and the shape of facilities. Regarding the facilities which are strongly related to leachate management, such as "leachate collection/drainage facility," "storage structure," "leachate treatment facility," we can find the design factors are the precipitation, landfill area and material of landfill waste. According to the landfill waste materials, the leachate quality, and the physic-chemical properties of landfill waste can be calculated or estimated by the leachate, which is calculated from precipitation and the landfill area.

In this section, we will mainly introduce the most essential "leachate management system" from the aspect of Japan's climatic characteristic.

Table 4.38 Design factors of the facilities configuring the landfill site

	Topography	Geology	Precipitation	Underground water	Landfill area	Landfill capacity	Material of landfill waste	Note
Construction	○	○	○	○	○	○		
Carry-in road	○	○						
Carry-in management facility	○	○				○		
Seepage control work	○	○		○	○		○	Protective work · Groundwater treatment
Collecting plant			○		○		○	
Landfill gas					○	○	○	
Storage structure	○	○	○	○		○		
Leachate treatment	○		○	○	○		○	Regulating plant for leachate
Rainwater drainage			○		○			
Disaster prevention facility	○	○	○		○			Flood regulating reservoir
Scattering prevention	○						○	
Monitoring facility			○	○	○	○	○	

4.6.1 Concept of the Leachate Treatment System

The semi-aerobic landfill is the center of the landfill technology. Japan builds facilities with the conscious of leachate management most since it is a country with high precipitation. The leachate does not retain in the disposal area, and the air is supply into the landfill layer through the water collecting pipe. For this purpose, leachate collection and drainage facility, landfill gas removal facility, storage structure, leachate adjustment facility, leachate treatment facility, and seepage control work must function organically. The relationship between each facility is shown in Fig. 4.86.

The capacity of the leachate treatment facility and the capacity of leachate adjustment facility are determined by calculating the water balance using precipitation data corresponding to the past maximum precipitation equivalent to the landfill period. On the other hand, leachate is removed from the landfill layer quickly to make the water level in the landfill layer to maintain the semi-aerobic landfill. It maintains a condition where water pressure is not applied towards the seepage control work in the bottom. Besides, with the underground water draining from the collection facility for underground water, the influence of water level raising on seepage control work is avoided.

However, due to the Guerrilla rainstorms and abnormal climatic condition in recent years, the precipitation exceeds the annual precipitation set before. In this instance, the leachate that exceeds the capacity of the leachate treatment facility and the adjustment facility is generated, and the untreated leachate may be discharged to the public water area. Therefore, it is necessary to store the leachate, which exceeds the capacity in the landfill layer temporarily to prevent the outflow. It also needs to make sure of the height of the storage structure.

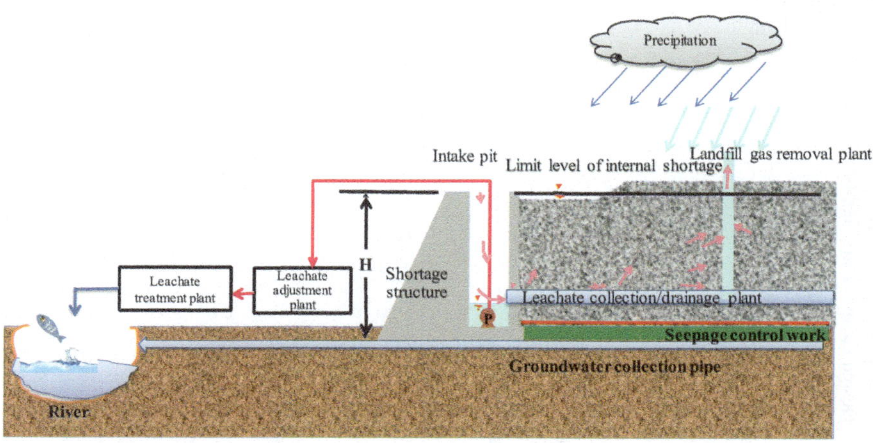

Fig. 4.86 Conceptual diagram of leachate management system

All the facilities are closely related to each other in the series of the collection, storage, and treatment for leachate. In other words, the "leachate management system" is composed of all the facilities. And consequently, it is necessary to take care of the function of the "leachate management system" properly when we plan, design, construct, and manage the MSW landfill site.

4.6.2 Leachate Collection/Drainage Facility

The leachate collection/drainage facility is one of the most critical facilities to calibrates the MSW landfill sites, and it maintains the semi-aerobic landfill by draining the leachate promptly and supplying air to the landfill layer from the outlet of the water collecting pipe. Since its functions, etc. are described in Sect. 2.6 "Semi-aerobic landfill," we will omit them in this section.

Perforated synthetic resin pipes and perforated fume pipes are used for the collection/drainage facility, and pipe diameter is preferably above 200 mm in consideration of air supply. The pipe diameter of the main line is more than 400 mm as usual, and the pipe diameter of the branched line is more than 200 mm as usual (Japan Waste Management Association 2010).

The branch lines are set at intervals of 10–20 m towards the main line (Japan Waste Management Association 2010).

In recent years, the proportion of incineration residues in landfill waste increases. Because of the unreacted slaked lime, etc. in incineration residues, the high pH and the consolidation leads the obstructions of ventilation and water flow in the landfill layer. With the utilization of the landfill gas removal facility as the vertical collection/drainage facility and the ventilation facility, the value of pH is reduced, and the water remains are avoided by preventing the perched water in landfill layer of the leachate. The role of the vertical collection/drainage facility becomes more critical in the landfill sites, which is mainly composed of incineration residues.

It is recommended to set more than one vertical collection and drainage facility at 2,000 m^2 (Japan Waste Management Association 2010). Recently, in order to complete functions of the leachate collection/drainage facility, crushed stones are used in the protective soil for seepage control work, and the number of landfill sites which the collecting function is enhanced is increased. The example is shown in Fig. 4.87.

4.6.3 Landfill Gas Removal Facility

The purpose of the landfill gas removal facility is to remove methane gas and hydrogen sulfide gas generated by the decomposition of organic substance. At MSW landfill sites, the content of organic substance becomes less, and the generation volume of the gas decreases among the landfill waste by the sophistication of the

Fig. 4.87 Example of a full-capacity collection layer consists of vertical collection facility and single grain crushed stone

intermediate processing facility. Although a similar tendency can be seen at landfill site of industrial waste, the methane gas and hydrogen sulfide gas are generated, and the necessity of landfill gas removal facility is also seen.

In order to make the landfill gas removal facility will not hinder the landfill operation, referring to Fig. 4.88, the gas removal facility is mainly the slope installation type, and the vertical type which connects vertical landfill gas removal facility with the main line or branch line of water collection/drainage facility is also be used. Besides, the vertical collection/drainage facility is combined used at many landfill sites as a countermeasure against the high density of landfill layers and worse water-permeable waste such as incineration residues, etc. Referring to Fig. 4.87, polyethylene pipes with a diameter of 75 to 600 mm and vinyl chloride pipes rolling with a nonwoven fabric or crushed stone are widely used. The interval in the case of slope type is set as 10–30 m, and it is 10–75 m in vertical type (Japan Waste Management Association 2010)

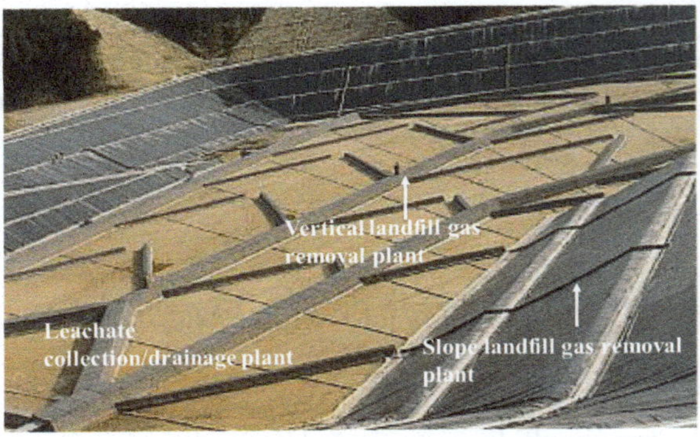

Fig. 4.88 Landfill gas removal facility

4.6.4 The Capacity Decision of the Facility According to the Leachate Management System

Figure 4.89 shows the flow sheet of the capacity decision of the facility according to the leachate management system (the capacity of the leachate treatment facility, the capacity of leachate adjustment facility, the height of storage structure).

① Landfill plan

In the case of determining the facility capacity, according to the leachate management system, it is necessary to pay attention to the landfill plan (landfill area, numbers of landfill region, landfill order, landfill method, landfill capacity, types of target landfill waste,etc.). A case study is performed, and we will explain the results in the following. This study is carried out at an open type MSW landfill site in a city with a population of 100,000 people. The landfill area is 9,000 $m^{2,}$ and the landfill capacity is 90,000 m^3. The landfill period is 15 years. There are two landfill regions which are 4500 m^2 for each. Referring to Fig. 4.90, The landfill order is from the downstream block to the upstream block. The landfill is completed in the downstream block, and it is under landfilling in the upstream block.

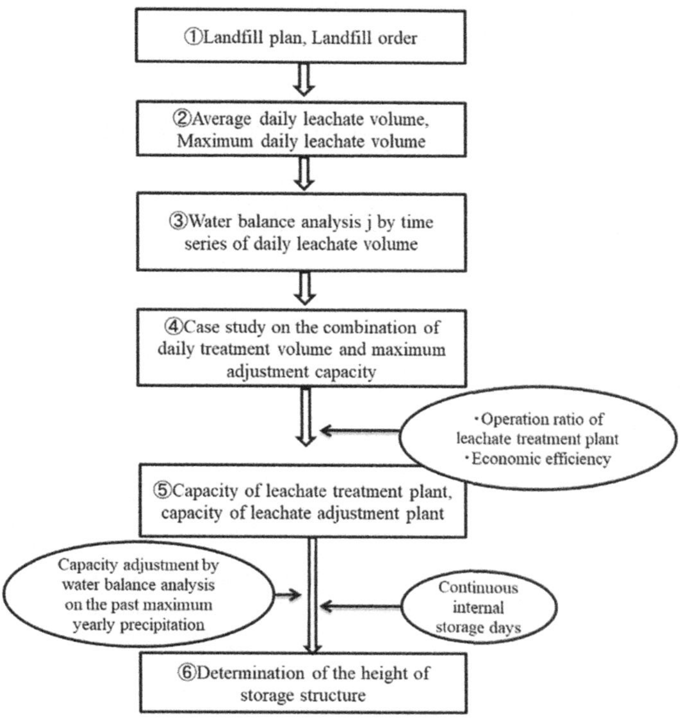

Fig. 4.89 The flow of capacity decision of the facility according to the leachate management

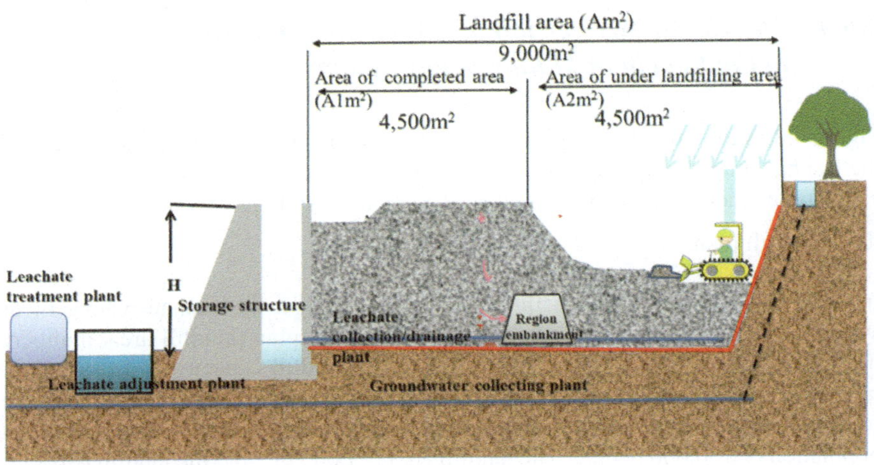

Fig. 4.90 Example of landfill plan

- Average daily leachate volume, Maximum daily leachate volume

The leachate volume is calculated by the rational formula. Based on the data in Table 4.39, the average daily value of the precipitation over the last 15 years is used as the average daily leachate volume.

The conversion value of the monthly maximum precipitation over the previous 15 years is set as the maximum daily leachate volume. The precipitation of MSW

Table 4.39 Leachate volume calculated by the rational formula

$Q = 1/1000 \cdot (C1 \cdot A1 + C2 \cdot A2) \cdot I$

Q : Leachate volume 〔m^3/day〕

A1 : Area of under landfilling area 〔m^2〕

A2 : Area of the region which directly removing surface water in the already landfilled region 〔m^2〕

C1 : Seepage coefficient of A1 region (Precipitation-Evaporation)

C2 : Seepage coefficient of A2 region ($C2 = C1 \times 0.6$)

I : Daily precipitation 〔mm/day〕

When calculating the average volume of leachate, the average daily precipitation 〔mm/day〕 is used by I.

When calculating the maximum volume of leachate, daily conversion value of maximum monthly precipitation is used by I.

As a rule, the data of precipitation is to investigate the past precipitation data over the same period as the landfill period. For instance, when the landfill period is 15 years, the precipitation over at least the last 15 years is investigated.

Besides, when the landfill period is less than 10 years, it is desirable to investigate the data over the last 15 years.

landfill site over 15 years is shown in Table 4.40. The average annual precipitation is 1,702 mm(4.7 mm/day). The monthly maximum precipitation occurred in July 2009, and it is 620 mm, and the daily conversion value is 620/31 = 20 mm/day. C1 is the region where is under landfilling and C2 is the region where the landfill is completed. The seepage coefficient is 0.63 in C1, and it is calculated from C1 = Precipitation-Evaporation amount, it. And the seepage coefficient is 0.38 in C2, it is calculated from C2 = C1 × 0.6. Based on the Blaney Criddle method (referring to Sect. 2.5 "Sea surface landfill site"), the monthly evaporation amount is obtained from the average value over 15 years and the average value over 15 years is taken as the seepage coefficient.

The calculation results are shown in Table 4.41.

③ Water balance analysis by time series of daily leachate volume

The capacity of the leachate treatment facility is set within the range of the average leachate volume and maximum leachate volume, and the water balance analysis is conducted by using the time series of the daily precipitation. The calculation flow sheet is shown in Fig. 4.91. The used time series of daily precipitation uses the year which the precipitation is the maximum over 15 years, the year which the precipitation is equal to average precipitation, the year which the precipitation amount is the maximum and the year which monthly precipitation is the maximum to find the maximum leachate adjustment capacity of each. Using the time series of precipitation amount, which is equivalent to each precipitation, the daily leachate volume is calculated and stored in the leachate adjustment facility.

The seepage coefficient used for calculating daily leachate volume provides monthly. The volume of leachate, which is the capacity of the leachate treatment facility set from the leachate adjustment facility, is extracted and sent to the treatment facility. The volume in the leachate adjustment facility fluctuates and the maximum adjustment capacity throughout the year is set as the maximum leachate adjust capacity. When the adjustment capacity becomes lower than the capacity of the leachate treatment facility, the process of extraction will be stopped. Water supply is carried out after the recovery of the adjustment capacity. The ratio of the number of days which water is sent and 365 days per year is set as the operating ratio.

④ Case study

The capacity of the management facility is changed within the average leachate volume, which is 20 m^3/day, and the maximum leachate volume which 90 m^3/day and the maximum leachate adjustment volume is calculated. Table 4.42 shows the results.

⑤ The capacity of the leachate treatment facility, Capacity of leachate adjustment facility

In Table 4.42, 10 cases of the capacity of the treatment facility are set within the range of maximum leachate volume from the average leachate volume, the annual precipitation and the maximum adjustment capacity are obtained.

Table 4.40 The monthly precipitation in last 15 years

	January	February	March	April	May	June	July	August	September	October	November	December	Annual precipitation
2001	134.5	94	84.5	53.5	117.5	490.5	311	78	211	148	145	75	1,942.5
2002	95.5	40.5	125.5	102	255.5	103.5	109.5	108.5	186	109.5	37	98.5	1,371.5
2003	50	74.5	103	144.5	110.5	200.5	444.5	232	69.5	11	124.5	36	1,600.5
2004	55	66	97	68.5	268.5	225	53	154.5	304.5	256.5	58.5	134.5	1,741.5
2005	47	91	93	36	62.5	15	301	73.5	115	18.5	109	58.5	1,020.0
2006	46	93	102	199.5	183.5	320	405.5	256	236.5	28	107	41	2,018.0
2007	47.5	61.5	81.5	74.5	64.5	40.5	381	178.5	72	88	13	92.5	1,195.0
2008	70	52	162	120	163.5	386	48	419	174	16.5	91	78.5	1,780.5
2009	62.5	86	46	99.5	45.5	219	619.5	94	83.5	105	181.5	50	1,692.0
2010	50.5	70	155	199.5	105	203	453.5	69.5	138.5	78.5	58	148	1,729.0
2011	106	48	60	37.5	277.5	409	172	322	85.5	127	166.5	38	1,849.0
2012	29.5	155	125	74.5	43.5	288.5	464	188.5	129	47.5	125	98.5	1,768.5
2013	57.5	81.5	56.5	108	37	268.5	134	501.5	133	227.5	119.5	77	1,801.5
2014	57.5	83	102.5	61	4	101	373	462.5	107	144.5	106	73.5	1,675.5
2015	83.5	42.5	94.5	239.5	120.5	222.5	266	319.5	146	83	140.5	109.5	1,867.5
2016	92	98.5	77.5	196	191.5	394.5	179.5	128	608.5	181	132.5	141	2,420.5
												Average	1,702.0
												Maximum	2,420.5

Table 4.41 Calculation result of average and maximum leachate volume

Daily average leachate volume

$$Q = 1/1000 \cdot (C1 \cdot A1 + C2 \cdot A2) \cdot I$$

$$= 1/1{,}000(0.63 \times 4{,}500\,m^2 + 0.38 \times 4{,}500\,m^2) \times 4.7\,mm/day$$

$$= 20\,m^2/day$$

Maximum leachate volume

$$Q = 1/1000 \cdot (C1 \cdot A1 + C2 \cdot A2) \cdot I$$

$$= 1/1{,}000(0.63 \times 4{,}500\,m^2 + 0.38 \times 4{,}500\,m^2) \times 20\,mm/day$$

$$= 90\,m^3/day$$

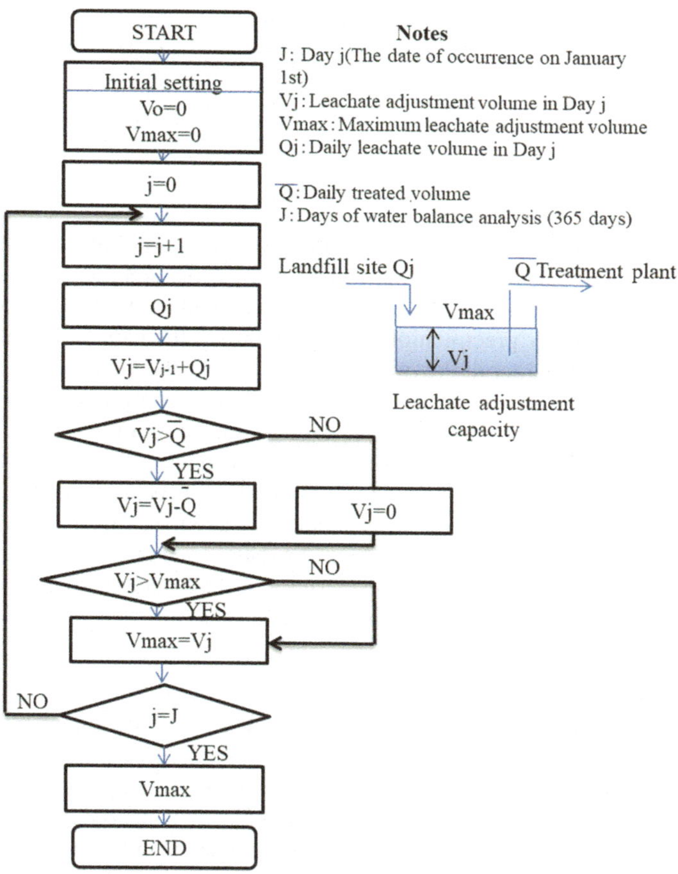

Fig. 4.91 The flow of water balance analysis

Table 4.42 Case study on water balance analysis

Case	The treated volume (m^3/day)	Adjustment capacity (m^3)				The average annual operation ratio (%)
		The year which the precipitation is the maximum over 15 years (2016) 2,420 mm	The year which the precipitation amount is the maximum (1980) 2,977 mm	The year which the precipitation is equal to average precipitation (2010) 1,729 mm	The year which monthly precipitation is the maximum (2009) 1,692 mm. 619 mm/month	
1	20	4,396	4,468	2,284	2,884	89
2	25	3,474	4,033	2,129	2,704	80
3	30	2,823	3,598	2,015	2,525	65
4	35	2,263	3,209	1,910	2,350	55
5	**40**	**1,933**	**2,844**	**1,805**	**2,175**	**46**
6	50	1,566	2,167	1,600	1,827	33
7	60	1,314	1,547	1,429	1,663	27
8	70	1,124	1,122	1,379	1,633	22
9	80	964	1,012	1,329	1,603	17
10	90	924	972	1,279	1,573	14

Based on the Revision of Guideline for Plan, Design, and Management of MSW Landfill Sites, the leachate adjustment facility uses the time series of the latest year which the precipitation is the maximum equivalent to the landfill period and the year which monthly precipitation is the maximum, the capacity of adjustment volume should be the larger of the two and will not cause the internal storage (Japan Waste Management Association 2010). Besides, considering the operation period, the economics of the leachate treatment facility, and the regional conditions, the capacity of the facility determines an appropriate planned inflow water volume of the leachate treatment facility (Japan Waste Management Association 2010). Therefore, the average operation ratio is set as about 50%, and the capacity of the leachate treatment facility is 40 m^3/day. Comparing the data in the maximum year, and the data in the monthly maximum year, the capacity of the leachate adjustment facility is set as 2,175 m^3 of the monthly maximum year.

⑥ Determination of the height of storage structure

The capacity of leachate treatment facilities and leachate adjustment facilities are determined on the condition without internal store at landfill sites in the year, which has maximum precipitation over 15 years. However, as the abnormal weather such as guerrilla rainstorm frequently occurs in recent year, it becomes more possible to exceed the set annual precipitation amount, and we should make preparations for it. As the countermeasures, it is common to store the leachate which exceeds the treatment capacity or adjustment capacity temporarily in the landfill layer and prevent the overflow of untreated leachate from storage structures. Therefore, it is necessary to secure the height of the storage structure to retain the untreated leachate. Figure 4.92 shows the association diagram.

From Fig. 4.92, the required internal storage volume (Sxi) is obtained by the difference between the leachate adjustment volume (Vi) and the necessary capacity from regulating reservoir (Vxi). Vxi in here is the adjustment capacity at the monthly maximum annual precipitation, and it is 2,175 m^3. Vi is the adjustment volume at the maximum precipitation over the past years, and it is 2,844 m^3 (referring to Table 4.42). And the required internal storage volume is

$$\begin{aligned} SXi &= Vi - VXi \\ &= 2,844\,m^3 - 2,175\,m^3 \\ &= 669 m^3 \end{aligned}$$

And based on Fig. 4.93, the possible internal storage capacity is

$$S = S_1 + S_2$$

S_1: Space volume below H line which is the height of storage structure (m^3)

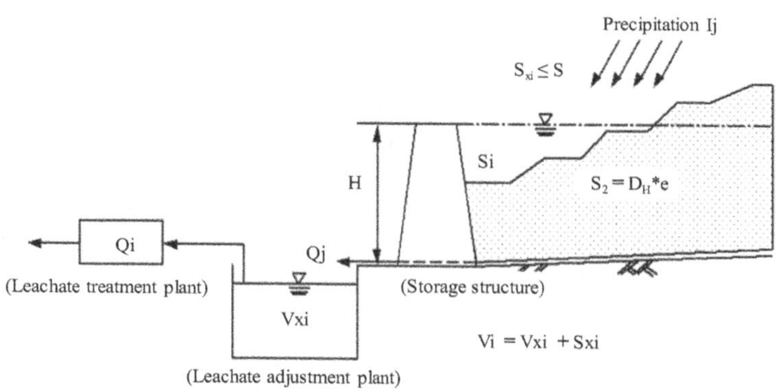

Leachate adjustment volume (Vi) = Required capacity of regulating reservoir (Vxi) + Required internal storage volume (Sxi)
$S = S1 + S2 = S1 + H\ (D_H \times e)$
S : Possible internal storage capacity
S1 : Space volume below H line which is the height of storage structure
S2 : Storage volume in the waste layer below H line which is the height of storage structure

Fig. 4.92 Association diagram of facility (Higuchi S et al. 1996)

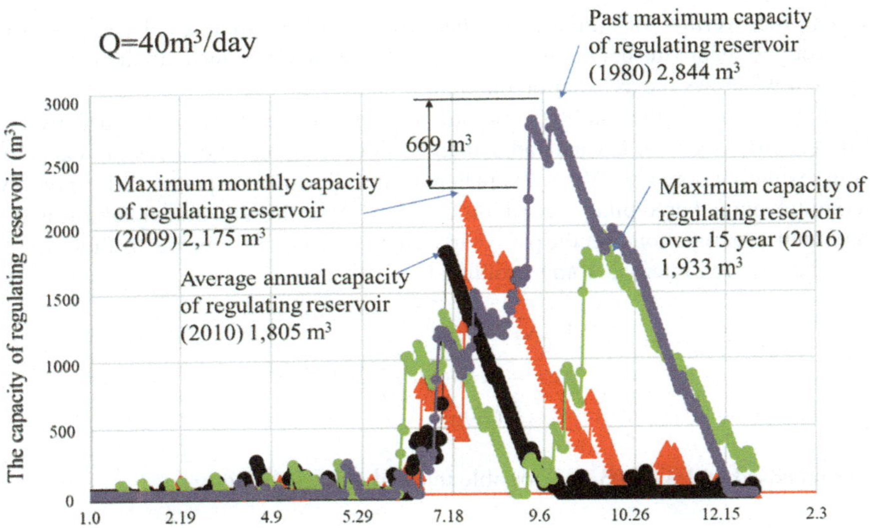

Fig. 4.93 The fluctuation diagram of the capacity of regulating reservoir of leachate according to the capacity of annual precipitation during $Q = 40\ m^3/day$

S_2:Volume of leachate which is possible to store in the waste space in the storage structure (m^3)=D_H (Waste volume in the storage structure m^3) × e (internal storage possible porosity ratio %)

e: The space ratio of the soil or waste layer is about 30%, it is already filled with precipitation during the internal storage, and the available space ratio of internal storage, e to the unsaturated zone becomes 5–10% (Higuchi n.d.).

$S_1 = 0$ and the height of the storage structure are set as 5 m, the capacity of waste up to the top of the storage structure is 5,000 m^3. When the available space ratio of internal storage is assumed as 7.5%, the available capacity of internal storage is S2 $= 5{,}000\ m^3 \times 0.075 = 375\ m^3$. In other words, $S = S_1 + S_2 = 375\ m^3$. The required internal storage amount, S_{xi}, is 669 m^3, it is bigger than S. 669 m^3–375 m^3 = 294 m^3. Therefore, the overflow occurred in the past maximum precipitation year.

When the height of the storage structure is 8 m, and the capacity of waste up to the top of the storage structure is 10,000 m^3, the available capacity of internal storage, S, becomes 750 m^3. It is possible to avoid the overflow even in the past maximum precipitation year. ($S_{xi} < S$)

Thus, the setting condition is as follows.

The capacity of the leachate treatment facility is 40 m^3/day

The capacity of the leachate adjustment facility is 2,175 m^3

The height of the storage structure is 8 m

Figure 4.93 shows the fluctuation diagram of the capacity of regulating reservoir of leachate according to the capacity of annual precipitation during $Q = 40\ m^3/day$.

Water balance is composed of the height of storage structure, the capacity of the leachate treatment facility, and the capacity of the leachate adjustment facility. Both the leachate collection/drainage facility which can remove leachate quickly and the landfill gas removal facility plays a significant role in maintaining the semi-aerobic landfill and conducting the early stabilization of landfill waste. In addition, internal storage to keep the water balance applies hydraulic pressure to seepage control work. With the increase in the risk of water leakage, the inflow of air into the landfill layer is cut, and the anaerobic has occurred. And they become the factor causing the deterioration of water quality and gas generation. Therefore, it is necessary to limit the number of continuous internal storage days, especially in summer, when the water temperature rises. The major facilities of MSW landfill site are closely linked to each other, and they become constituent elements of the leachate management system.

4.6.5 *Quality Prediction on Leachate*

It is no exaggeration to say that the accurate prediction of leachate quality is not possible at the present time. The reason is that the leachate quality differs depending on the landfill waste materials, landfill method, and weather conditions such as precipitation. In particular, the landfill waste materials vary depending on the treatment method and the treatment level of intermediate processing. For example, in the case of the incineration facility, depending on the different treatment methods of exhaust gas and based on the different effluent standards, the used amount of medicine, the types of medicine and the methods of stabilizing fly ash are different. In regards to the resource recovery facility, it varies according to the sorting methods during the time of collection, the crushing methods and the selection accuracy, etc. Therefore, both the landfill waste materials of each final landfill site and the leachate quality are different. In the Revision of Guideline for Plan, Design, and Management of MSW Landfill Sites, the setting of leachate quality is according to "The inflow water quality differs depending on the landfill waste and landfill methods. It is difficult to set the planned inflow water quality uniquely. It is usually to make an investigation on the water quality of the other MSW landfill site where the landfill waste is similar. The decision is made with the consideration of the difference of landfill structure, landfill operation, intake area, etc. between it and the target planned landfill site (Japan Waste Management Association. 2010)." The amount of planned inflow water quality of the landfill waste which is mainly incineration residue and noncombustible waste is shown in Table 4.43.

In Japan, with the improvement of the intermediate processing facility in recent year, the percentage of incineration residues in the landfill waste is increasing, and the leachate shows a high salinity. For this reason, the number of cases which process Ca^{2+} and Cl^- is increasing. However, the range of the standard of water quality in Table 4.43 is large, and there isn't any rational method for setting at the present time. In this part, we will introduce an example of Cl^- prediction method proposed by the authors.

As stated previously, there isn't any unscientific thing for the setting of the raw water quality in the leachate treatment plan. It is difficult to set the Cl^- concentration,

Table 4.43 Amount of planned inflow water quality (when the landfill waste is incineration residue and noncombustible waste) (Japan Waste Management Association 2010)

Item	Unit	Standard of water quality
BOD	mg/l	50–250
SS	mg/l	100–200
COD	mg/l	50–200
T-N	mg/l	50–100
Ca^{2+}	mg/l	500–3,000
Cl^-	mg/l	2,000–20,000

most of the cases are set it based on the precedent of other cities. The factors that affect water quality are described in Fig. 2.11 in Sect. 2 Issues of high concentration inorganic salts in Chap. 2. It fluctuates depending on the Cl content in fly ash or bottom ash which changes with the concentration of exhaust gas treatment from the incineration facility. Besides, as the Cl^- in leachate does not react with other substances in the landfill layer, it leaches out by advection and diffusion. Therefore, it also fluctuates mostly depending on the factors of landfill operation such as the thickness of the landfill layer, the dilution area in the landfill site, and the landfill order, etc.

4.6.5.1 The Influence of Cl Content in Incineration Residues on Peak Concentration of Cl^- in Leachate

As Cl easily dissolves in water, the higher the content is, the higher the concentration becomes. A simulated landfill with a diameter of 30 cm, a sufficient height of 1.0 m is filled with incineration residues, and then an artificial water sprinkling is conducted. Figure 4.94 shows the relationship between the initial peak concentration of leachate and the Cl content in the incineration residue in that condition. Because the incineration residue filled in the simulated landfill is discharged from different

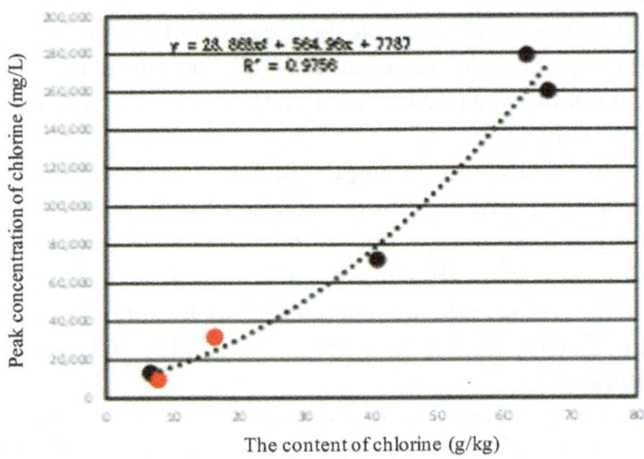

Fig. 4.94 The content of Cl and the peak concentration of Cl (Tameda K et al 2010)

incineration facilities, the concentration of HCl gas in the exhaust gas is 50 ppm. Besides, two points with a low peak concentration are only filled with bottom ash. Based on these results, we can indicate that the higher the content of Cl is, the higher the peak concentration of Cl^- becomes.

4.6.5.2 The Landfill Height of One Layer and the Peak Concentration

Cl cannot be decomposed during the infiltration process and hardly reacts with other substances, so it moves from the upper to the bottom while dissolving in water, that is the advection. Therefore, the higher the landfill height, the more amount dissolves with the leachate, the higher the concentration is. On the other hand, the lower the landfill height is, the lower the Cl^- concentration becomes.

Incineration residues are filled into the simulated landfill layer with a diameter of 20 cm which is shown in Fig. 4.95, the landfill height is 50 cm (run 1), 1.0 m (run 2) and 2.0 m (run 3). Then, artificial water sprinkling is performed. The aging of Cl- concentration in leachate is shown in Fig. 4.96. We can find the higher the landfill height is, the higher the peak concentration of Cl- in the leachate tends to be.

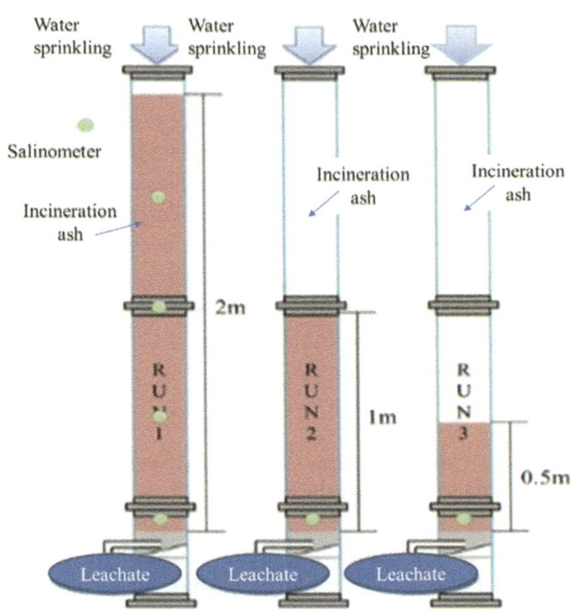

Fig. 4.95 Simulated landfill layer with different thickness (Yahata A and Higuchi S 2012)

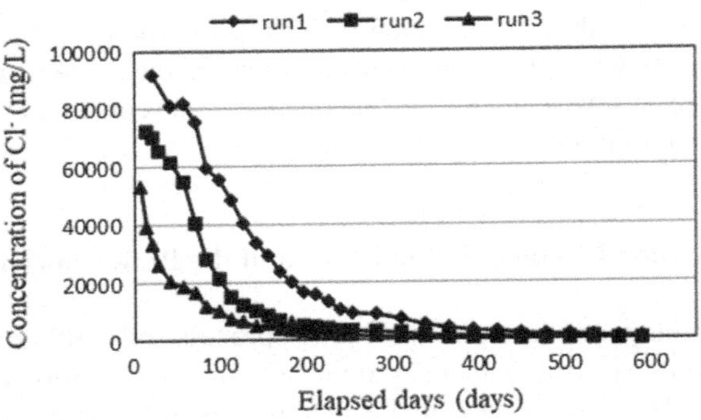

Fig. 4.96 Difference in the thickness of layer and the concentration of Cl-in leachate (Yahata A and Higuchi S 2012)

4.6.5.3 Concentration-Inhibition by Landfill by Stacking up

The Cl in the previously landfill layer is washed out as leachate, and its content decreases. When a newly waste layer with a high content of Cl is landfilled above the landfill layer where the Cl content is reduced, the lower layer becomes a filter layer. Because of the diffusion mixing of Cl^- and the delay of leaching, it is known that the value will never exceed the peak concentration in the landfill of the lower layer (Higuchi 1996). When the landfill by stacking up is conducted in the simulated tank which is shown in Fig. 4.95, the aging of Cl- concentration is shown in Fig. 4.97. There are four layers which the thickness is 50 cm, two layers which the thickness is 1 m, and 1 layer which the thickness is 2 m. Comparing with the landfill in these three conditions, the landfill by stacking up should be conducted at the timing of the Cl^- concentration in the leachate becomes lower than 1,000 mg/l. Based on the data, it will not exceed the peak concentration at the initial stage in adding either the thickness of the layer, and it indicates that the concentration can be inhibited. It is possible to inhibit the concentration of Cl^- by incorporating the landfill by stacking up to the landfill operation.

4.6.5.4 Concentration Control by Securing the Dilution Zone in the Landfill Site

The Cl^- concentration in leachate from the reclaimed part shows a high concentration, it is 10^5 mg/l at a peak period, while the precipitation in un-reclaimed part becomes dilution water, and the Cl- concentration in leachate is reduced. For this reason, the Cl- concentration in leachate changes according to whether the dilution area is existing or not.

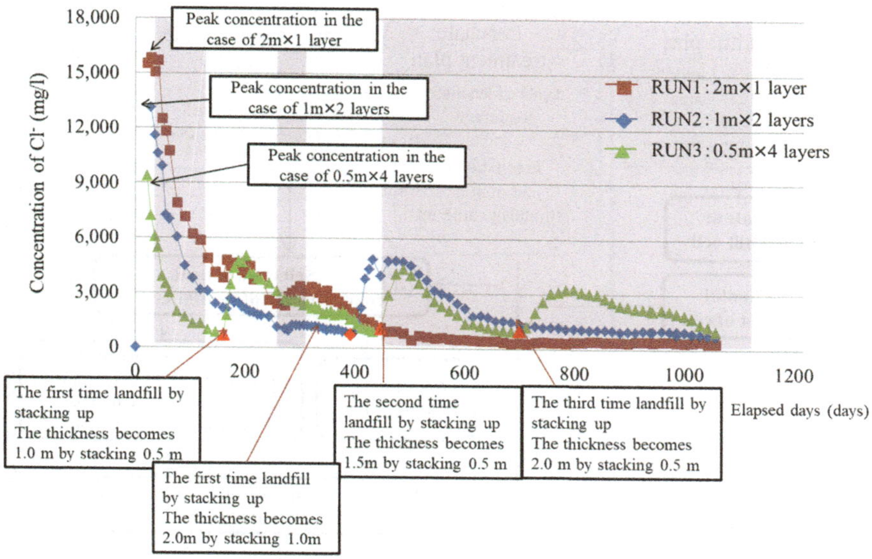

Fig. 4.97 The effective of inhibiting the concentration of Cl-by landfill by stacking up (Yahata A and Higuchi S 2012)

4.6.5.5 Concentration Control of Cl^- by Landfill Order

It is possible to make the concentration control of Cl- in leachate by methods for landfill operation, such as landfill region, the area of a landfill cell, landfill direction, landfill height of one layer, numbers of adding layers, etc.

4.6.5.6 Setting Method of Cl^- Concentration

Through the elution characteristics of Cl^- in leachate, which was mentioned above, Fig. 4.98 shows the setting flow of the Cl^- concentration which using the landfill cell model. The explanation of each process is as follows.

① Plan for landfill

Based on the plan for landfill, the landfill cell is set as one unit, and a model of the landfill site, which is assumed the landfill order is set.

The landfill cell model is shown in Fig. 4.99. The cell increases in both the horizontal direction and the vertical direction during the progress of landfill. The positions of the cell are expressed as Hi to Hn in the X direction, Vi to Vn in the Y direction, and Di to Dn in the Z direction. For example, the cell-A is expressed as (Hα, Vβ, Dγ).

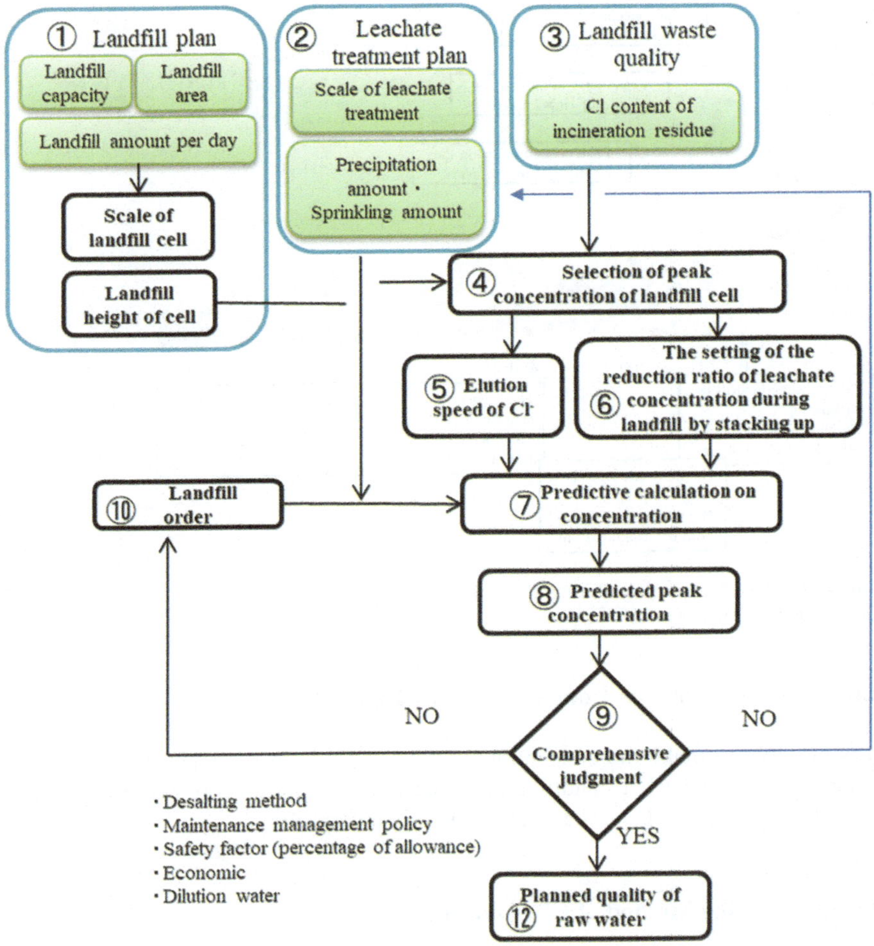

Fig. 4.98 The setting flow of the Cl-concentration by using the landfill cell mode (Tameda et al. 2017) revision

A. The calculation of landfill capacity per day

The landfill capacity per day is calculated from the landfill capacity, the landfill area, the landfill period, all of them are based on the MSW Disposal Plan.

B. The setting of the cell scale

The daily landfill capacity of the target landfill site is calculated and the capacity and size (length × width × height) of the Landfill Cell Model are set from the landfill capacity per day. The capacity per cell is about 10 to 30 days, and the landfill height of a cell is about 1.0 to 2.0 m.

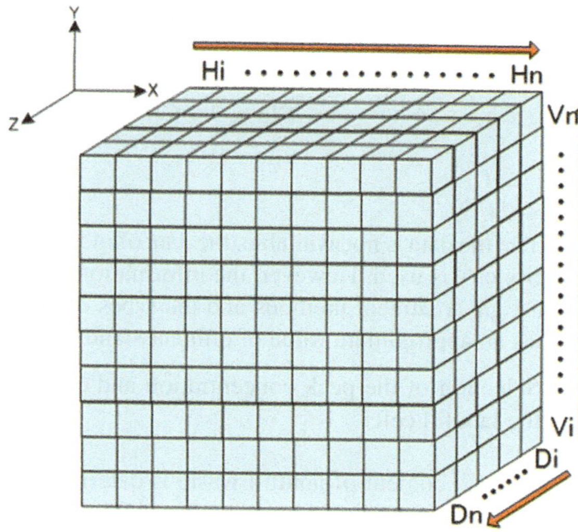

Fig. 4.99 Landfill cell model (Tameda and Higuchi 2016; Tameda et al. 2017)

C. The decision on the shape of the landfill sites

The number of cells until the end of the landfill is set according to the landfill area, the disposal capacity, and the cell capacity, and it is divided into the number of cells in the horizontal direction and the number of cells in the vertical direction. In other words, the number of cells in the X direction (Hi to Hn), the Y direction (Vi to Vn) and the Z direction (Di to Dn) of a cell model of landfill sites and the shapes of landfill sites are determined.

① Plan for the leachate treatment

A. The capacity of the leachate treatment facility

Based on the plan for leachate treatment, the capacity of the leachate treatment facility is set. The setting method is referring to Chap. 5. (4) The capacity determination of facility according to the leachate management system.

B. The setting of daily precipitation and water sprinkling volume

The daily precipitation at open type landfill site and the daily water sprinkling volume at closed type landfill site are set. In the case of open type landfill site, the precipitation used for calculating the capacity of the leachate treatment facility is taken as the standard.

A. The material of landfill waste (Cl content of landfill waste)

Cl content tests for each landfill waste are conducted. The Cl content is calculated from the target waste and its landfill ratio. For instance, when the landfill waste is bottom ash, fly ash, shredded noncombustible residue, and their landfill ratio is 7:2:1, their Cl content is Cl_b, Cl_f, Cl_n. The Cl content of landfill waste, $C1_l$ is represented by the following formula:

$$Clb \times 7 + Clf \times 2 + Cln \times 1$$

$Cl_l = Cl \bullet$
$7 + 2{+}1$
When the data is not available, the data of incineration facilities in another similar-capacity city is used. However, the information is limited to the facilities where the exhaust gas treatment methods and the types of desalinization agents are the same and has an approximate value of effluent standard of hydrogen chloride gas.

④ Selection of the peak concentration and concentration attenuation formula of the landfill cell

When the Cl content of landfill waste is determined, the peak concentration of Cl- in landfill cell is predicted by using Fig. 4.94.

⑤ Determination of Cl^- elution speed

Then, the concentration attenuation formula is set. Based on the daily changes of Cl^- concentration in leachate from the simulated tank filled with incineration residue with a Cl content of 67 g/kg, the prediction formulas are shown in Fig. 4.100. The landfill periods are divided into the early period ($Y = -3.31X + 114{,}971$), the middle period ($Y = -0.52X + 44{,}550$), the late period ($Y = -0.041X + 8{,}940$), and the stable period ($Y = 495$). Since the basic data of concentration attenuation formula is small, when the emission value of HCl gas treatment of the incineration facility is about 50 ppm, and the height of the landfill cell is 1.0 m, the formulas in Fig. 4.100 are used.

⑥, ⑦, ⑧ The calculation of leaching concentration of Cl- (the setting of decreasing rate) and concentration prediction during the adding process

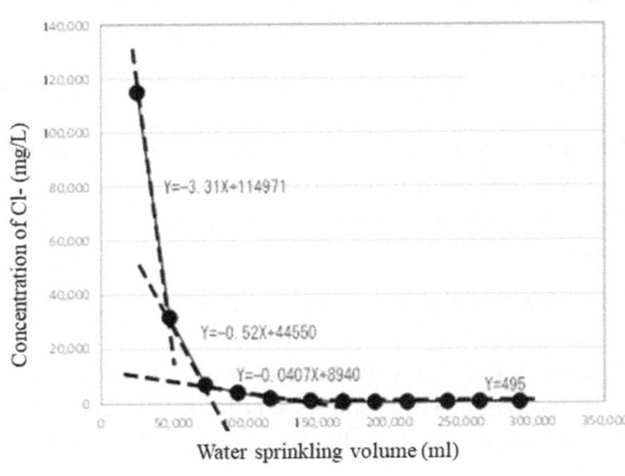

Fig. 4.100 Prediction formulas on peak concentration of landfill cell (Tameda K et al. 2017)

A. Migration properties of leachate concentration αi (mg/l) of Day ni cell is calculated by multiplying migration rates and αi (mg/l). When a migration rate for the case (Vm) in which Day ni cell is landfilled at the m-th cell from the basement is defined as M%, the rate of migration from Day ni cell to the cell immediately below it is 100%, and those to the second and subsequent cells below it are preset as M-1%,...$M_{3\,\%}$, $M_{2\,\%}$, $M_{1\,\%}$. (Figure 4.101)

B. Leachate concentration at the m-th cell from the basement on Day ni is determined according to the location (Vm) of Day ni cell in the y-direction.

The location (Vm) of Day ni cell in the y-direction gives the following calculations:

Leachate concentration of all Day ni cells at the basement of the landfill site, V1, is expressed by the following equation: αi (mg/l) × 100%,

Migration of Day ni cell at Vm starts {(m−1) × n1}days after Day ni, that is, at Day Ni = ni + {(m−1) × n1}, and it ends at Day Ni + n1 × j (j = 1, 2, 3, 4...)

The leachate concentration of Day n_i cell on Day N_i is expressed by the following equation: α_i (mg/l) × M%.

Based on the leachate ratio M, the leachate concentration of Day n_i cell on Day N_i + ni is expressed by the following equation. Day n1 is the number of days required to landfill one cell, which was set in Step ①.

When $50 < M \leqq 100$ αi(mg/l) × (100-M)%

When $0 < M \leqq 50$ αi(mg/l) × M%

The calculation mentioned above is repeated and the total leachate ratio ΣM in Day ni is 100-ΣM<M. The leachate concentration of the cell in Day ni in the Day Ni + n_1 × j is αi(mg/l) × (100−ΣM)%.

C. The calculation described above in "B" is performed for all cells in the landfill area in this Landfill Cell Model and the total leachate concentration.

D. As shown in Figs. 4.102 and 4.103, the following is an example of the calculation described above of the case where four cell layers are stacked up:

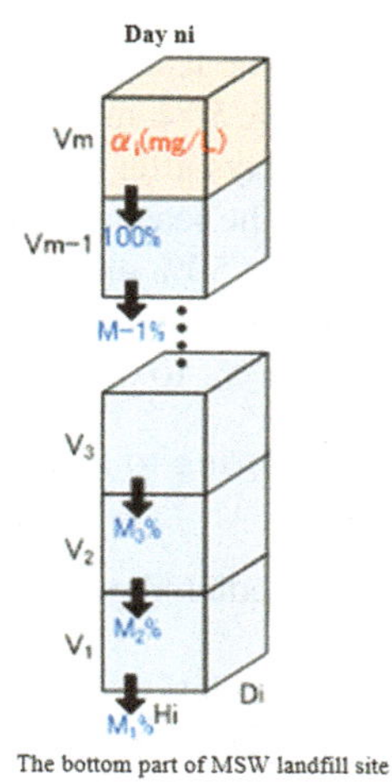

Fig. 4.101 Conceptual diagram of reduction ratio of Cl-during adding (Tameda K et al. 2017)

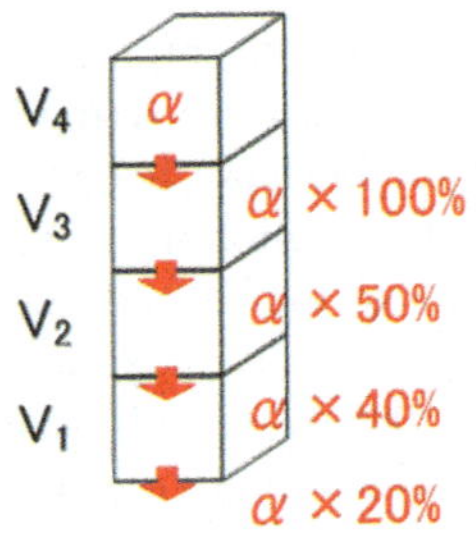

Fig. 4.102 Example of reduction ratio of Cl-when adding four layers (Tameda K et al. 2017)

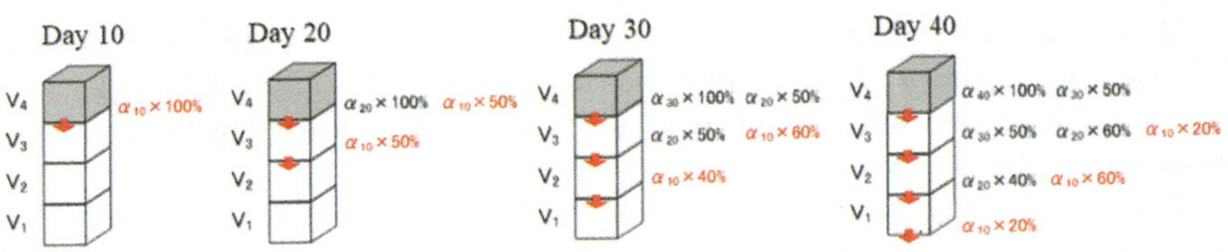

Fig. 4.103 Transition concept of Cl-leachate of landfill cell model (Tameda K et al. 2017)

- Landfill is advanced from V_1 to V_4.
- Only the migration property from V_4 is considered.
- The leachate concentration from V_4 is α_i mg/l (i represents the number of days; i = 10, 20, 30,…).
- The migration rates from V_4 in the y-direction: first layer, 100%; second layer, 50%; third layer, 40%; and fourth layer, 20%.
- 10 days after the landfill: The leachate quality $\alpha_{10} \times 100$ migrates from V_4 to V_3.
- 20 days after the landfill: The leachate quality $\alpha_{20} \times 100\%$ migrates from V_4 to V_3 and the leachate quality $\alpha_{10} \times 50\%$ migrates from V_3 to V_2.
- 30 days after the landfill: The leachate quality $\alpha_{30} \times 100\%$ migrates from V_4 to V_3, the leachate quality $\alpha_{10} \times 60\%$ migrates from V_3 to V_2, the leachate quality $\alpha_{20} \times 50\%$ migrates, the leachate quality $\alpha_{10} \times 40\%$ migrates from V_2 to V_1.
- 40 days after the landfill: The leachate quality $\alpha_{40} \times 100\%$ migrates from V_4 to V_3, the leachate quality $\alpha_{20} \times 50\%$ migrates from V_3 to V_2, the leachate quality $\alpha_{30} \times 50\%$ migrates, the leachate quality $\alpha_{10} \times 40\%$ migrates from V_2 to V_1, the leachate quality $\alpha_{20} \times 40\%$ migrates, and the leachate quality $\alpha_{10} \times 20\%$ migrates from V_1 to outside of the landfill site.

Referring to Fig. 4.104, the subsequent calculations are conducted in the same way.

⑧ Predicted peak concentration (Calculation examples)

The calculation of the predicted peak concentration is produced by the landfill cell model. The landfill capacity is 40,000 m^3, including the soil cover, the landfill area is

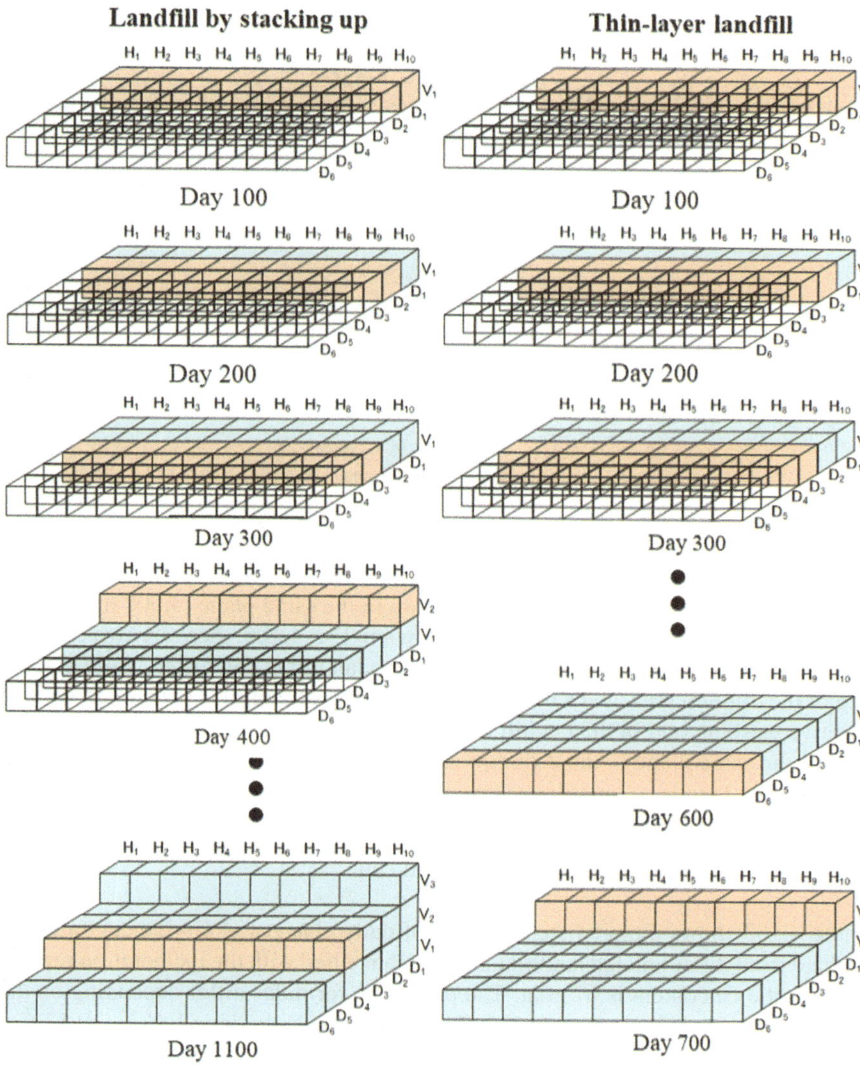

Fig. 4.104 Landfill methods (landfill by stacking up, thin-layer landfill) (Tameda K et al. 2017)

4,000 m^2, the height of landfill is 10 m on average and the landfill period is 15 years. In regard to the capacity of the leachate treatment facility, it is 30 m^3/day calculated from the meteorological data of City A in the past 15 years (from 2001 to 2015).

A. Plans for landfill and leachate treatment

The landfill period of one landfill cell is 10 days (landfill capacity: 40,000 m^3 $\div$ 15 years $\div$ 365 days $\times$ 10 days = about 73 m^3), and the total number of cells is

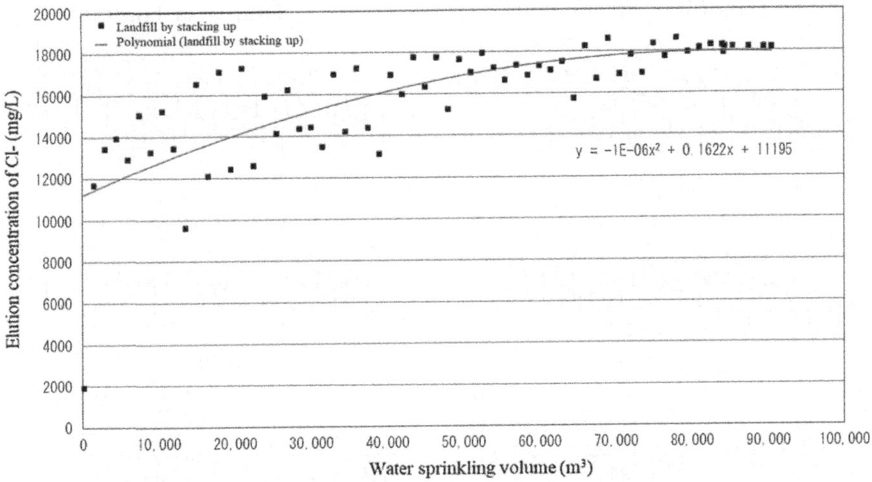

Fig. 4.105 Peak concentration of Cl-in leachate (landfill by stacking up)

548. The height of one cell is 1 m, 55 cells are in the same plane (4,000 m^2/73 m^2), and 10 steps in the vertical direction. As shown in Fig. 4.105, there are two cases on landfill methods, a landfill by stacking up which landfill is conducted while making sure of the dilution area, and a thin-layer landfill which landfill is performed with an effective washing out. In the landfill by stacking up, a landfill is carried out on three rows of the same plane in the Z direction and then restacking to the upper stage. In the thin-layer landfill, a landfill is carried out on the entire row of the same plane in the Z direction and after then restacking to the upper stage.

B. Peak concentration formula of a landfill cell

Figure 4.100 shows the predicted formula of daily changes of Cl- concentration in leachate from the simulated landfill tank, which is filled with the incineration residue in which the Cl content is 67 g/kg. The predictive formulas differ depending on the period of landfill. It is $Y = -3.31X + 114,971$ in the early period, $Y = -0.52X + 44,550$ in the middle period, $Y = -0.041X + 8,940$ in the late period and $Y = 495$ in the stable period of landfill.

C. Predictive calculation of the concentration

The peak concentration of Cl^- in leachate is set according to the conditions mentioned above. The results are shown in Figs. 4.105 and 4.106. The concentration seldom fluctuates in the case of the thin-layer landfill from the early period of the landfill, and a gradual increase can be seen. In the case of the landfill by stacking up, the concentration fluctuates intensely from the early period of the landfill and shows a sharp rise than the condition in the thin-layer landfill. It increases to about 17,000 mg/l

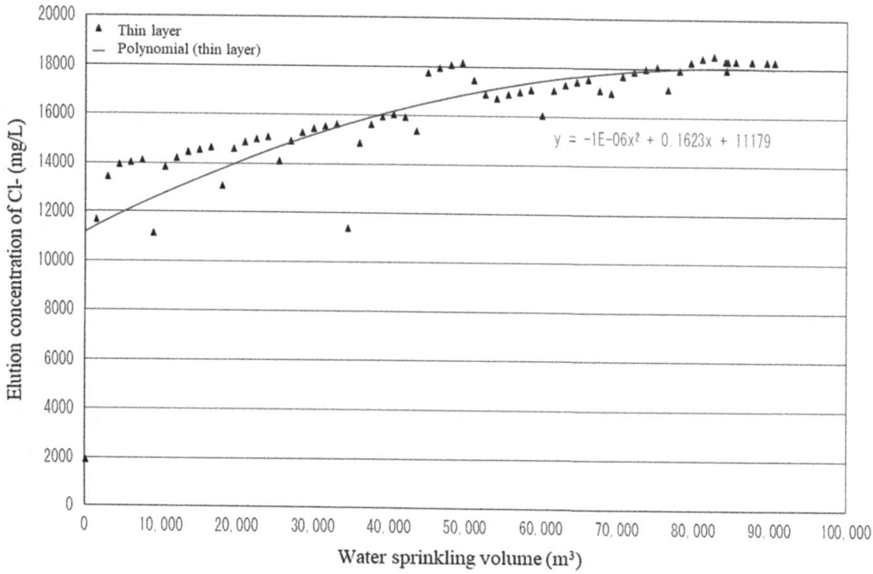

Fig. 4.106 Peak concentration of Cl-in leachate (thin-layerlandfill)

when the volume of water sprinkling is 15,000 m^3. In the late period of the landfill, the concentration fluctuation shows a coverage tendency and a gradual increase can be seen.

As a result, the predicted peak concentration of Cl- is around 18,000 mg/l for both the landfill by stacking up and the thin-layer landfill.

⑨ Comprehensive determination

Based on the result of concentration prediction, the peak concentration of Cl^- is about 18,000 mg/l in either the landfill by stacking up or thin-layer landfill. However, since the capacity of desalination treatment equipment of the leachate treatment facility is determined by the inflow load, it is safe to set the peak concentration as the raw water quality, but it is not economical. Therefore, it is necessary to set the raw water quality with the overall conscious of the following points.

A. Desalination treatment method

The desalination methods are the electrodialysis membrane method and the reverse osmosis membrane method. As the effects of raw water quality, the concentration of Cl^- and concentrated salt (concentration factor) are listed. Generally speaking, the concentration factor in the electrodialysis membrane method is higher than that in the reverse osmosis membrane method. Therefore, it is better to use the electrodialysis membrane method when the Cl^-concentration in raw water exceeds 3%. In this case, there is no problem by using either method.

B. Maintenance management policy

In this case, there is no difference in the peak concentration by either the landfill by stacking up or the thin-layer landfill, however, the load amount applied to the desalination treatment equipment varies depending on the period of the landfill in which the peak concentration appears. The period is either the prior period or the late period.

When the peak concentration appears in the prior period of the landfill, the leachate amount is small, so it is likely to be within the allowable load amount of the desalination treatment equipment. On the other hand, when the peak concentration appears in the late period of the landfill, the leachate amount is large, it may exceed the allowable load amount of the desalination treatment equipment. Therefore, a stable treatment is more possible performed when the peak concentration is induced in the prior period of the landfill. Thus, it is desirable to control the concentration according to the policy of the landfill operation and to set the concentration of raw water quality.

C. Economy efficiency

With the setting of the concentration of raw water quality becomes higher, it is possible to make a safe and reliable treatment. On the other hand, the construction cost becomes higher and causes deterioration in the economy. The period in which the load amount is maximum is short, therefore, it is necessary to set the raw water quality in consideration of these.

D. Safety

In regard to designing the facilities or the equipment, playing safety as countermeasures against unexpected issues. And it is desirable to estimate a margin of 10 to 30% for the set raw water.

⑩ Planned raw water quality

Considering the load amount of water quality, the setting flow of the planned raw water quality is shown in Fig. 4.107.

In the case of open type, first of all, setting the capacity of leachate treatment facility (Qop) from the water balance analysis which using the time series of daily precipitation (maximum daily precipitation). Then, the predicted peak

concentration (P) using the cell model is multiplied by the leachate amount on the day when the peak concentration is presented (Vop) to obtain the maximum load amount (Cmax). After that, Cmax is divided by the capacity of the leachate treatment facility (Qop) to obtain the planned raw water quality. In the case of closed type, the capacity of the leachate treatment facility (Qcp) is calculated from the landfill capacity and the liquid-solid ratio. The maximum load amount Cmax is calculated from the capacity of the leachate treatment facility (Qcp), the peak concentration (P), and the volume of water sprinkling on that day (Vcp) (the volume of water sprinkling Vcp $\times$ seepage coefficient 0.6–0.8), then, it is divided by the capacity of leachate treatment facility (Qcp) to obtain the planned raw water quality.

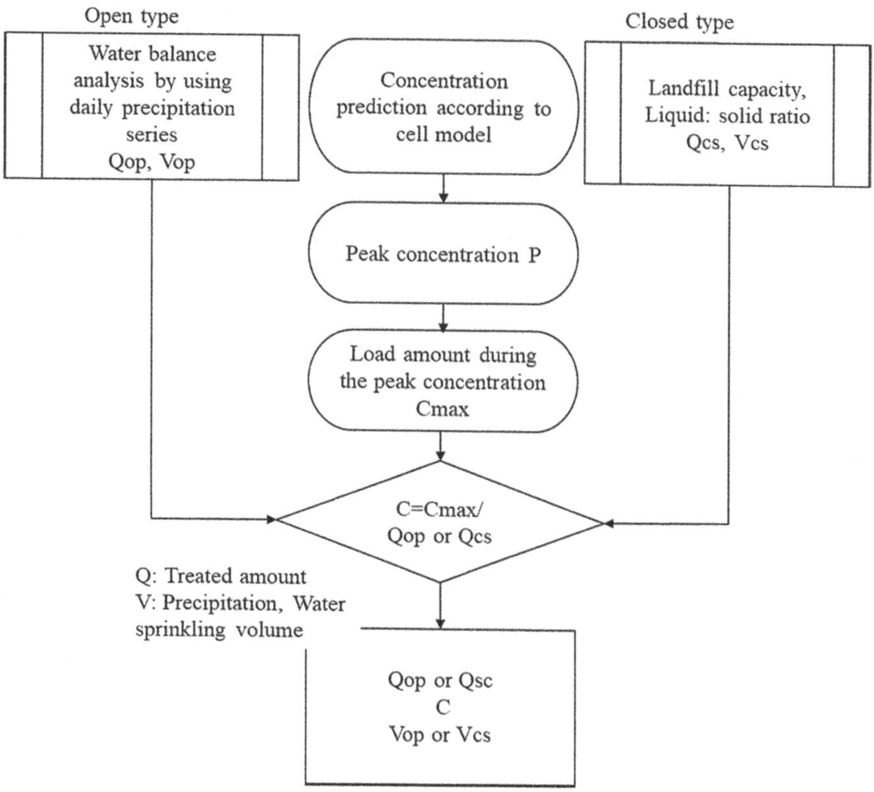

Fig. 4.107 Setting flow of planned raw water quality by considering the load amount of the

Figure 4.108 shows the predicted peak concentration of Cl^- and the load amount of Cl which calculated from the open type landfill by stacking up by landfill cell model.

Based on the data in Fig. 4.108, we can find the peak concentration of Cl^- is 18,600 mg/l, according to the water balance analysis of leachate, the leachate amount on the day which the peak concentration appears is 15.3 m^3/day. The maximum load amount is obtained from multiplying the peak concentration and the leachate amount.

The maximum load amount is 285 kg/day (18,622 mg/l × 15.3 m^3/day=285 kg/day).

The capacity of the leachate treatment facility is calculated from the water balance analysis, and it is 30 m^3/day.

The planned raw water quality is 9,497 mg/l. It gets from the maximum load amount÷the capacity of the leachate treatment facility, that is 285 kg/day÷30 m^3/day=9,497 mg/l.

Therefore, the planned raw water quality is estimated to be 10,000 mg/l. In this case, both the maximum load amount and the peak concentration appears on the

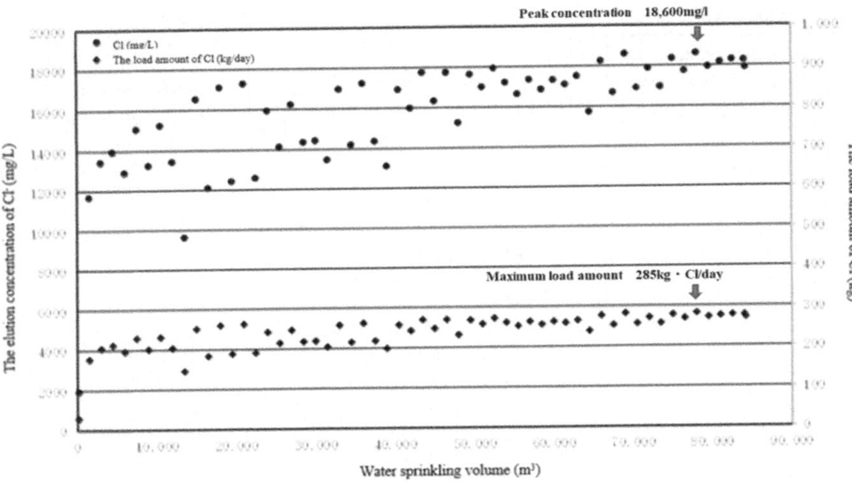

Fig. 4.108 Predicted peak concentration of Cl-, Load amount of Cl

same day. However, these two values may show on a different day in other cases, so we need to pay attention to this point.

Though the capacity of the leachate treatment facility is once determined and the planned raw water quality is calculated, the capacity of the leachate treatment facility is determined from the capacity of the leachate adjustment facility and the facility operation rate of the leachate treatment facility. The capacity of the leachate treatment facility is set as 30 m^3/day from the range of 10–40 m^3/day in this case, while there are some cases which the capacity is determined according to the C. Economy in ⑨Comprehensive determination. For example, since the construction cost of the leachate treatment facility is high, there is also the other option to set the capacity of the leachate treatment facility as 20 m^3/day and the planned raw water quality as 12,900 mg/l (2285 kg/day–20 m^3/day).

The water quality prediction by the landfill cell model must be connected with the landfill method because the landfill method affects the maximum load amount. It is essential to provide a margin to handle the changes in the landfill methods because the unpredicted things are open happens in the actual field.

4.6.6 Seepage Control System

The seepage control system is composed of not only seepage control works such as sheets and cohesive soil, but also collection/drainage facility, seepage control work, monitoring, etc. The basic is to prevent leachate from retaining in the landfill site as much as possible. The penetration of the leachate is hindered by seepage control work, and the leachate is removed promptly out of the landfill site by the collection/drainage facility.

The seepage control system has been developed rapidly in the past 20 years, and many landfill sites have introduced it excessively. For this reason, there are few new technologies to be presented anew, and we will make an introduction about the basic technical structure of the seepage control work.

4.6.6.1 Basic Requirements of the Seepage Control

Among all the facilities that consist of the MSW landfill site, the effluent quality of the leachate and the seepage control performance are the only facilities determining the performance of the facilities quantitatively. The seepage control performance is defined by the standard prescribed by the Ministerial Ordinance, and the outline is as follows.

① Impermeable layer

According to the Ministerial Ordinance, as an unnecessary ground condition, the seepage control work exists in a continuous stratum of above 5 m when the permeability coefficient is 1 × 10–5 cm/s.

② The basic structure of seepage control (Japan Waste Management Association 2010)

The seepage control structure is divided into surface seepage control work and vertical seepage control work. The structure of surface seepage control work is

- A seepage control sheet is laid on the surface such as clay, etc. The permeability coefficient is 1 × 10–6 cm/s, and the thickness is more than 50 cm.
- A seepage control sheet is laid on the surface such as water seal asphalt concrete, etc. The permeability coefficient is 1 × 10–7 cm/s, and the thickness is more than 5 cm.
- A double seepage control sheet is laid on the surface such as nonwoven fabric, etc. In order to prevent the upper and the lower sheet from being simultaneous damaged, the nonwoven fabric, etc. is laid between the double seepage control sheet.
- As exceptional rules, when the slope gradient is more than 50% (1:2), there isn't any possibility of storage leachate in the slope part. It is fine to lay a seepage control sheet or a rubber-asphalt on the mortar spraying, etc.

On the other hand, when the impervious geology is existing in the whole underground of the landfill site, the structure of vertical seepage control work is

- The solidified substance which lugeon value is less than 1 by injecting chemical, etc. until to the impervious geology.

- A continuous wall with a permeability coefficient of 1 × 10–6 cm/s and a thickness of less than 50 cm is provided until to the impervious geology.
- A steel sheet pile which is provided until to the impervious geology, or the wall of sheet method, etc.

4.6.6.2 Structure of Seepage Control

The basic structures and the main types of seepage control are listed in Fig. 4.109.

For the surface seepage control, it is selected depending on the ground properties of the landfill site. In the case of the sheet, the landfill site is constructed in a steep gradient area in the mountainous area based on the characteristic of the MSW landfill site in our country, and the low-elastic type and the medium-elastic type and fiber composites are frequently used in the covered type waste landfill site, etc. In regard to the soil-based seepage control, bentonite mixed soil (500 mm) is commonly used, while in recent years, bentonite crushed stones (100–150 mm) has been used and a method of improving performance and economy has been developed. Figure 4.110 shows a conceptual diagram of surface seepage control.

In the case of vertical seepage control, there is a permeable layer in the lower of the landfill layer. If there is an impervious ground in a relatively shallow position, a vertical seepage control work such as water stop concrete, etc. will be installed up to the impermeable ground. Though it depends on the topography, the groundwater is also removed and treated as leachate. Figure 4.111 shows a conceptual diagram of vertical seepage control.

4.6.6.3 The Standard of Surface Seepage Control

① Impermeable layer

As mentioned above, the Ministerial Ordinance requires that, as an unnecessary ground condition, the seepage control work is a continuous stratum of more than 5 m. The permeability coefficient is 1×10^{-5} cm/s. The schematic diagram is shown in Fig. 4.112.

② Cohesive soil and seepage control sheet

Figure 4.113 shows the schematic diagram of a combination of cohesive soil and seepage control sheet. The thickness of the sheet is more than 1.5 mm, and it is more than 3.0 mm in the case of asphalt. The slope gradient is more than 1: 2. When the sheet is exposed to solar light for an extended period, the shading nonwoven fabric is laid.

Surface seepage control work	Seepage control sheet	Synthetic rubber based/Synthetic resin-based	Low-elastic type	EPDM PVC, etc.
			Medium-elastic type	TPO-PP.-PE TPU LDPE,LLDPE EVA META TPE FPA, etc.
			High-elastic type	HDPE MDPE
			Fiber reinforced type	EPDM-R PVC-R
		Asphalt-based	Sheet type	Impregnation asphalt Layered asphalt
			Spraying type	Single spraying Fiber spraying
		Fiber composite based	Bentonite-based (GCL) Polymer-based (GHPL)	
		Composite type	Sheet bentonite composite seepage control material Fiber · bentonite composite seepage control material	
	Soil-based seepage control	Cohesive soil-based		
		Cement-based		
		Bentonite mixed soil		
		Other mixed soil		
	Asphalt concrete			
Vertical seepage control work	Injection solidifying method	Permeable injection method High-pressure injection method		
	Diaphragm wall method	Continuous diaphragm wall method Soil cement solidifying wall method		
	Steel sheet pile method	Steel sheet pile method Thin steel sheet water stop method		
	Other methods	Vertical seepage control sheet method Thin wall continuous wall work method		

Fig. 4.109 Structures and types of seepage control (Japan Waste Management Association 2010)

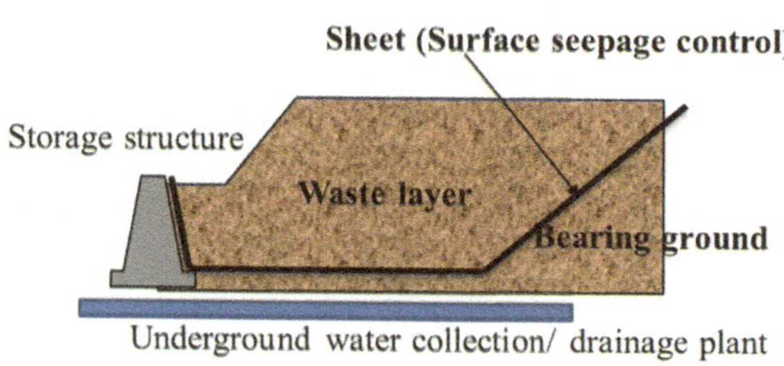

Fig. 4.110 Conceptual diagram of surface seepage control

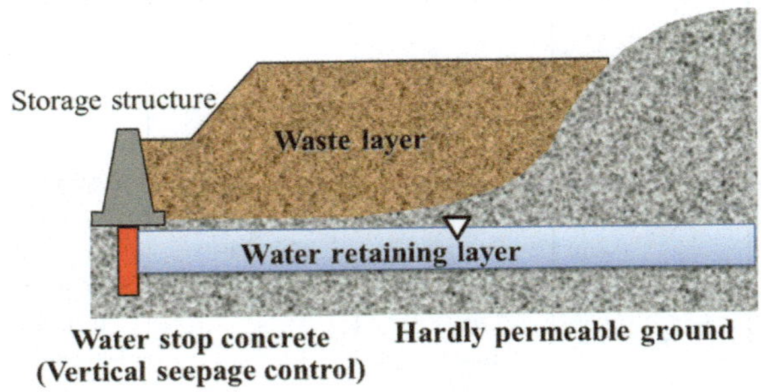

Fig. 4.111 Conceptual diagram of vertical seepage control

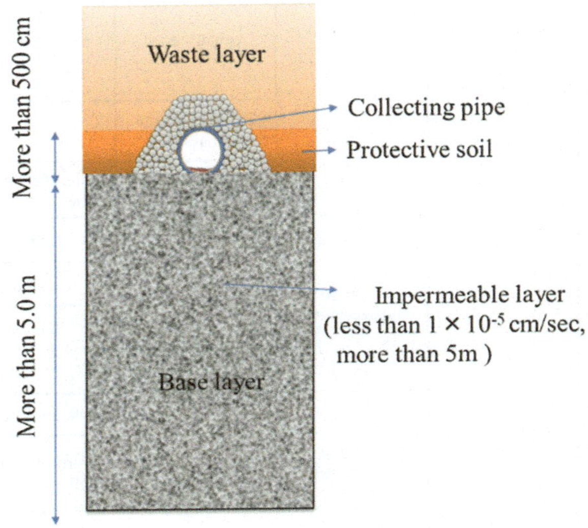

Fig. 4.112 The case that does not require seepage control work

③ Asphalt concrete and seepage control sheet

Figure 4.114 shows the schematic diagram of the combination of asphalt concrete and seepage control sheet. The thickness of the sheet is more than 1.5 mm, and it is more than 3.0 mm in the case of asphalt.

④ Double seepage control sheet

Figure 4.115 shows a conceptual diagram of the double seepage control sheet. The thickness of the sheet is more than 1.5 mm, and it is more than 3.0 mm in the case of asphalt. The slope gradient is more than 1: 2. When the sheet is exposed to solar light for an extended period, the shading nonwoven fabric is laid.

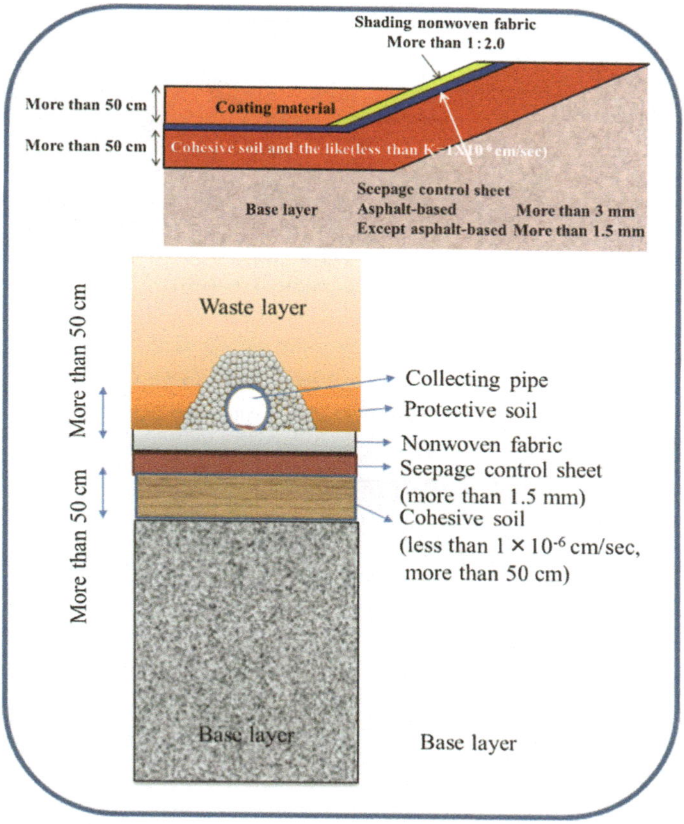

Fig. 4.113 Case of cohesive soil and seepage control sheet

⑤ Seepage control work with the slope gradient less than 1:2

The slope gradient is basically above 1: 2. When it is less than 1: 2, as shown in Fig. 4.116, the seepage control sheet is laid on the mortar spraying. In this case, the internal storage of leachate cannot be performed.

4.6.6.4 Seepage Control Standard in Foreign Countries

① Seepage control standard in German

From the viewpoint of the quality of the landfill waste, Germany has a seepage control standard, which is relatively close to the quality of landfill waste in Japan. In Germany, the seepage control standards are set for each landfill waste materials. As the seepage control standard of MSW landfill site for "Waste generated from

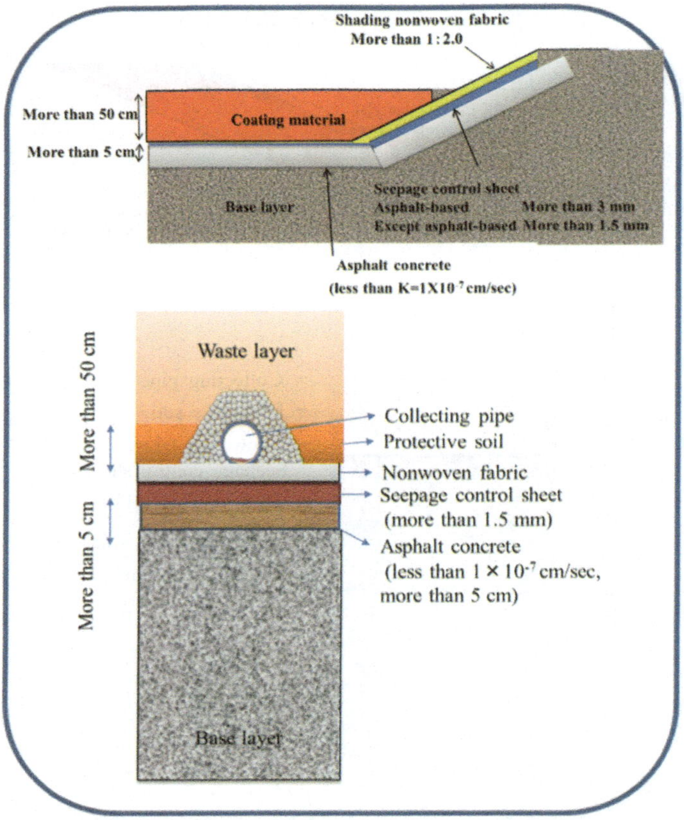

Fig. 4.114 Case of asphalt concrete and seepage control sheet

the residential area" which equals to general waste in Japan, the MSW landfill site where the waste of Class I (the ignition loss is less than 3%) is landfilled is shown in Fig. 4.117, and the MSW landfill site where the waste of Class II (the ignition loss is less than 5%) is landfilled is shown in Fig. 4.118. Besides, the site where the waste of Class III (the ignition loss is less than 10%) is landfilled is shown in Fig. 4.119. The waste of Class III is equivalent to the industrial waste in Japan.

The seepage control standard of the site in Class I and Class II is close to the standard of MSW landfill site in Japan, while Germany makes a stricter value towards seepage control performance, such as the permeability coefficient of cohesive soil or the thickness of the sheet, etc. The most differences from Japan are that it is based on the soil-based seepage control cohesive soil, and a rule is provided for the collection/drainage layer, the entire bottom of the landfill site becomes the collection layer, the clogging preventing is considered. Since gravel (single-sized crushed stone) is used for the collection/drainage layer, a weight of nonwoven fabric installed on the upper part of the sheet is more than 2,000 g/m^2 is to protect the sheet. Besides,

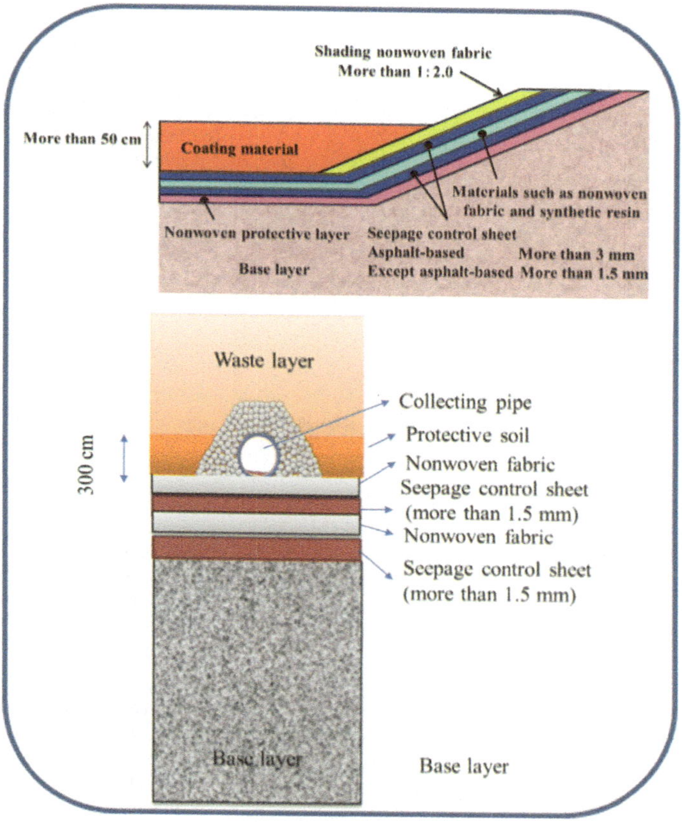

Fig. 4.115 Case of double seepage control sheet

Fig. 4.116 Case of slope gradient

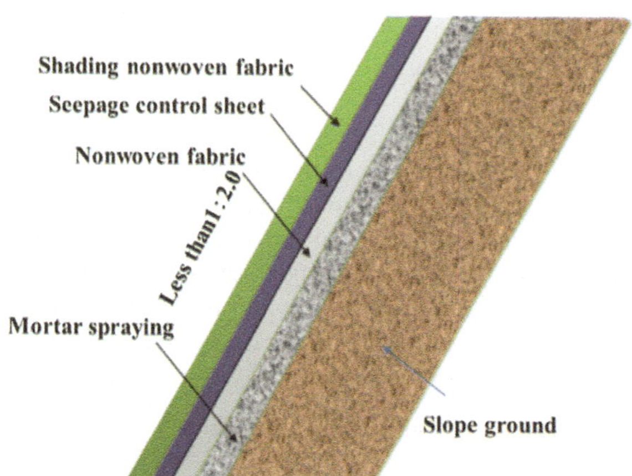

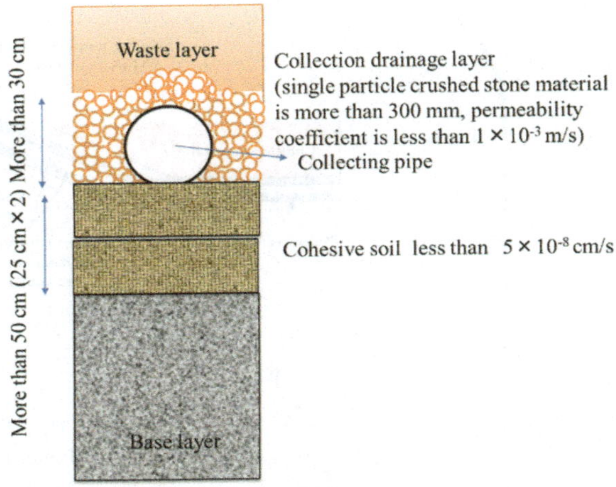

Fig. 4.117 Germany: Seepage control standard of ClassI (Higuchi S 2005)

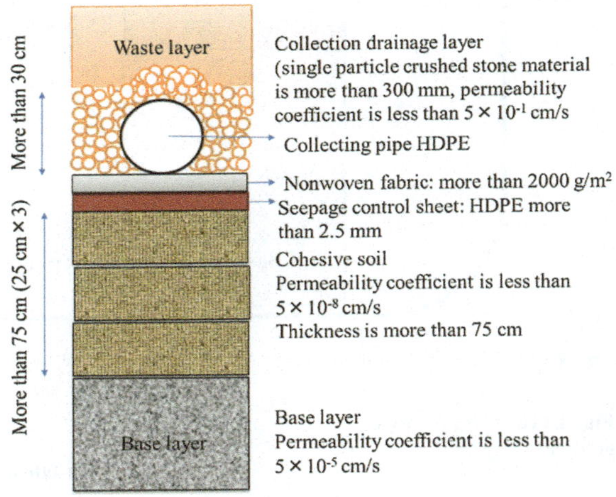

Fig. 4.118 Germany: Seepage control standard of ClassII (Higuchi S 2005)

since the entire bottom of the landfill site is a collection layer, the diameter of the collecting pipe is 300 mm as basic. In Japan, the sheet covering member is usually soil material such as typical soil, water collecting function is not required much. The diameter of the collecting pipe is recommended to be more than 600 mm in the main line. It purposes to supply air into the landfill layer through the collecting pipe. In recent years, even in Japan, the cases in which crushed stones are used as the sheet covering member to promote the water collecting function are increasing.

② Seepage control standard in China

The construction of incineration facilities has been carried out rapidly in China. The quality of landfill waste is basically mainly composed of organic substance. Under

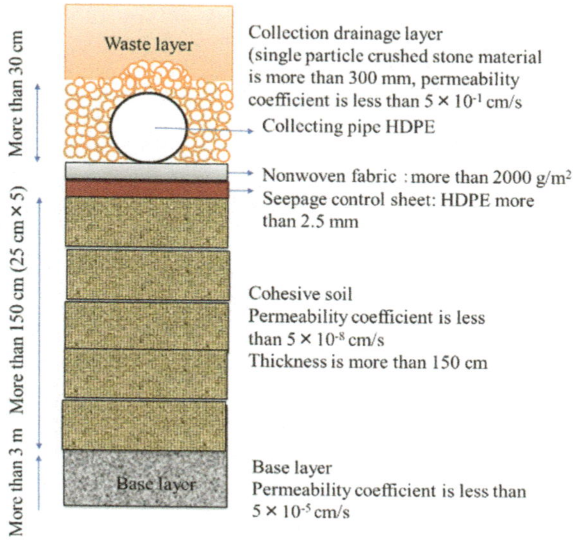

Fig. 4.119 Germany: Seepage control standard of ClassIII (Higuchi S 1996)

such circumstances, the seepage control standard defined in 2007 is as shown in Figs. 4.120, 4.121, 4.122, 4.123 and 4.124.

The seepage control standard in China is the combination of cohesive soil and seepage control sheet (HDPE). Although the seepage control performance in China is looser than in Germany and in Japan, the Chinese lay emphasis on the function of

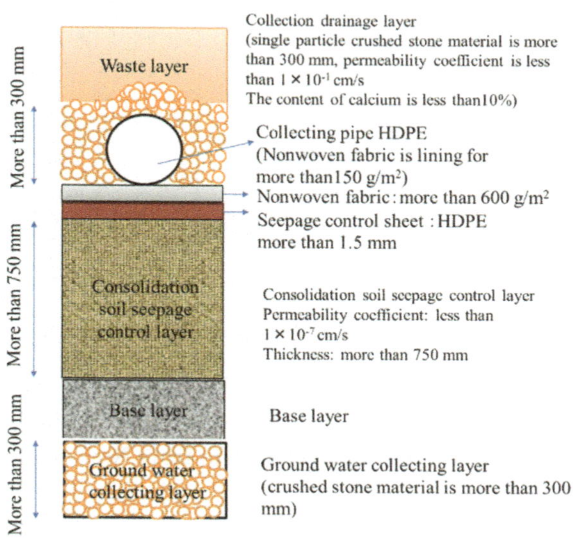

Fig. 4.120 Seepage control standard in China (Cohesive soil + HDPE) (Yahata A and Higuchi S 2012)

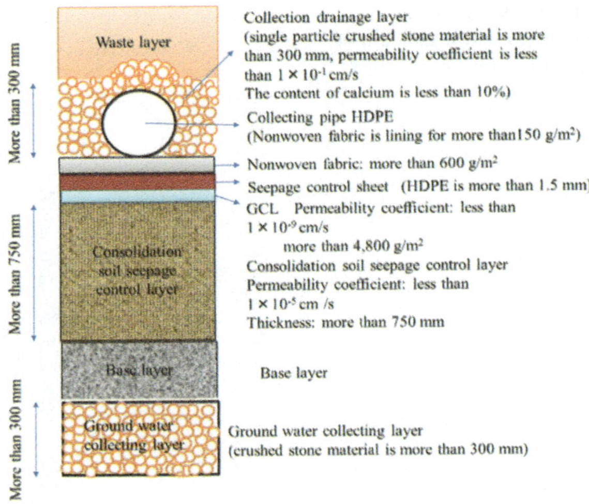

Fig. 4.121 Seepage control standard in China (HDPE + GCL + Cohesive soil) (Yahata A and Higuchi S 2012)

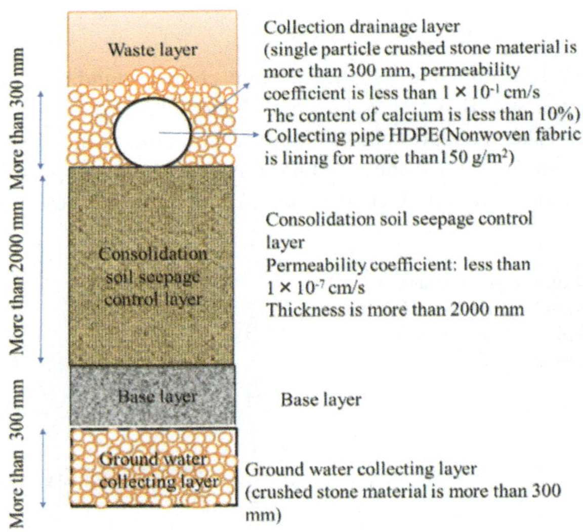

Fig. 4.122 Seepage control standard in China (Cohesive soil) (Yahata A and Higuchi S 2012)

collection/drainage layer. And the basis of the seepage control system is in conformity with the standard in Germany.

4.6.6.5 Seepage Control System from Now on

As stated previously, the seepage control system can achieve its aim not only by the seepage control work but also by the function of the collection/drainage facility

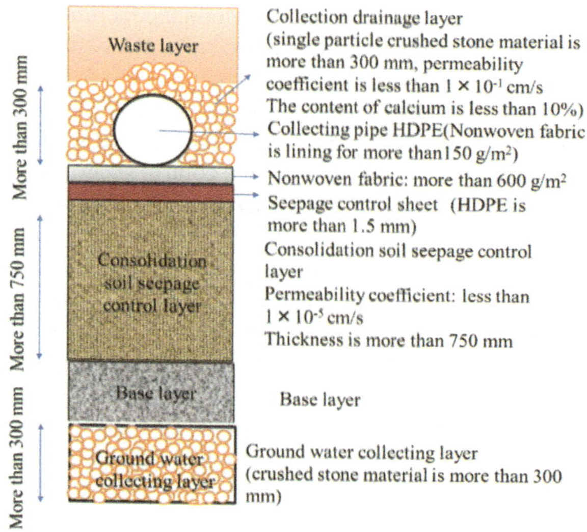

Fig. 4.123 Seepage control standard in China (HDPE + Cohesive soil2) (Yahata A and Higuchi S 2012)

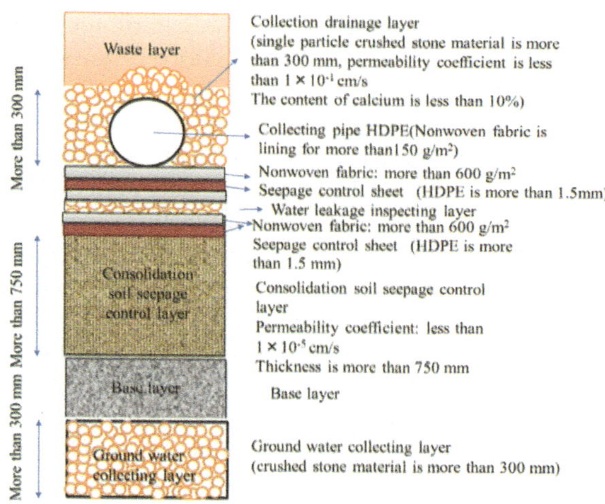

Fig. 4.124 Seepage control standard in China (HDPE + HDPE +Cohesive soil) (Yahata A and Higuchi S 2012)

and monitoring, etc. Besides, among all the previous seepage control systems, the seepage control work has the symptomatic treatment and the post-treatment, and these advanced cases have been adopted in other landfill sites as actual results. As the advanced instances are obtained from the typical consents of surrounding residents, the previous seepage control work becomes one rank higher, and it is upgraded gradually. As a result, the excessive facilities are the present state which double seepage control has passed, and triple and quadruple stoppages have become ineffective. Therefore, it is necessary to make efforts to let the surrounding residents understand the importance of the seepage control system, which is not only consisting of seepage control work but also collection/drainage facilities. Among all the foreign instances introduced in this section, the idea of collection/drainage layer can compensate for the shortcomings of the seepage control covering material in Japan. For example, in Japan, the soil material is used as the covering material on the upper part of the seepage control sheet. Since the soil material was used as the covering material of the landfill site where the seepage control sheet was first introduced, the subsequent sites follow this precedent. Therefore, it exists until now. When a large amount of precipitation occurs before the landfill, the soil material will flow to the gradient direction, the thickness of the covering layer will fluctuate. In this condition, when heavy machinery, especially caterpillar type heavy machinery travels, it will reach the seepage control sheet during the twisting moment, and it is judged as the reason causing the damage of the sheet. With it changing to gravel, the bottom of the landfill site will become an entire collection layer, the drainage will be performed promptly to the leachate adjustment facility or the leachate treatment facility without the leachate retaining in the landfill layer, and it is possible to reduce the risk of the damage of the seepage control work. The landfill management with an emphasis on collection/drainage function is carried on even in Japan, through the covering material layer is used as a collection layer for gravel, the effectiveness of early stabilization of the landfill layer is high as well as the function of seepage control. Therefore, it is necessary to increase the basis weight of nonwoven fabric between the seepage control sheet and the coating layer and to prevent the damage of seepage control sheet by gravel, etc. Besides, in many foreign countries, the seepage control sheet is mainly cohesive soil-based seepage control sheet. In Japan, since the thickness of cohesive soil is required, there is also an aspect that thin sheet-based seepage control has been widely used to secure the landfill capacity even by a small amount. Natural materials such as cohesive soil perform their functions semi-permanently as seen in pyramids, etc. The debate about sheet-based seepage control that "How many years does a sheet can be used?" is still existing even at the present day. Generally speaking, when the waste is landfilled on the sheet and without the effect of the ultraviolet rays, the function can be lasted for an extended interval, sometimes semi-permanently. However, it has been only decades since the seepage control sheet was laid at the landfill site, and its verification will be entrusted to the next generation. Recently, fiber composite material in which bentonite or polymer materials are incorporated into fibers have been widely used, and a method of constructing bentonite crushed stone on a thin layer (Narushima 2016) has also been developed. By using these effects, it is possible to make a rational and economy seepage control work. It is desirable to construct a seepage control work which suits the land conditions such as topography or the landfill waste materials in Japan.

4.6.7 *Multi-function System*

4.6.7.1 What Is the Multi-function System

Based on Japan's waste management policy, the state and technology of future MSW landfill site are described in this chapter. It is an MSW landfill site where "Preservation of surrounding environment" and "Early stabilization and abolishment" is possible.

① Pre-treatment. (inert matter and factors which inhibit the stabilization is removed in advance to secure the ventilation and water flow of the landfill layer)

The need for pre-treatment depends on the quality of the carry-in waste. In regard to industrial waste, it depends on the WDS (Waste Data Sheet). By the pre-treatment, the effect of decomposing the organic substance or washing the inorganic substance can be enhanced, and it is possible to promote the reduction or the neutralization of the leachate quality, the decline of the gas concentration and the amount of gas.

② Landfill operation (a landfill method which can control the volume and the quality of the leachate)

By the method of landfill operation, the concentration of the leachate can be controlled, and it is possible to make a stable operation of the leachate treatment facility.

③ Maintenance the function of the leachate collection/drainage facility or the landfill gas removal facility

As stated previously, the leachate collection/drainage facility removes the leachate from the landfill site quickly. It prevents the water from remaining in the landfill layer. By air inflow to landfill layer, it has a function of decomposing the organic substances and making a pH neutralization. Besides, by the cooperation with the vertical gas removal facility, a further effect can be achieved. In order to maintain this function, it is necessary to prevent the sediment deposit in the facility and to prevent clogging in the hole of water collecting pipe. Therefore, a high-pressure nozzle with monitoring by a television camera and a mobile camera as needed for washing inside the pipe is effective.

④ Post-treatment (the function that can remove the inhibiting factors of stabilization and abolishment after the completion of landfill)

Even if landfill pretreatment, appropriate landfill work, or management of water collection pipes, etc. are performed, the perched water still generates in the wide and deep landfill layer and inhibits the ventilation and water flow. Because of this, the decomposition and the washing out of the organic substance are delayed in this place, and the leachate quality is deteriorated. In order to solve this problem, as needed, it is necessary to make an investigation on this place, make the refractory organic easily decomposed by ventilation, water flow or oxidizing agent, etc. after the completion of landfill. Therefore, it is desirable to incorporate a structure that allows these measures during the construction of the MSW landfill site.

⑤ Leachate management (landfill and leachate management that can control the volume and the quality of the leachate)

With the sophistication of the incineration facility, the treatment of inorganic substances such as Cl^-, etc. become the biggest problem in leachate treatment in recent years. Especially in the case of Cl^-, the increasing of concentration causes the increasing of load amount and inhibits the desalination treatment. This problem can be avoided by setting the concentration of Cl^- in raw water quality to a high value, while the equipment cost becomes higher. On the other hand, the period of the peak concentration of Cl^- is short and the concentration decreases gradually. Generally speaking, the period in which the peak concentration is shown is almost similar to the period in which the peak salt load amount (salt concentration × leachate volume) of leachate is exhibited. However, the leachate volume at this period is smaller than the capacity of the planned leachate treatment, although it depends on the landfill order (Tameda et al. 2017). For this reason, the setting of the planned raw water quality of the open type landfill site is as follows.

$$Cl^- = \frac{Lc}{Q}$$

Cl^- = Chloride ion concentration in planned raw water quality (mg/l)

Lc	=	Cmax	×	Qmax
Peak load by landfill cell model, etc. (Cl-g/day)		Peak water quality by landfill cell model, etc. (mg/l)		Daily leachate volume at the peak water (m^3/day)

Q: Capacity of leachate treatment plant set by water balance analysis (m^3/day)

In the case of the covered type landfill site, as the leachate volume is calculated from the solid-liquid ratio of water sprinkling, not the water balance analysis, the capacity of the leachate treatment facility is smaller. For this reason, the planned raw water quality is close to the peak quality, and it is larger than the open type. Therefore, in the case of the covered type landfill site, it is necessary to consider the scale of the facility including the management of the landfill operation such as a more careful landfill by stacking up and the addition of dilution water.

⑥ Monitoring

The monitoring for the facilities configuring the landfill site to function for the preservation of the surrounding environment fully, early stabilization and abolishment are necessary. And the facilities for providing countermeasures in an emergency is also required. In general, there is leakage detection system for seepage control work, and the monitoring facility for surrounding underground water, etc., the functional maintenance of water collecting pipe and landfill gas removal facility is also essential to the early stabilization. Furthermore, the ventilation and the water flow of the landfill waste is also becoming important.

⑦ Minimization of by-product from the landfill site

The MSW landfill site is a landfill site for waste as the literal interpretation. However, sludge is generated from landfill gas or leachate treatment, and in recent year, by-product salt (concentrated salt) is produced from desalination treatment. There isn't any proper disposal method for these by-product salt, at the current state, it is dried and filled into the flexible containers and stored, or it is disposed on consignment outside. This kind of landfill site isn't a veritable landfill site. Therefore, it is necessary to make an effort to let the amount which is brought out from the landfill site become zero. The dried by-product salt can be used as road antifreeze in some facilities. The authors are searching methods to produce sodium hypochlorite from by-product salt and use it as a disinfectant in a final sewage treatment facility or wastewater treatment facility and developing technology to produce acid and alkali by using bipolar membrane. In this way, it is desirable to apply the technology that is possible at the present time to recycle by-products generated from landfill and aim for a true landfill site. "Multi-function system" is a project that many functions of the landfill site can be utilized, an early stabilization can be carried out safely, and the local environment can be preserved (Lv 2017).

4.6.7.2 Examples of Multi-function System

① Pre-treatment

As facilities for the pre-treatment of the landfill, the reforming by waste washing or by granulated solidification mentioned in Sect. 4.4 "The early stabilization" is listed. We will introduce the pre-treatment for the assumed cases where waste with high organic load, such as unavoidably organic sludge needs to be landfilled. Figure 4.125 shows the conceptual diagram of it. Figure 4.126 provides the installation at MBP facility.

A pre-treatment area incorporating ventilation pipe is provided previously, the waste with high organic load is placed, ventilation is carried out for 2 to 3 weeks, the waste is disposed of after the decomposition of the organic substance being performed (Fig. 4.127).

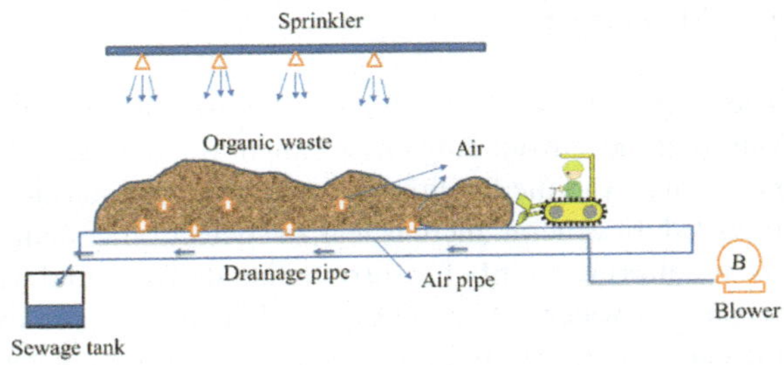

Fig. 4.125 Example of the conceptual diagram of pre-treatment

Fig. 4.126 Example of ventilation before landfill

② Landfill operation

It becomes possible to control the leachate quality by the method of the landfill by stacking up. The method of the landfill by stacking up is another example. The concentration of leachate increases when the landfill height of waste increases, and it decreases when the height decreases. Because of this characteristic, the landfill height of one layer is lowered, and after stabilization progresses by washing out, etc., the lower layer becomes a filter by adding new waste. Comparing with the peak concentration when the landfill is completed at once with high height, its peak concentration is low. The tendency is particularly remarkable for Cl- which is not biodegradable (Higuchi S 1998).

Figure 4.127 explains its conceptual diagram.

Figure 4.128 shows the temporal changes of Cl^- concentration and Ca^{2+} concentration by filling the incineration residue (the content of Cl is 28,000 mg/kg) into a simulated landfill layer with a diameter of 10 cm and conducting the landfill by stacking up. The adding dump is carried out eight times the height is 25 cm. When the Ca^{2+} concentration becomes less than 100 mg/l, the landfill by stacking up is carried out. As a result, the maximum Cl^-concentration by the landfill by stacking up was 9,220 mg/l for the third layer (the landfill height is 75 cm), and the Cl^- concentration

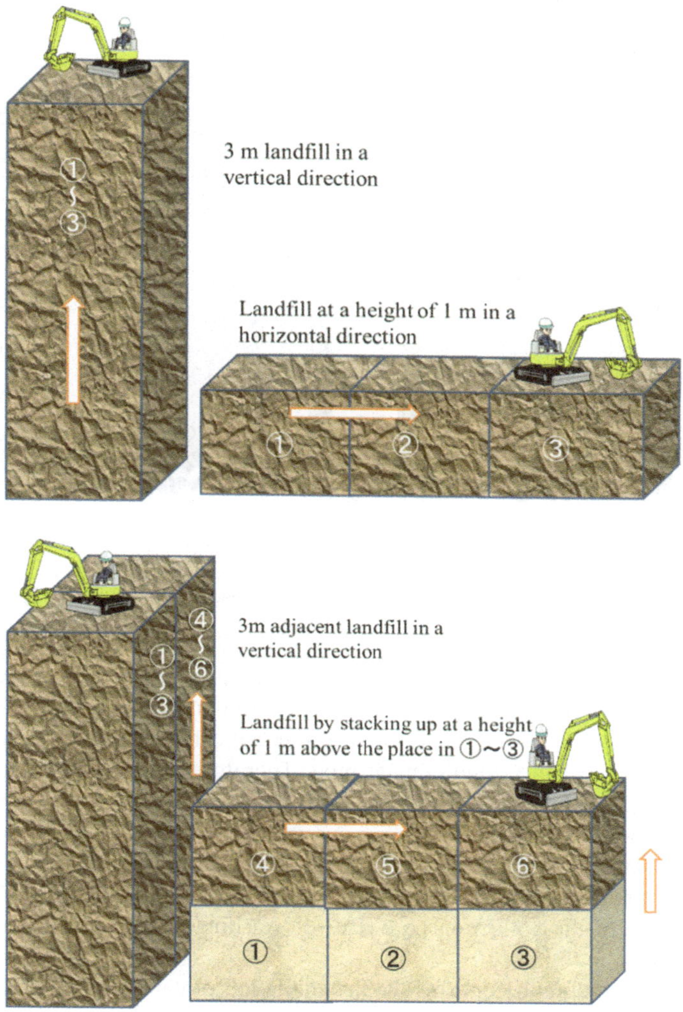

Fig. 4.127 Conceptual diagram of landfill by stacking up

did not exceed the maximum value from the fourth layer, it was 1/4 to 1/2 of the maximum value.

③ Maintenance function of leachate collection/drainage facilities and landfill gas removal facilities

The leachate collection/drainage facility and the landfill gas removal facility are one of the most important facilities to promote ventilation and water flow of the landfill layer and to stabilize the landfill waste. Their functions will be reduced,

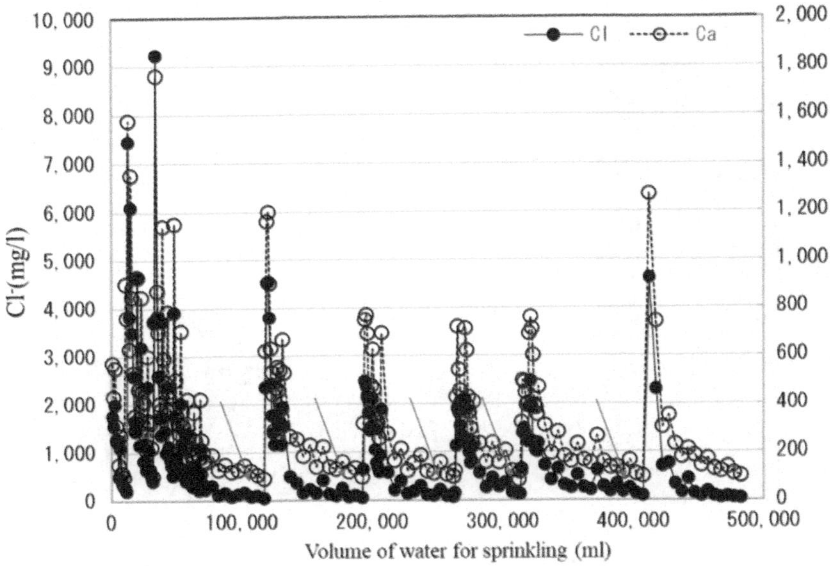

Fig. 4.128 Landfill by stacking up and Cl-concentration in leachate (Tameda K et al. 2017)

and the stabilization will be delayed when the facilities are blocked by earth and sand or Ca scale. For this reason, it is necessary to conduct a function test on the collection/drainage facility regularly and to remove earth and sand when they are accumulated in the collection/drainage pipe. Therefore, design or construction is required so that a mobile camera with a high-pressure washing can be inserted during the construction.

Figure 4.129 shows a conceptual diagram of the washing and inspection of the collection/drainage facility.

The interior of the collection pipe and the washing example by high-pressure water are shown in Fig. 4.130.

The upper photograph confirms the accumulation of earth and sand. The lower photograph shows the condition where the earth and sand in the bottom of the pipe, the scale substances on the pipe wall and water collecting hole are washed by high-pressure water installed on the side of a mobile camera. Through these photographs, it can be visually confirmed that water collection and ventilation have been secured.

In addition, at the terminal section of the collection pipe, it necessary to measure the wind velocity in the pipe regularly to confirm the semi-aerobic function. Incidentally, the wind velocity from the terminal section of the collection pipe of the landfill site where the photo mentioned above is taken to the landfill layer is 0.2–0.4 m/s, the air capacity calculated from the cross-section area of the pipe is 8.6–26.8 m^3/min (12,400–38,600 m^3/day), it indicates the semi-aerobic function is excellently maintained (Higuchi et al. 2005).

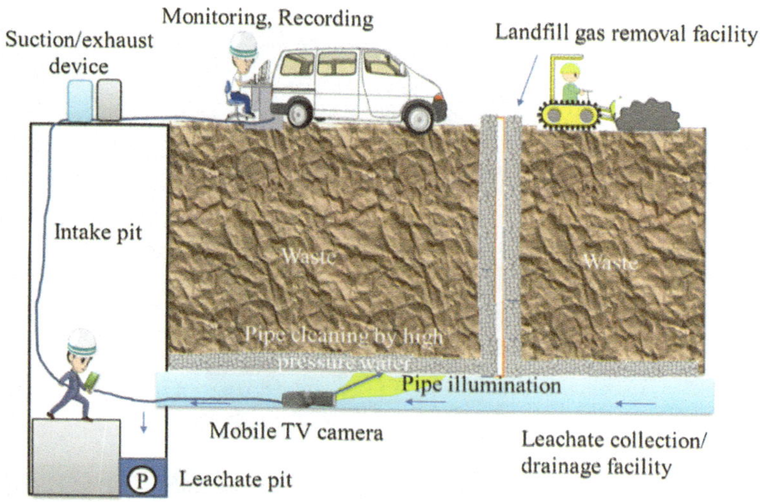

Fig. 4.129 Conceptual diagram of washing and inspection of collection drainage

④ Post-treatment (Miyata et al. 2007, 2008; Yoshinaga S et al. 2008, 2009)

Even after the pre-treatment or an appropriate landfill operation, the waste may generate perched water or heat due to the consolidation of the landfill layer or chemical reaction, etc. A post-treatment is required to solve these matters. In general, large-scale equipment, as introduced in Sect. 4.4.2 is required for the post-treatment. By incorporating equipment which can be used for post-treatment into the landfill structure from the beginning, measures towards stabilization can be taken without adding equipment. We will make an explanation of the methods, etc.

Figure 4.131 presents a plan to install the pipe to make the oxidizing agent can be injected when ventilation and water flow are previously conducted in the soil covering layer.

In order to promote ventilation and water flow, crushed stones (it is preferred to use single crushed stone or grain crushed stone) are used for the intermediate cover soil, and a perforated pipe is laid. This pipe can supply air, water, oxidizing agent (hydrogen peroxide water, ozone water, etc.) to the landfill layer through a switching valve. Water and oxidizing agent diffuse in the crushed stone layer and they are supplied to the lower landfill layer. Air is supplied from the blower and compressor to upper landfill layer. In the case of ventilation, it is more effective to place it in the impermeable protective layer at the bottom of the landfill site. In order to perform ventilation and water flow with this pipe, the position where ventilation or water flow is performed is judged by landfill layer monitoring. The temperature and the gas (CH_4, H_2S, O_2, CO, CO_2) in the landfill layer are used as the index.

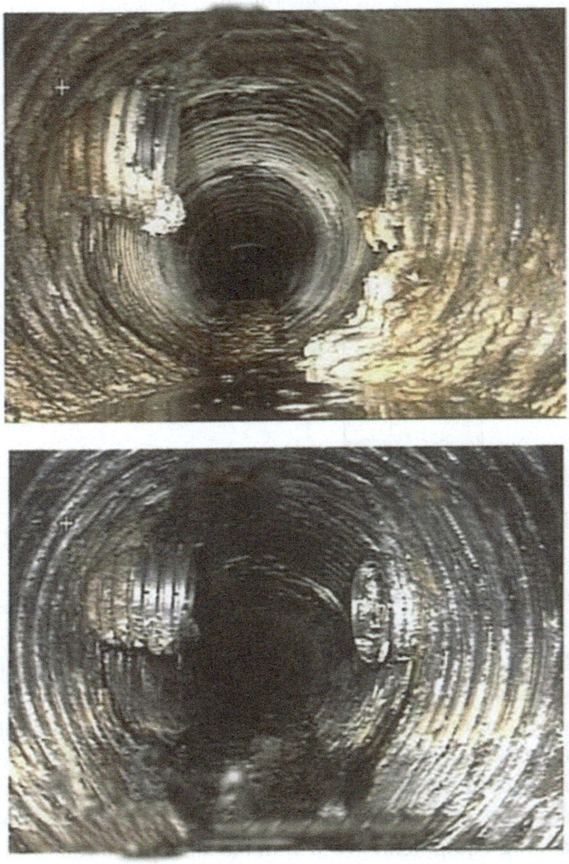

Fig. 4.130 Photograph of leachate collection drainage pipe

⑤ Minimization of by-product from leachate treatment and landfill site

In recent years, leachate treatment facilities in Japan have increased desalination treatment from landfill waste and introduced electrodialysis membranes or reverse osmosis membranes. Referring to Fig. 4.132, when introducing electrodialysis membrane into the desalination treatment, the sludge without Ca, the flocculated and sedimented sludge, and the by-product salt come out as by-products. While introducing the reverse osmosis membrane, as shown in Fig. 4.133. the sludge without Ca and the by-product salt comes out. At present, the sludge is disposed of, and the by-product salts are dried and filled in flexible containers, then stored or outsourced. Aside from the landfill disposal at own MSW landfill site, the landfill site where storage or outsourcing until the appropriate treatment is established is not a veritable

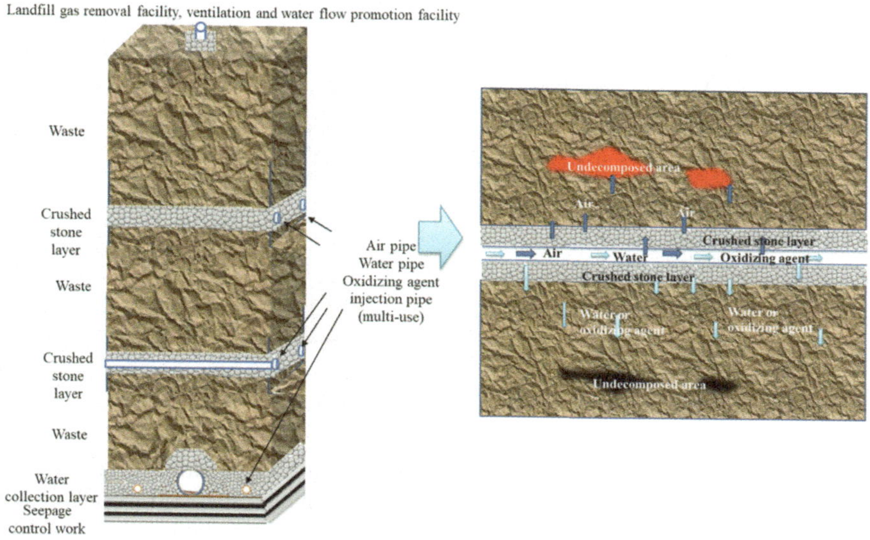

Fig. 4.131 Conceptual diagram of built-in landfill layerin the post-treatment

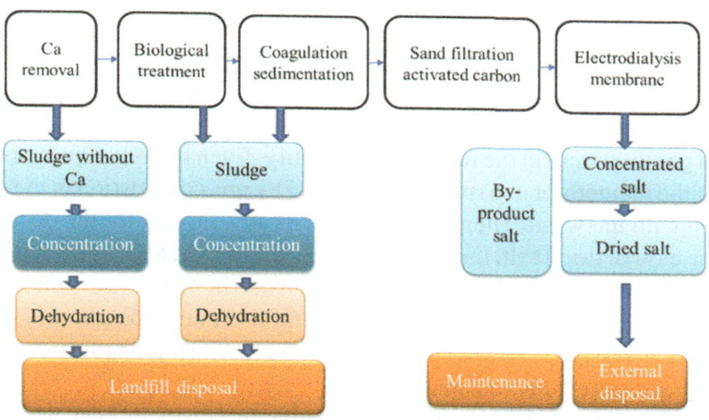

Fig. 4.132 By-products from electrodialysis membrane

one. The function of the landfill site cannot be performed if the waste is carried out from the landfill site and disposed of outside.

Regarding the recycling of by-product salt, it is often used as an anti-freezing agent. Recently, a method for producing sodium hypochlorite (eco-hypo) from by-product salt and using it as a disinfectant at the sewer final treatment plant was started in Matsuyama City (Matsuyama method). It is called Matsuyama method. In

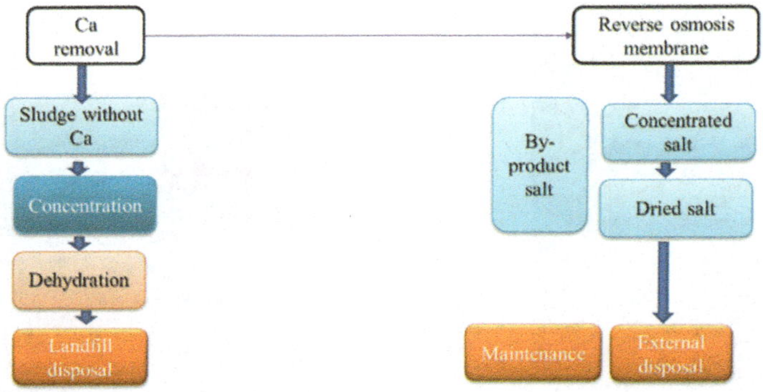

Fig. 4.133 By-products from reverse osmosis membrane

the future, it is necessary to promote recycling by using as an antifreezing agent or by the Matsuyama method. The author conducts a formation study on eco-alkali and eco-acid by using bipolar membrane electrodialysis as one method of the recycling of by-product salt and obtains the eco-alkali (NaOH + KOH) is 6–8%, and the eco-acid (HCl + H_2SO_4) is 10–15% (Han J and Higuchi S 2007; Han J et al. 2018). The concentration is sufficient for use as a neutralizing agent. As the concentration of desalted water is less than Cl-300 mg/l, it also can be used as reused water (Han J and Higuchi S 2007; Han J et al. 2018). (Referring to Fig. 4.134).

Based on the data in Fig. 4.135, since Ca sludge is almost entirely composed of $CaCO_3$, it becomes CaO by burning. CaO can be used as a dechlorinating agent during the exhaust gas treatment in the incineration facility. Even if it is not burnt, put into the waste pit of the incinerator and oxidized into CaO by pre-combustion in the process of incineration with the waste Then it can be used as an adjuvant for the dechlorinating agent. In the case of dried salt, the by-product salt can generate 0.3–0.5% of saltwater, while in the case of concentrated salt, 0.1–1.0% of eco hypochlorous can be produced directly by the non-diaphragm electrolysis method. Stable operation is possible to be provided when the Ca^{2+} in saltwater is about 50 mg/l, and the pre-treatment is not necessary to perform if the leachate treatment facility has a Ca^{2+} removal process. When the eco hypochlorous is generated by the diaphragm method by using saturated salt solution, it is possible to produce 5–8% of eco hypochlorous by pre-treating Ca^{2+} by using 200 μg/l of chelate agents (Takimoto et al. 2016). When the eco-alkali or eco-acid is generated by the previously mentioned bipolar membrane electrodialysis method, it is necessary to pre-treat the Ca^{2+} in 4 N saltwater or the concentration liquid be less than 1 mg/l (Han and Higuchi 2007).

⑥ Monitoring

Various kinds of monitoring are required in order to perform the function of the multi-function system.

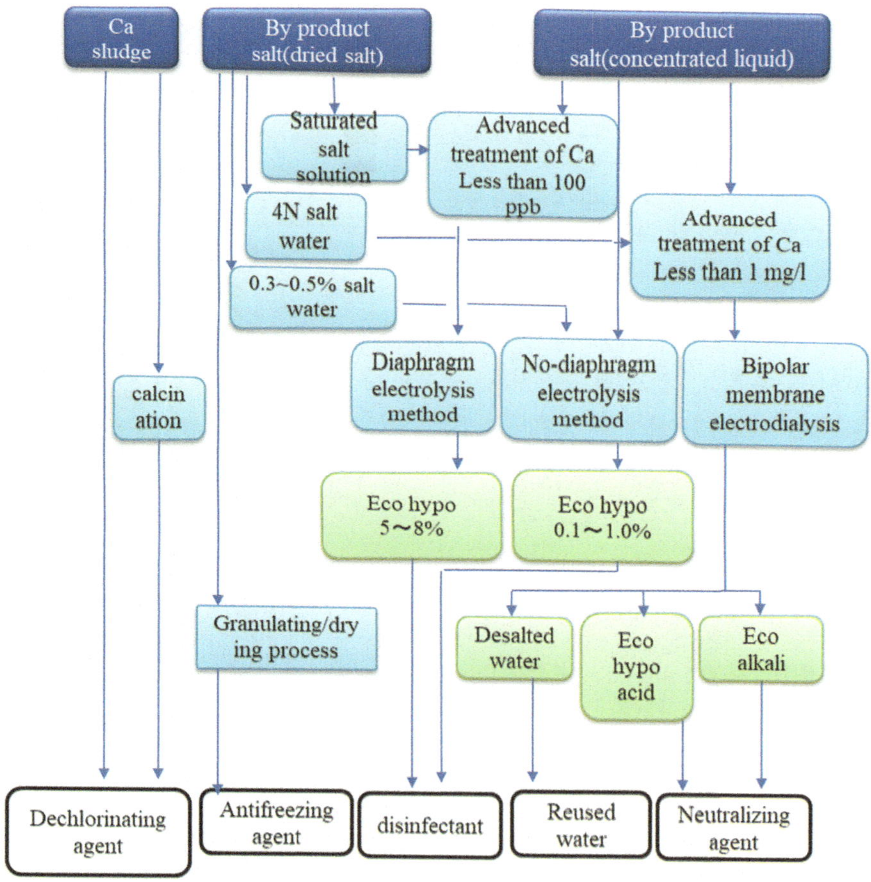

Fig. 4.134 Recycling of by-products accompanied with leachate treatment

- Carrying-in waste: Conduct the elution test and composition analysis regularly. The kinds of carrying-in waste, especially industrial waste changes according to the changes in the social situation and economic situation, it is necessary to investigate newly accepted waste and waste with a large quantity.
- Monitoring the seepage control function: Monitoring by the leakage detection system, groundwater monitoring, etc. The hexa-diagram of leachate, surrounding river water, and groundwater is also effective.
- Leachate: Interlocking with carrying-in control, leachate quality monitoring is essential. Besides, analysis of bacterial flora can be used as an index of the process of stabilization.
- Treated water: It is necessary to analyze regularly whether the stabilization of leachate can be performed or not.

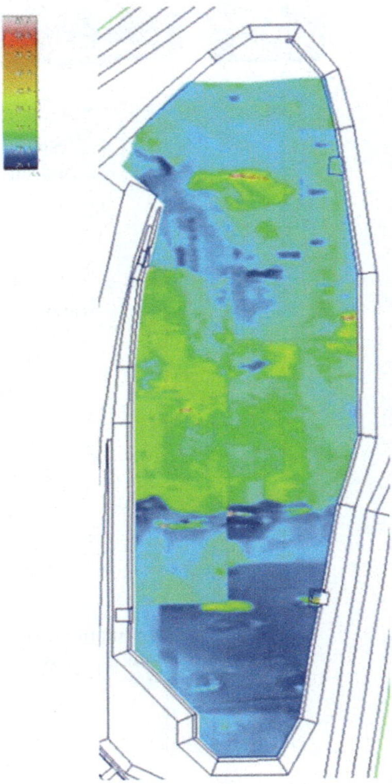

Fig. 4.135 Ground temperature monitoring during landfill operation

- Landfill gas: Conduct landfill gas such as methane and hydrogen sulfide by depth. Conduct temperature control, which is described later in combination.
- The function of collecting pipe: Monitoring the wind velocity, etc. at the inlet of the pipe mentioned in ③ regularly.
- Surrounding groundwater: One of the essential monitoring as verification of seepage control function.
- The temperature of the landfill layer: The temperature rises in the landfill layer is one process of stabilization. Since it is related to gas concentration, leachate quality, it is desirable to carry out routinely among all the monitoring items. In recent years, the surface temperature can be measured easily by a thermal infrared camera. In the case of abnormally high temperature, it is possible to identify the heat source by using underground temperature gradient and specific heat (Higuchi S 2005). Fig. 4.135 shows an example by the infrared camera at the time of excavation of the waste layer, and Fig. 4.136 shows an example of the surface temperature distribution of the whole landfill site.

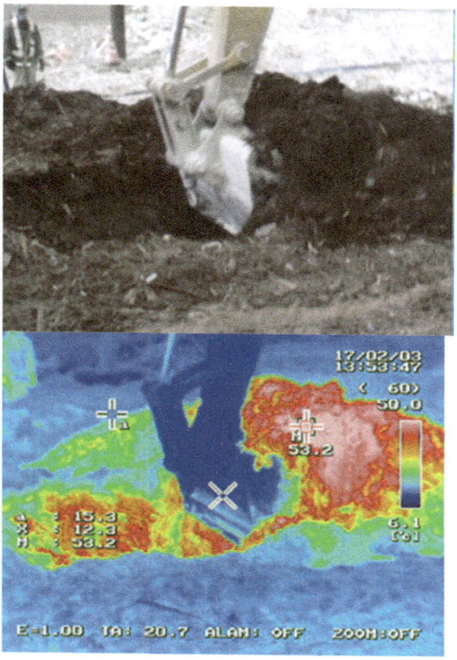

Fig. 4.136 Example of surface temperature distribution of the whole landfill site

4.6.7.3 Case of a Multi-function System

We will introduce the case of a landfill site built based on the multi-function system.

① Outline

Public intervention type landfill site for industrial waste

Form: A covered type landfill site

Landfill area: About 44,000 m^2

Landfill capacity: About 840,000 m^3

Leachate capacity: 60 m^3/day

Leachate treatment method: Desalination (concentrated salt is dried storage and recycled in the future) is conducted by way of Ca removal + Biological treatment + Flocculation sedimentation + Sand filtration · Activated carbon · Chelating adsorbent + Electrodialysis membrane. Figure 4.137 shows the view inside and outside the facilities.

② Characteristics

It fits all the multi-function system mentioned above and conducts facilities and maintenance management for early stabilization, including structure safety and pre-treatment. In particular, it secures the semi-aerobic function and has an aerobic function of discharging the air from the water collecting pipe to the outside by forced

Fig. 4.137 A complete view inside and outside the landfill site

ventilation. Also, we used crushed stone in the intermediate covering soil, and it has multiple functions since we installed the water flow pipes and air pipes among it. It can be said to be the state-of-the-art landfill site in Japan as well as in the world.

③ Safety structure

Figure 4.138 shows the schematic diagram of the bottom seepage control work. In addition to the three-layer structure of the seepage control work with different materials, it becomes a five-layer structure, including a self-repairing sheet (GCL)

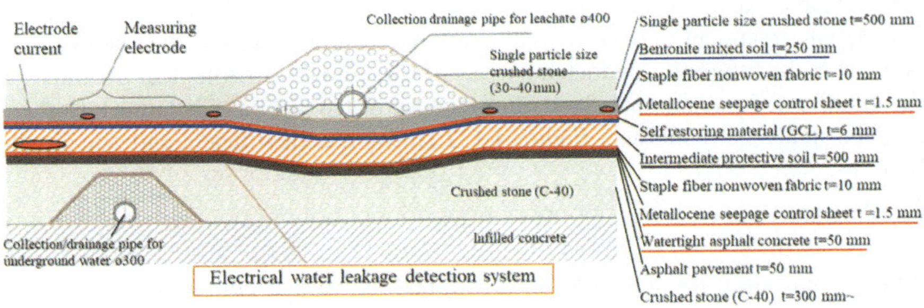

Fig. 4.138 Multiple structures see page control

and 250 mm of bentonite mixed soil. Also, it has an electrical water leakage detection system.

④ Pre-treatment

As shown in Figs. 4.139, and 4.140, a policy of pre-treatment is made towards the carrying-in waste to make an early stabilization. As explained in Fig. 4.141, the

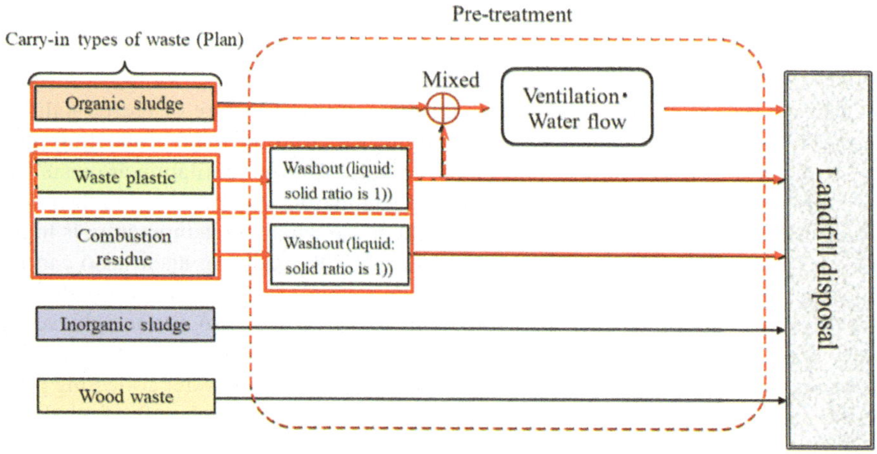

Fig. 4.139 Landfill pre-treatment policy

Fig. 4.140 Diagram of pre-landfill treatment

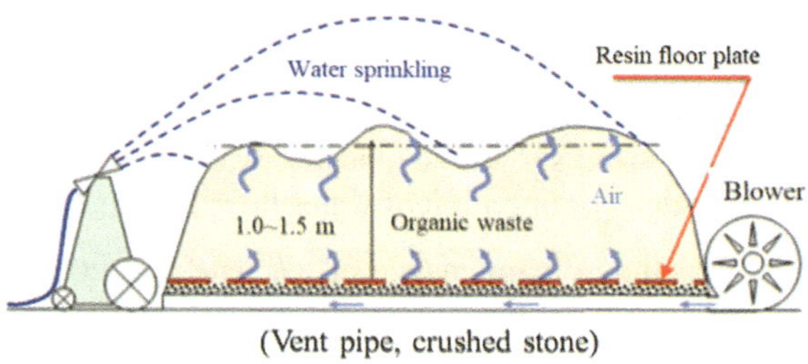

Fig. 4.141 Landfill pre-treatment area

yard for the pre-treatment is installed in the landfill site. It is possible to perform ventilation and artificial sprinkling by the controllable inverter blower.

⑤ Landfill operation

Soil material (shirasu) protects the slope area in contact with the seepage control work, and the soil covering material uses crushed stone to secure ventilation and water flow. The landfill site is divided into meshes and managed so that the management of landfill waste can be achieved.

⑥ Treatment during or after the completion of the landfill

Except for the semi-aerobic landfill, the most significant feature of these facilities is that it has an aerobic function of discharging air from the water collecting pipe to the outside by forced ventilation. Pipes for water flow or ventilation are installed in intermediate covering soil. Water flow and ventilation are carried out in the place where high concentration gas is generated, the place where the temperature is high, and the place where the washing out is delayed. The oxidizing agent also can be injected into the landfill layer as needed. In other words, not only the pre-treatment, the treatments during or after the completion of the landfill also aim to achieve an early stabilization. A schematic diagram is shown in Fig. 4.142. At all times, the semi-aerobic is maintained, and the air can be discharged from water collecting pipe to the outside by four forced fans as appropriate. It is aerobic when the forced fan is in operation. Also, pipes for water flow, ventilation, or oxidizing agent injection are installed in the intermediate covering soil where the crushed stone is used. The ventilation pipes and the water flow pipes are divided into 30 m $\times$ 30 m blocks, and the pipes are rising to the final cover for each block. Based on the monitoring results, it is possible to perform water flow, ventilation, and oxidizing agent injection of the block at the high-temperature places or the place where high concentration gas generated.

Figure 4.143 shows the pipes for water flow and ventilation. Figure 4.144 shows the situation confirmation of water flow. Figure 4.145 shows the arrangement plan.

⑦ Functional maintenance of leachate collection/drainage facilities, landfill gas removal facilities

Except for the function confirmation by the TV camera of the water collecting pipe, since the operation of the forced ventilation fan changes the movement of the gas, and the control of the landfill gas performs. Also, in the final section of the water collecting pipe, the speed of the air flowing into the landfill layer is measured regularly.

⑧ Monitoring

The monitoring includes soil temperature, gas, leachate quality, and surrounding underground water, etc. Besides, the leachate collecting pit and groundwater collecting pit are in the deep part of the underground, the underground inspection passage is set up in two places, and it becomes possible to measure the air quantity

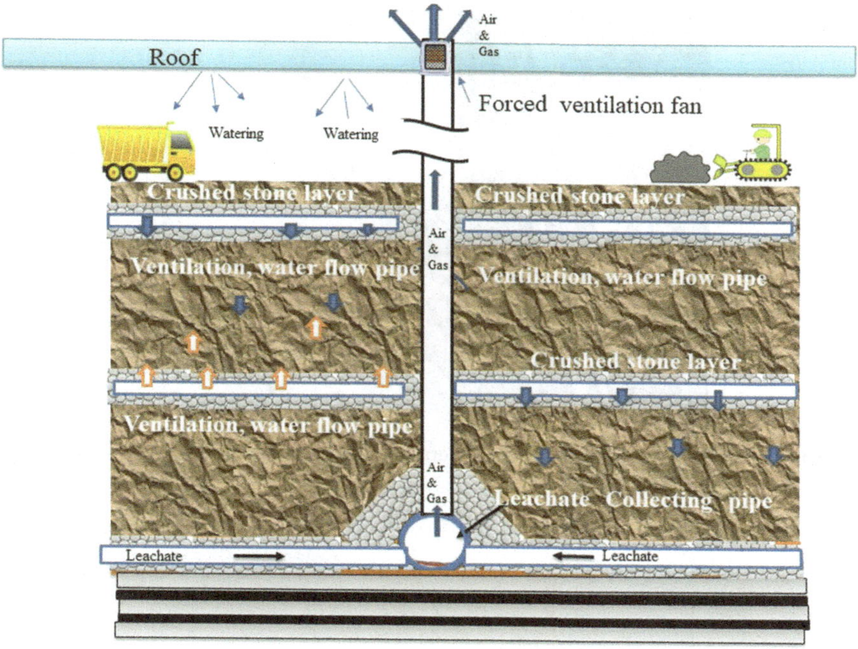

Fig. 4.142 Exanple of ventilation and water flow during or after the completion of landfill (Yokoyama R M 2015)

Fig. 4.143 Installation work for water flow pipes and ventilation pipes in the crushed

through the collecting pipe and monitor the underground water. Figure 4.146 shows the picture of the underground inspection passage. Figure 4.147 shows the functional overview of the landfill site.

Fig. 4.144 The situation confirmation of pipe arrangement and water flow

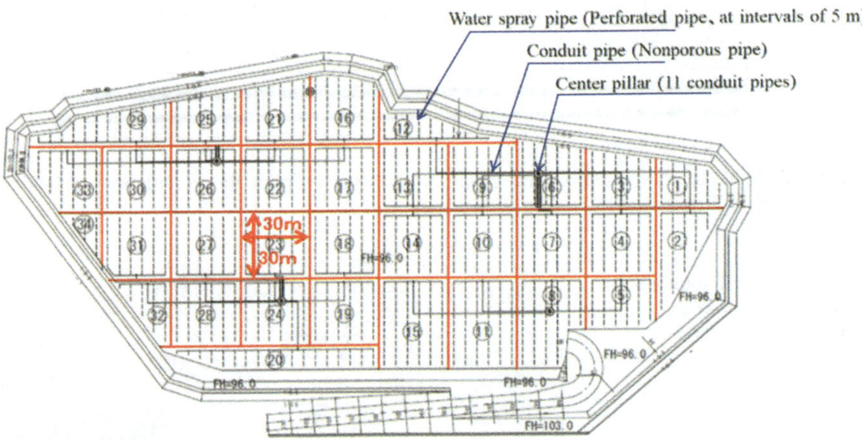

Fig. 4.145 Block arrangement of ventilation and water flow

4.7 Landfill According to the Waste Management System

Japan's landfill technology has been formed and developed through trial and errors. It is towards symptomatic treatment and post-treatment. The lessons from business or research through these processes is that the landfill site must be considered as one of the "waste management total system." For instance, there was no concept of the landfill waste materials for a long time. The primary function of the landfill site is to landfill the carry-in waste safely and to process the leachate without retaining in the landfill layer and discharged it with the prescribed water quality. With the obligations to establish the exhaust gas treatment facilities due to the spread of

Fig. 4.146 Theunder ground inspection passage

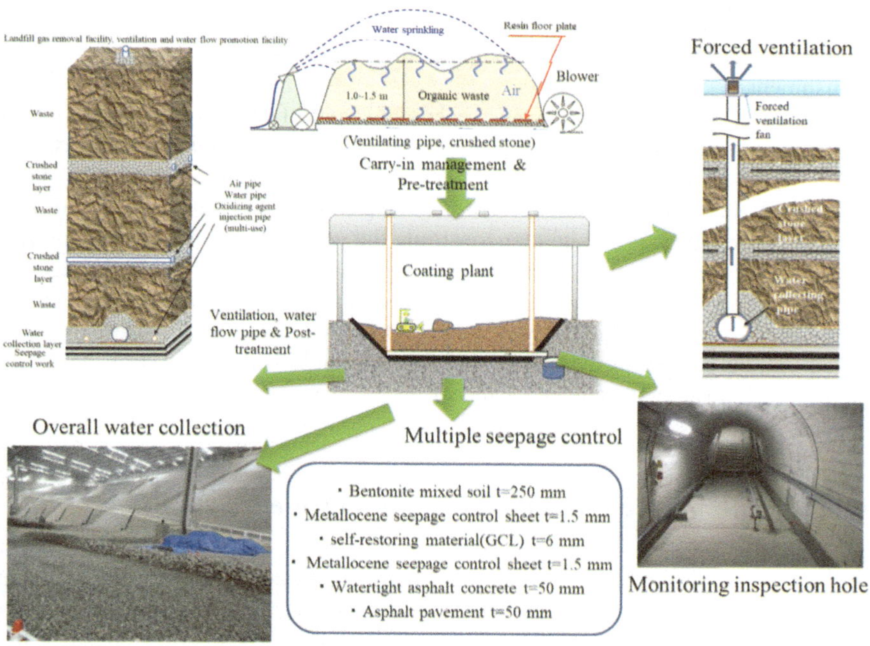

Fig. 4.147 Functional overview of Kagoshima Eco-park

incineration facilities and nitrogen oxide and hydrogen chloride gas regulation, the lime, etc. is blown in. Since the 1980s, Ca scale has adhered to the stirrers and pumps of leachate treatment facilities, and the pH of leachate has increased. Furthermore, the Cl^- concentration has increased, and salt damage such as agricultural damage has occurred at the destination of treated water. As a result, calcium removal equipment, neutralization equipment, and demineralization equipment have been installed in the leachate treatment facilities. Besides, by the influence of the chelating agent, the fly ash, which is designed as the specially controlled waste has caused problems in nitrogen treatment and COD treatment at the leachate treatment facilities.

It is because the landfill site did not notice changes in landfill waste materials.

The performance requirements for the incineration facilities are to meet the ignition loss of incineration residue and the exhaust standard and to clear the landfill standard by the stabilizing treatment of fly ash. For this purpose, agents such as desalination agents and chelating agents are used. With these agents, the required performance can be satisfied. However, the effects of these agents on the landfill disposal were not considered. Naturally, they did not transmit to the landfill site. It caused the trouble in the leachate treatment at the landfill site, the prolongment of maintenance period after the completion of the landfill, and the delay of the abolishment.

Due to this reason, the maintenance cost of the landfill site has increased. And the total cost of waste treatment has increased. Although the minimum cost system has been selected in each facility, the total cost has still increased.

In Japan, the technology for incineration, the technology for crushing and sorting, and technology for landfill site that constitute the waste treatment system is almost mature at present, and it is considered that there will be no significant changes in the future. Therefore, it is desirable to establish a total system as a waste management system now. In other words, many problems of the landfill site cannot be solved without cooperation with the incineration facilities. This section describes the desirable waste management system from the viewpoint of landfill.

4.7.1 Issues of the Current Waste Management System

Figure 4.148 shows the general waste management system at present. Combustible waste is incinerated at an incineration facility. At that time, unique lime or highly reactive lime is frequently used as a desalting agent for exhaust gas treatment. Besides, the organic chelating agents are commonly used to stabilize the fly ash. The incineration residue is disposed of at the landfill site. Incombustible waste and oversized waste are crushed and sorted at the recycling center, and the shredded incombustible waste is landfilled at the landfill site. Leachate generates in the landfill site, the leachate quality is high pH, high salt and high alkali due to the influence of fly ash, and it is treated and discharged according to the utilization situation of the destination of the treated water. When nitrogen removal is incorporated into leachate treatment, the influence of the residual chelates (organic chelating agent) causes nitrification

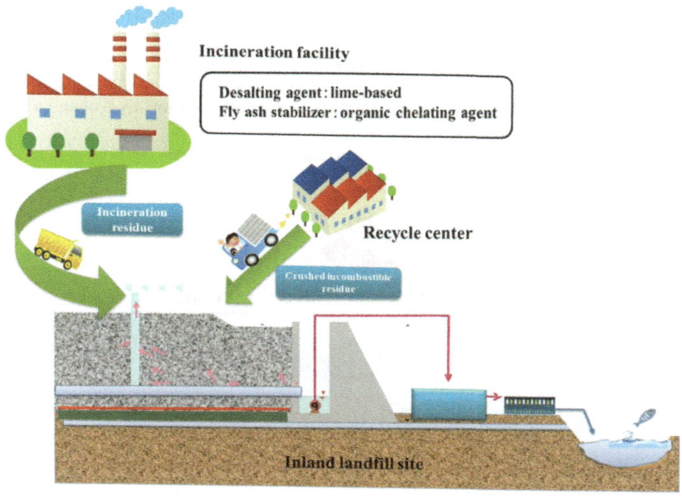

Fig. 4.148 Waste management system at the present time (Inland landfill site)

inhibition. Besides, the landfill site will be abolished based on the abolition standards after the completion of landfill. However, when strict standards are established in the abolition standards, the reduction of nitrogen is slow, and it takes time to abolish. The nitrogen source is considered to be organic nitrogen derived from residual chelates. Besides, when the landfill site is constructed at the sea surface, as the seawater forms an immersion condition, it becomes an anaerobic condition in the landfill layer, there will be no supply of oxygen and carbon dioxide, and the high pH continues, those are the most significant factors of discontinuance. Furthermore, it is challenging to remove nitrogen from leachate (retaining water) as same as the condition in the inland landfill site. It is due to the use of a lime-based desalination agent and the residual chelate (organic chelating agent).

4.7.2 The Waste Management System and MSW Landfill in the Future

4.7.2.1 Inland Landfill Site

Figure 4.149 shows the waste management system of the inland landfill site. By changing the desalination agent for exhaust gas treatment in incineration facilities from calcium-based to sodium-based (sodium hydroxide, sodium hydrogen carbonate or dolomite hydroxide), the pH problem has occurred. By changing the stabilizing agent from organic chelates to inorganic chelates (phosphorus), the

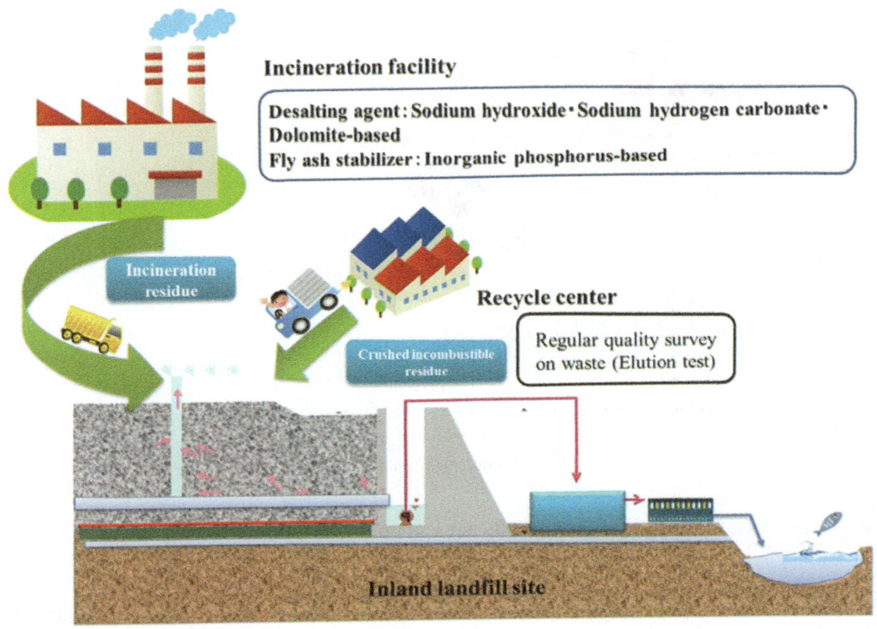

Fig. 4.149 A desirable waste management system (Inland landfill site)

hindrance of the nitrification inhibition and the nitrogen treatment derived from the residual chelates can be solved during the process of leachate treatment.

And it becomes possible to conduct a type of landfill site with an early stabilization and a rapid abolishment. In the case of the shredded incombustible waste, it is desirable to do a periodic elution test on the shredded incombustible waste, since lead and mercury derived from toy base and fragments of the fluorescent tube which mixing in incombustible waste may elute. With this method, it is possible to stabilize early and to abolish rapidly and to reduce the cost of waste disposal.

4.7.2.2 Sea Surface Landfill Site

Sea surface landfill sites exist only in Japan and Singapore, and the number may increase with the spread of the incineration facilities in the future. Although the sea surface landfill site is most suitable as a landfill of high salt and low organic substance, which is a feature of incineration residues, the biggest issue of the sea surface landfill site is that the value pH does not decrease. The reason for high pH is the influence of calcium-based chemicals used for exhaust gas treatment (desalination agent) of the incineration facilities. Furthermore, without the airflow in the landfill layer, there is no supply of oxygen or carbon dioxide. Since the decrease of pH has to rely on the dilution by precipitation, it takes a long period of the units of 100 years to decrease

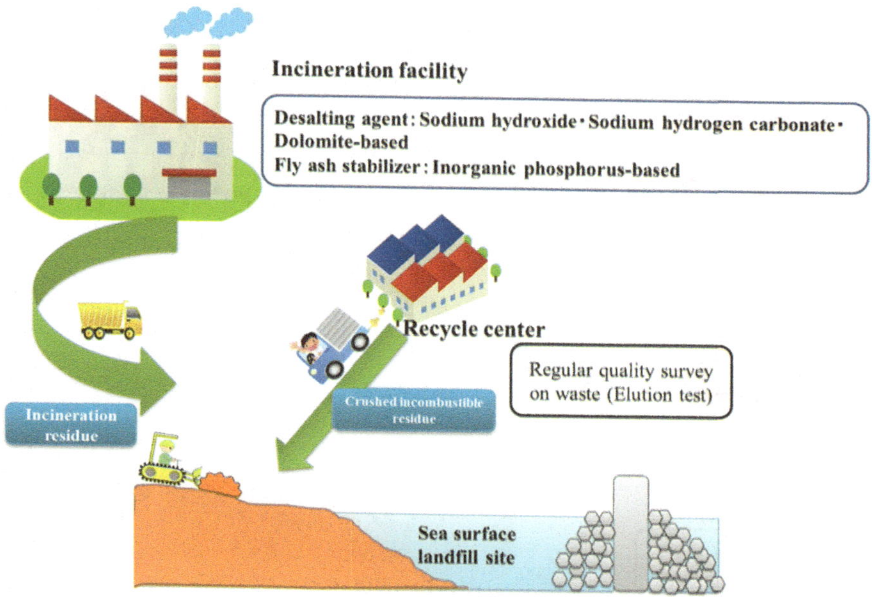

Fig. 4.150 A desirable waste management system (Sea surfacel and fill site)

the value. For this reason, the fundamental solution is to change the exhaust gas treatment chemical from calcium-based to sodium-based, such as sodium hydroxide or sodium hydrogen carbonate. Furthermore, through changing the organic chelate used for stabilizing treatment agent for fly ash to an inorganic chelate, the problems of nitrogen and COD can be solved. Although the sodium-based chemical is more expensive than the calcium-based one, the cost will be offset by the maintenance and management cost of the landfill site on the units of 100 years. As a result, the total cost, including the incineration facilities and final disposal, is reduced, and it is improved economically. Besides, the use of hydroxide dolomite (a mixture of lime and magnesium) in desalination agent can lower the pH. In the case of shredded incombustible waste, the monitoring of landfill waste materials is required, it is the same as in inland landfill site (Referring to Fig. 4.150).

4.7.2.3 Waste Management System Presupposed Ash Recycling

Incineration residue is recycling as a part of cement manufacturing substances. Cement is an industry of calcium whose primary raw material is lime. Therefore, the incineration residue used the lime-based chemical in exhaust gas treatment chemical is suitable to be used as the cement raw material. However, since cement is under chloride control, and incineration residues contain high concentrations of salts, it is necessary to remove the salt in the incineration residues. From these facts, the

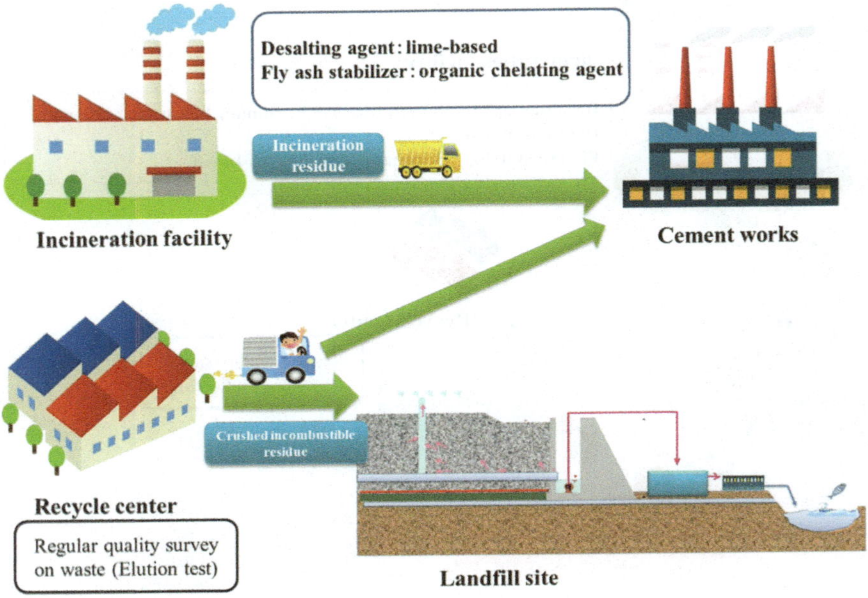

Fig. 4.151 A waste management system which resource is advanced

waste management system at present is suitable for cement recycling of incineration residue.

Figure 4.151 shows an example of a waste management system presupposed cement recycling. In other words, the waste management system at present (stabilizing fly ash by using a lime-based exhaust gas treatment chemical and organic chelate) is most suitable for cement recycling. Since the cement kiln is burned at a high temperature of 1450 °C, the stabilization treatment of fly ash is not required in the case of the precondition that the cement kiln of incineration residues is thrown. However, as the incineration residue contains a high concentration of salts, it is necessary to conduct a desalination treatment on the salts exceeding the allowable value of the cement raw material.

4.7.2.4 A Waste Management System Which Resource Is Advanced Used

In recent years, the number of facilities where the desalination treatment of leachate is carried out in final disposal system for pre-treatment, undischarged of treated water, cement recycling of incineration residue, etc. When the desalination treatment is conducted, by-product salts are generated, facilities for recycling the by-product salts are also established. When the recycling system of incineration residue and the recycling system of by-product salt are functioning, the landfill site can be

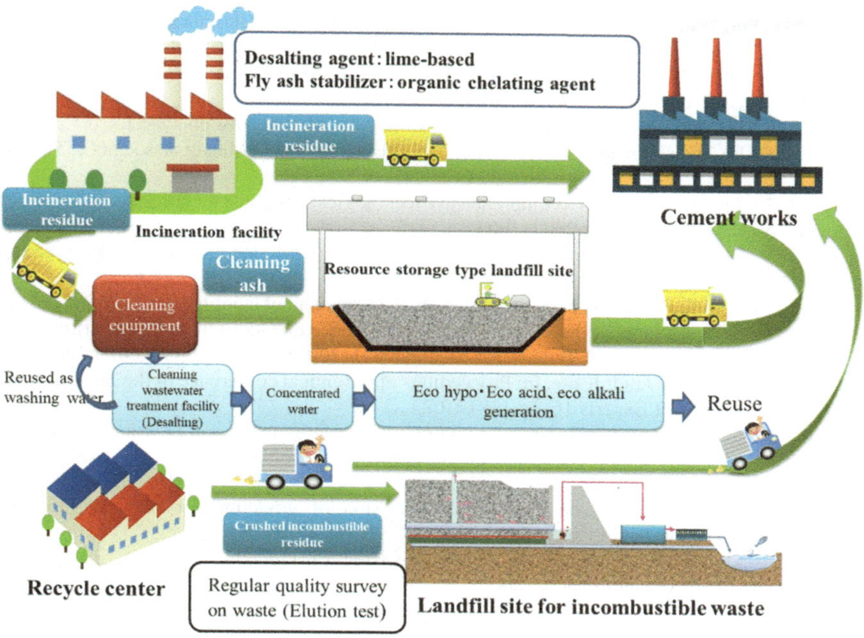

Fig. 4.152 Example of a waste management system which resource is advanced

utilized as a facility where the resource is advanced used, and the landfill disposal of shredded incombustible residue is enough. As shown in Fig. 4.152, the incineration residue is desalination-washed and stored temporarily, and then taken out to the cement facilities. On the other hand, the by-product salt is generated by ash washing. Then, by electrolysis, eco-hypochlorous is generated and reused as a disinfectant for effluent treatment. Or by bipolar membrane electrodialysis, eco-alkali and eco acid are produced and reused as neutralizing agents, etc. The example of waste management treatment system presupposed advanced resource is shown. Eco-hypochlorous has already been put into use. The fundamental researches on eco-alkali and eco-acid have already been completed, and the goal of practical application has been set.

References

Han J, Ito T, Higuchi S (2018) Study on recycling of by-product salt accompanied by desalination treatment of leachate (No. 3). J Jpn Waste Manag Assoc 39:277–279

Han J, Lv H, Higuch S, Uchida M (2017) Study on recycling of by-product salt accompanied by desalination treatment of leachete. J Jpn waste Manag Assoc 8:242–244

Higuchi S (1994) Construction and management of MSW landfill sites based on the leachate treatment system. J Jpn Waste Manab Assoc 47(198):23–28

Higuchi S (1996) Study on high concentration inorganic salts in waste landfill sites. Thesis for Kyushu University, pp 76–87

Higuchi S (1998) Study on high concentrations of inorganic salts in MSW landfill sites. The Doctoral Dissertation of Kyushu University, pp 76–87

Higuchi S (1999) Planning and construction of landfill site. Daily Report, pp 42–52

Higuchi S (2003) The abolishment of MSW landfill site. Japan Waste Manag Assoc 56(255):428–433

Higuchi S (2005) Development of a comprehensive tecnological system to reclain landfill. Scientific research on waste treatment in 2002–2004

Higuchi S et al (1996) Leachate management system of solid waste landfill sites. In: Proceedings of ISWA International Congress, pp 259–260

Higuchi Y et al (2001) Research on an in-situ density measurement of landfill waste at MSW landfill site. The 36th Proceedings of the Japan National Conference on Geotechical Engineering

Higuchi S, Yamada T, Tameda K, Hanashima M, Tanaka H (2005) Research on functional recovery of the stabilized MSW landfill site. In: Proceedings of the annual conference of the Japan Society of Waste Management Experts 16, pp 1081–1083

Ikeda M et al (2003) Demonstration test on volume reduction by atatic compression method and its application to early stabilization Proceedings of the annual conference of the Japan Society of Waste Management 26, pp 465–466

Japan Waste Management Association (2010) Revision of guideline for plan, design, and management of MSW landfill sites, 329

Japan Waste Management Association (2010) Revision of guideline for plan, design, and management of MSW landfill sites, 328

Japan Waste Management Association (2010) Revision of guideline for plan, design, and management of MSW landfill sites, 409

Japan Waste Management Association (2010) Revision of guideline for plan, design, and management of MSW landfill sites, 412

Japan Waste Management Association (2010) Revision of guideline for plan, design, and management of MSW landfill sites, 349

Japan Waste Management Association (2010) revision of guideline for plan, design, and management of MSW landfill sites, 363

Japan Waste Management Association (2010) Revision of guideline for plan, design, and management of MSW landfill sites, 364

Japan Waste Management Association (2010) Revision of guideline for plan, design, and management of MSW landfill sites, 238

Japan Waste Management Association (2010) Revision of guideline for plan, design, and management of MSW landfill sites, pp 242–246

Kubo S et al (1988) Identification of heat generation source and load inside object based on the boudary integral. Transactions of the JSME(A), 54(503):1329–1334

Lei X (2014) Study on pre-treatment of incombustible shredded waste before reclamation. Graduate school of engineering, recycling and eco-technology, Fukuoka University. A Master's Thesis for graduate school in 2014

Lei X et al (2014) Study on landfill characteristic of incombustible Shredded Residue (II). In: Proceedings of the annual conference of the Japan Waste Management Association 35, pp 209–211

Minakata M et al (2003) Waste washing by ejector washing facility. In: Proceedings of the annual conference of the Japan Society of Waste Management Experts 14, pp 503–505

Misumi F et al (2001) An elution study on organic substance and inorganic salts from washing ash and non-washing ash. In: Proceedings of the annual conference of the Japan Society of Management Experts 12(2), pp 993–995

Misumi F et al (2003) An elution study on organic substance and inorganic salts from washing ash and non-washing ash. In: Proceedings of the annual conference of the Japan Society of Management Experts 14, pp 1014–1016
Misumi F et al (2004) Investigation on the influence of washing operations as a pre-treatment of landfill on leschate. In: Proceedings of the annual conference of the Japan Society of Waste Management Experts 15, pp 1211–1213
Miyata Y, Higuchi S, Hanashima M, Tameda K (2007) Study on development on the early stabilization landfill system. In: Proceedings of the annual conference of the Japan Society of Waste Management Experts 18, pp 793–795
Miyata Y, Higuchi S, Yoshinaga S, Hanashima M (2008) Study on development on the early stabilization landfill system (No. 2). J Jpn Waste Manag Assoc 29:209–211
Narushima S (2016) The suggestion of cut off water that paid its attention to pure bentonite in clay liner. In: Proceedings of the annual conference of Japan Society of Material Cycles and Waste Management 27, pp 389–390
Reference to TA-Abfall (1991)
Reference to TA-Siedlunsg fall (1993)
Representation researcher, Higuchi S (2005) Development of a comprehensive technological system to reclaim landfills. Waste Manag Res 16(3):132–141
Takeuchi A (1983) Groundwater survey method by measuring landslide underground temperature. Published by Yoshi Bookstore
Takimoto T, Horii Y, Higuchi S (2016) Study on recycling of desalted concentrated water by diaphragm electrolysis (No. 2). In: Proceedings of the annual conference of Japan Society of Material Cycles and Waste Management 27, pp 439–440
Tameda K et al (2010) Technology for early stabilization of MSW landfill site by the injection of dry fog oxidizing agent (II). In: Proceedings of annual conference Japan Society of Material Cycles and Waste Management Experts 21, pp 529–530
Tameda K, Higuchi S (2016) Study on the development of the concentration prediction method of Cl^- in leachate. In: Proceedings of the annual conference of the Japan Society of Material Cycles and Waste Management Experts 27, pp 431–432
Tameda K, Song Y, Ito T, Lee N, Higuchi S (2017) Study on the development of the concentration prediction method of Cl^- in leachate. J Jpn Waste Manag Assoc 70(339):101–107
Tanioka T et al (2002) Waste washing by spiral wet washing facility. In: Proceedings of the annual conference of the Japan Society of Waste Management Experts 13, pp 978–980
Uryo N et al (2005) Experiment for promoting stabilization of landfill waste by RP method. In: Proceedings of the annual conference of the Japan Society of Waste Management Experts 16, pp 1084–1086
Uryo N et al (2006) Experiment for promoting stabilization of landfill waste by PR method. In: Proceedings of the annual conference of the Japan Society of Waste Management Experts, 17
Wada Y (2016) Densification of landfill waste at sea surface MSW landfill sites. In: Proceedings of the annual conference of the Japan Waste Association 37, pp 258–260
Yahata A, Higuchi S (2012) Study on salts elution control system in MSW landfill sites (II). J Jpn Waste Manag Assoc 3, pp 251–253
Yamatani K et al (1995) Idenfication of poin source locations for poisson equation using harmonic functions. Trans Jpn Soc Industr Appl Math 5(1):87–99
Yoshizaki K et al (2006) Study on early abolishment and stabilization promotion for waste landfill site. In: Proceedings of the annual conference of the Lapan Society of waste Management Experts 17, pp 959–961
Yokoyama R (2015). General outline of eco park kagoshima, an industrial waste management type landfill site by public intervention. Mater Cycles Waste Manag Res 26(1)

Yoshinaga S, Higuchi S, Miyata Y, Hanashima M, Tameda K (2008) Study on development on the early stabilization landfill system (No. 3). In: Proceedings of the annual conference of the Japan Society of Waste Management Experts 19, pp 775–777

Yoshinaga S, Higuchi S, Miyata Y, Hanashima M, Tameda K (2009) Study on development on the early stabilization landfill system (No. 4). J Jpn Waste Manag Assoc 30:231–233

Chapter 5
Conclusion

Technology for Japan's final disposal is towards symptomatic treatment and post-treatment; it has been formed and developed through trial and errors. Japan is considered to be one of the countries with the most advanced technology even by international standards.

Personally, when I was a senior at college, I entered a laboratory for research in waste. I worked in a consultant company for nearly 30 years, gaining a lot of experience in planning and designing the landfill site of local government, and then changed my job to a university. It has been 19 years since I worked in the university, I have carried out student education and made research on waste focusing on the MSW landfill site. During this time, both in consultant companies and in universities, I have solved the problems one by one until now.

The contents contained in this book are focusing on the results of research conducted to address the issues related to the MSW landfill site. The lesson from business and research through these processes over the years is that we must consider the MSW landfill site as a part of "waste management system."

The "waste management system" means a complete process of waste management starting from waste collection and transportation. It is generally considered as an ideal form by everyone even from old times.

Unfortunately, the reality is that it is only an armchair theory.

Since Japan is small in size and it is difficult to secure an area for MSW landfill site, Japanese have taken a half-century to make a study on the intermediate processing such as incineration, ahead of other countries. They have solved various kinds of problems such as exhaust gas problem, dioxin problems, and the like until now. When we realize that about 90% of the landfilled waste becomes the residues produced through the intermediate processing, the adverse effects are appearing in the MSW landfill site. As described in this book, the adverse effects are issues that cannot be solved at the MSW landfill site, and it is a problem that must be considered as an adverse effect of the "waste management system."

S. Higuchi, *Municipal Solid Waste Landfill Technology in Japan*, Environmental Science and Engineering, https://doi.org/10.1007/978-981-16-2734-7_5

Several factors are impeding the "waste management system," among all the elements, the authors list the "the inheritance of technology," "the correspondence to residents" and "the vertical society."

"The inheritance of technology" is improved and established by the predecessors, and we do not have the opportunity to reestablish the completed technologies. As one of the reasons why it could not be done, the local governments and companies must always respond to the residents who are strict to the "waste management system".

For example, the waste disposal facilities at present are focusing on the construction based on comprehensive evaluation type ordering and the maintenance management with performance guarantee based on long-term comprehensive contracts. Management prioritizes technologies and environment measures over costs is a sound system, aiming to protect the living environment of the residents and the surrounding environment, with the risk-sharing between the public and the private sectors. However, due to the problems in "final disposal," the system changes. It is rejected because of the contractor's responsibility called "performance guarantee," even if the improvement is required in "incineration." Besides, in the case of "vertical organization" of administrations and companies, even if in the same environment department, less of them consider "collection and transportation," "incineration," "crushing sorting" and "final disposal" as a section.

As a countermeasure, it is necessary to have a department or organization that consistently checks low environment load, high economic, and more intelligent technology throughout the waste management system. It is the role of the planning department in consultants and administrations with expertise.

Through the whole situation, it is necessary to have the courage to carry out improvement and risk communication with the related residents in regard to proving a better measure than the current situation towards economic efficiency and environment conservation and the like.

As mentioned above, in the past 50 years, "waste management system" has been formed by the development of problem-solving techniques. It is desirable to conduct a technology assessment to integrate environmental, economic, and technical, that is, to secure the total system.

Index

S. Higuchi, *Municipal Solid Waste Landfill Technology in Japan*, Environmental Science and Engineering, https://doi.org/10.1007/978-981-16-2734-7

The manufacturer's authorised representative in the EU is Springer Nature Customer Service Centre GmbH, Europaplatz 3, 69115 Heidelberg, Germany. If you have any concerns regarding our products, please contact ProductSafety@springernature.com

Printed and bound by CPI Group (UK) Ltd, Croydon, CR0 4YY
30/09/2026
02225294-0013